Microelectromechanical Systems—Materials and Devices II

MATERIALS RESEARCH SOCIETY
SYMPOSIUM PROCEEDINGS VOLUME 1139

Microelectromechanical Systems—Materials and Devices II

Symposium held December 1–2, 2008, Boston, Massachusetts, U.S.A.

EDITORS:

Srikar Vengallatore
McGill University
Montreal, Canada H3A 2K6

Jörg Bagdahn
Fraunhofer Center for Silicon Photovoltaics
Halle (Saale), Germany

Norman F. Sheppard Jr.
MicroCHIPS, Inc.
Bedford, Massachusetts, U.S.A.

S. Mark Spearing
University of Southampton
Southampton, United Kingdom

Materials Research Society
Warrendale, Pennsylvania

CAMBRIDGE
UNIVERSITY PRESS

University Printing House, Cambridge CB2 8BS, United Kingdom

One Liberty Plaza, 20th Floor, New York, NY 10006, USA

477 Williamstown Road, Port Melbourne, VIC 3207, Australia

314-321, 3rd Floor, Plot 3, Splendor Forum, Jasola District Centre, New Delhi - 110025, India

79 Anson Road, #06-04/06, Singapore 079906

Cambridge University Press is part of the University of Cambridge.

It furthers the University's mission by disseminating knowledge in the pursuit of education, learning and research at the highest international levels of excellence.

www.cambridge.org
Information on this title: www.cambridge.org/9781605111117

Materials Research Society
506 Keystone Drive, Warrendale, PA 15086
http://www.mrs.org

First published 2009
First paperback edition 2012

Single article reprints from this publication are available through University Microfilms Inc., 300 North Zeeb Road, Ann Arbor, MI 48106

CODEN: MRSPDH

A catalogue record for this publication is available from the British Library

ISBN 978-1-605-11111-7 Hardback
ISBN 978-1-107-40839-5 Paperback

CONTENTS

*Invited Paper

MICRO/NANOMECHANICS

MEMS RELIABILITY AND TRIBOLOGY

*Invited Paper

PREFACE

Over the past fifteen years, Microelectromechanical Systems (MEMS) have transitioned from occupying a technology niche to having major industrial significance. The worldwide market for MEMS is now approximately $10 billion, and the total value of systems enabled by MEMS is several orders of magnitude higher than this figure. Initially, commercially successful MEMS utilized pre-existing materials and processes derived from conventional silicon-based semiconductor microelectronics. As the market has grown the material and process sets have broadened and departed from their semiconductor roots. The opportunities created by this broadening have generated a vibrant research community working on new materials and processes. In addition, during this period, MEMS and microfabrication have become important tools for the development and characterization of materials in general. Beginning in 1998, a series of Materials Research Society symposia has documented these trends. This proceedings volume reports on research presented at the latest of these symposia, Symposium GG, "Microelectromechanical Systems—Materials and Devices II," which was held December 1–2 at the 2008 MRS Fall Meeting in Boston, Massachusetts.

The topics covered by the symposium and in these proceedings provide an accurate reflection of the breadth of topics currently under investigation in this field. Many novel materials and accompanying processes are discussed, as well as detailed analyses of more conventional materials and processes. A consistent theme in previous symposia has been the need to conduct accurate material property assessment at the relevant length scales and the need for suitable metrology tools to support the introduction of new materials. These topics are well represented in the present proceedings. We also note the increasing trend towards the inclusion of papers in the proceedings that demonstrate the close coupling between the materials, processes and the MEMS they have been developed for. The growth in the number of papers with this character is a positive indication of the highly interdisciplinary nature of the field and also the extent to which researchers in the community have embraced the need to address system design issues as well as fundamental material science.

There is every indication that the continued growth of MEMS as an important area of technology will continue to provide a strong motivation for the accompanying development of materials and processes. We fully expect that the MRS symposium will also continue to provide a record of these developments.

Srikar Vengallatore
Jörg Bagdahn
Norm Sheppard
S. Mark Spearing

March 2009

MATERIALS RESEARCH SOCIETY SYMPOSIUM PROCEEDINGS

Volume 1108 — Performance and Reliability of Semiconductor Devices, M. Mastro, J. LaRoche, F. Ren, J-I. Chyi, J. Kim, 2009, ISBN 978-1-60511-080-6

Volume 1109E —Transparent Conductors and Semiconductors for Optoelectronics, J.D. Perkins, T.O. Mason, J.F. Wager, Y. Shigesato, 2009, ISBN 978-1-60511-081-3

Volume 1110E —Theory and Applications of Ferroelectric and Multiferroic Materials, M. Dawber, 2009, ISBN 978-1-60511-082-0

Volume 1111 — Rare-Earth Doping of Advanced Materials for Photonic Applications, V. Dierolf, Y. Fujiwara, U. Hommerich, P. Ruterana, J. Zavada, 2009, ISBN 978-1-60511-083-7

Volume 1112 — Materials and Technologies for 3-D Integration, F. Roozeboom, C. Bower, P. Garrou, M. Koyanagi, P. Ramm, 2009, ISBN 978-1-60511-084-4

Volume 1113E —Low-Cost Solution-Based Deposition of Inorganic Films for Electronic/Photonic Devices, D.B. Mitzi, D. Ginley, B. Smarsly, D.V. Talapin, 2009, ISBN 978-1-60511-085-1

Volume 1114E —Organic and Hybrid Materials for Large-Area Functional Systems, A. Salleo, A.C. Arias, D.M. DeLongchamp, C.R. Kagan, 2009, ISBN 978-1-60511-086-8

Volume 1115 — Physics and Technology of Organic Semiconductor Devices, M. Baldo, A. Kahn, P.W.M. Blom, P. Peumans, 2009, ISBN 978-1-60511-087-5

Volume 1116E —Reliability and Properties of Electronic Devices on Flexible Substrates, J.R. Greer, J. Vlassak, J. Daniel, T. Tsui, 2009, ISBN 978-1-60511-088-2

Volume 1117E —Materials Science for Quantum Information Processing Technologies, M. Fanciulli, J. Martinis, M. Eriksson, 2009, ISBN 978-1-60511-089-9

Volume 1118E —Magnetic Nanostructures by Design, J. Shen, Z. Bandic, S. Sun, J. Shi, 2009, ISBN 978-1-60511-090-5

Volume 1119E —New Materials with High Spin Polarization and Their Applications, C. Felser, A. Gupta, B. Hillebrands, S. Wurmehl, 2009, ISBN 978-1-60511-091-2

Volume 1120E —Energy Harvesting—Molecules and Materials, D.L. Andrews, K.P. Ghiggino, T. Goodson III, A.J. Nozik, 2009, ISBN 978-1-60511-092-9

Volume 1121E —Next-Generation and Nano-Architecture Photovoltaics, V.G. Stoleru, A.G. Norman, N.J. Ekins-Daukes, 2009, ISBN 978-1-60511-093-6

Volume 1122E —Structure/Property Relationships in Fluorite-Derivative Compounds, K.E. Sickafus, A. Navrotsky S.R. Phillpot, 2009, ISBN 978-1-60511-094-3

Volume 1123 — Photovoltaic Materials and Manufacturing Issues, B. Sopori, J. Yang, T. Surek, B. Dimmler, 2009, ISBN 978-1-60511-095-0

Volume 1124 — Scientific Basis for Nuclear Waste Management XXXII, R.B. Rebak, N.C. Hyatt, D.A. Pickett, 2009, ISBN 978-1-60511-096-7

Volume 1125 — Materials for Future Fusion and Fission Technologies, C.C. Fu, A. Kimura, M. Samaras, M. Serrano de Caro, R.E. Stoller, 2009, ISBN 978-1-60511-097-4

Volume 1126 — Solid-State Ionics—2008, E. Traversa, T. Armstrong, K. Eguchi, M.R. Palacin, 2009, ISBN 978-1-60511-098-1

Volume 1127E —Mobile Energy, M.C. Smart, M. Nookala, G. Amaratunga, A. Nathan, 2009, ISBN 978-1-60511-099-8

Volume 1128 — Advanced Intermetallic-Based Alloys for Extreme Environment and Energy Applications, M. Palm, Y-H. He, B.P. Bewlay, M. Takeyama, J.M.K. Wiezorek, 2009, ISBN 978-1-60511-100-1

Volume 1129 — Materials and Devices for Smart Systems III, J. Su, L-P. Wang, Y. Furuya, S. Trolier-McKinstry, J. Leng, 2009, ISBN 978-1-60511-101-8

Volume 1130E —Computational Materials Design via Multiscale Modeling, H.E. Fang, Y. Qi, N. Reynolds, Z-K. Liu, 2009, ISBN 978-1-60511-102-5

Volume 1131E —Biomineral Interfaces—From Experiment to Theory, J.H. Harding, J.A. Elliott, J.S. Evans, 2009, ISBN 978-1-60511-103-2

MATERIALS RESEARCH SOCIETY SYMPOSIUM PROCEEDINGS

Prior Materials Research Society Symposium Proceedings available by contacting Materials Research Society

Materials and Processes for MEMS

Mater. Res. Soc. Symp. Proc. Vol. 1139 © 2009 Materials Research Society 1139-GG01-01

Commercial MEMS Case Studies: The Impact of Materials, Processes and Designs

Jack Martin
Analog Devices, Inc., Micromachined Products Division,
Cambridge, MA 02139, U.S.A.

ABSTRACT

Minimizing risk is an important factor in new product planning because high volume breakthrough products require tens of millions of dollars to develop and bring to market. Sometimes risk can be minimized by following the IC model: build new devices on an existing process – just change the mask set. This approach obviously has limits. Adoption of new materials and processes greatly expands the horizon for "disruptive" products. This paper uses a case study approach to examine how changes in masks, materials and unit processes were used, and will continue to be used, to produce MEMS products for high volume applications.

INTRODUCTION

New devices typically start with an idea followed by a lab scale investigation. The next step – development focused towards a commercial product – is considerably larger. What lessons can we learn from a review of past successes and failures in commercial MEMS products? How can we build on those lessons for future products?

The MEMS industry is almost as old as the IC industry. MEMS pressure sensors have been commercially available since the early '70s. Through the '70s and '80s, MEMS was described as being on the verge of explosive growth. For example, it was the cover story [1] of the April 1983 issue of Scientific American. Yes, that growth is occurring. However, it has taken much longer than expected. This paper starts by examining three promising MEMS product opportunities that did not succeed. It then moves onto a series of successful examples, and how they were affected by material properties, product architecture and market forces.

DISCUSSION

Technical excellence is not sufficient

The 1983 Scientific American article described a variety of MEMS devices that had been demonstrated at an R&D level such as accelerometers, inkjet print nozzles and pressure sensors that had active circuitry integrated on the chip. However, the focus was a Stanford University gas chromatograph which had injection and carrier gas ports, a capillary column, valves, detector, exhaust port and connecting capillaries all integrated on a 2-inch silicon wafer. By minimizing system volume, this lab-on-a-chip optimized one important chromatograph design goal. Unfortunately, other performance metrics such as column separation efficiency did not match conventional chromatographs. Standard chromatographs with discrete components performed better and cost less. The wafer-level chromatograph was an impressive technology, but the resulting product could not compete in a marketplace.

There is a Lesson here: MEMS integration is difficult. The example involved integration of mechanical and chemical components, but electronic integration poses similar challenges. This does not mean that integration is impossible. Indeed, Texas Instruments' DMD image projection products with two million independently-controlled MEMS mirrors requires on-chip electronic integration. What it does mean is that level of integration is a key design factor in MEMS products. This issue is discussed later in this report.

Optical cross-connects were a highly visible MEMS product class that failed when the internet bubble collapsed in 2001. Most long distance voice and data communications is carried by optical fibers. These messages are switched electrically, so signal routing requires that they pass through optical-electrical and electrical-optical converters. Telecommunication system traffic projections in the late 1990s showed that the converters would soon become a serious bottleneck. The proposed solution was to deploy an all-optical network based on arrays of addressable MEMS mirrors (Figure 1). New optical cross-connects would send signals from each fiber in a bundle directly to the appropriate fiber of another bundle, thus avoiding optical-electronic conversion. When the market collapsed in 2001, major corporations like AT&T, WorldCom, Lucent, JDS and Nortel lost 2.5 trillion dollars in valuation and shed over 500,000 jobs. The first MEMS cross-connects were just coming onto the market – a market that had disappeared. It has recovered slowly in the last seven years but the "competition" (electrical-optical converters) has also improved.

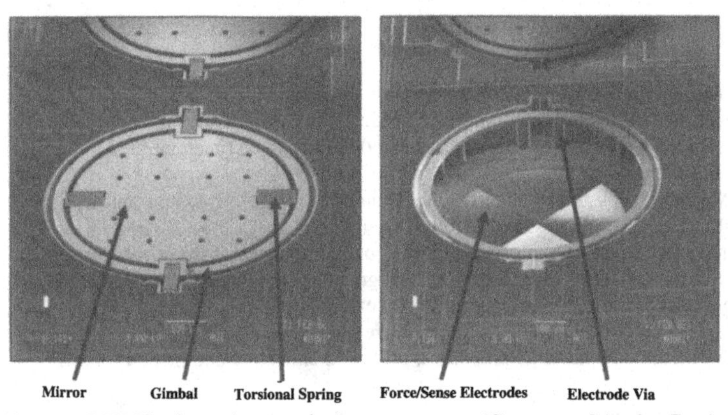

Figure 1: MEMS mirrors in an optical cross-connect. (Courtesy of Analog Devices, Inc.)

High performance RF MEMS switches were a less visible casualty of the 2001 economic turbulence. In the late 1990s, a Northeastern University consortium developed an electrostatically-actuated MEMS switch that extended early Foxboro Company R&D. These devices exhibited remarkable switch life. Analog Devices planned to produce them for automated test equipment (ATE) applications. Those plans were cancelled when collapse of the internet bubble devastated sales of semiconductor test equipment. Sample quantities had been incorporated in developmental defense systems so a license to Radant MEMS allowed them to supply the switches for limited defense applications [2]. Like optical cross-connects, there is renewed interest in MEMS switches. The challenge is to identify stable, high volume markets that place a value on switch capabilities that is consistent with manufacturing costs. ATE, handset, automotive and telecom applications have been proposed. This is not surprising, because MEMS switches offer unique capabilities with respect to architecture of small electronic systems. For example, different fab technologies can be integrated to re-partition products (example: high voltage bipolar with low voltage CMOS).

The optical cross-connect and MEMS switch experience offers a Lesson: There must be a Customer willing to buy the product at an acceptable price in sufficient quantity. Technology alone is insufficient.

AT&T Bell Laboratories developed integrated MEMS microphones in the 1980s [3-5]. However, they couldn't compete with electret microphones. Electret microphones are not particularly good; acoustic performance is mediocre and solder temperatures destroy them. However, they had one outstanding advantage over MEMS microphones: lower price. Knowles Acoustics broke the price barrier and made their first shipment of MEMS microphones in 2003. By the end of 2006, they had shipped 300 million foundry-manufactured microphones [6]. It took over 20 years but MEMS microphones are now a reality. We are now seeing competition from other MEMS suppliers, driven by acoustic performance as well as price.

The Lesson: In high volume products, price is paramount.

Even subtle material properties are critical in MEMS

In the mid-80s, The Foxboro Company introduced pressure and differential pressure transmitters based on piezoresistive pressure sensors. These instruments were used from -40 to +125°C in process control loops (refineries, paper mills, power plants, etc). The differential transmitter typically had a full scale of 20 kPa, but that measurement might be superimposed on a hydrostatic pressure of 20,000 kPa (3000 psi). Silicon pressure sensors are extremely sensitive to package stresses so they were anodically bonded to borosilicate glass (Figure 2). The glass provided a mechanically predictable, low stress mount to isolate the sensor from the metal transmitter housing. Unfortunately, performance tests showed that the measured differential pressure shifted as a function of hydrostatic pressure. The cause of the shift was traced to the glass. It was thermally matched to silicon. However, the bulk modulus of silicon and glass is different. As hydrostatic pressure increased, the difference in volumetric compression applied a torque to the silicon frame, causing it to rotate slightly. This rotation affected the output of piezoresistors implanted in the sensing diaphragm.

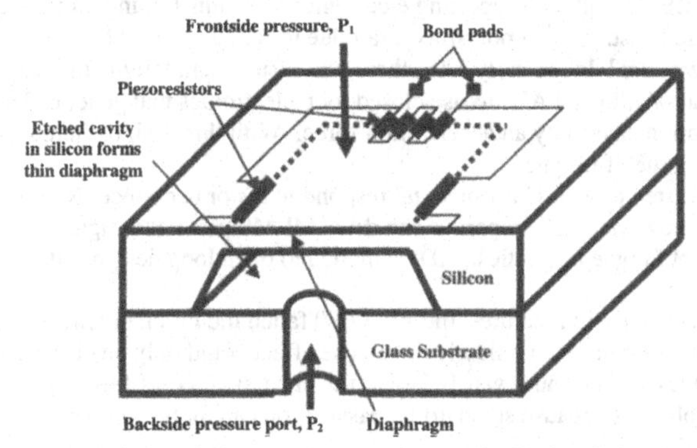

Figure 2: Piezoresistive pressure sensor anodically bonded to borosilicate glass.

To eliminate hydrostatic pressure sensitivity, the glass hard-mount was replaced by silicone rubber. Chemical interaction between this elastomeric mount and the hydraulic oil (a dimethyl silicone) was avoided by making the mounts from a fluorosilicone elastomer.

The Lesson: It is seldom possible to anticipate every effect of material properties on MEMS performance, yield and reliability. Thin film effects are particularly common.

The "Killer" Application

The MEMS literature has many references to "Killer" applications – high volume markets that place high value on a MEMS product. Reality is not that simple. As shown in Table 1, even companies that identify such applications typically wait about 15 years before achieving profitability.

Table 1
Historic MEMS Gestation Times
R&D to Break-Even

Company	Product	Time (years)
Texas Instruments	DLP	18 (27)
Analog Devices	Air Bag Sensor	13
Knowles	Microphone	14

In the late '80s, Analog Devices (ADI) recognized that automotive air bag sensors had the potential to be a killer application. The non-MEMS air bag sensors in use at the time were relatively unreliable so each installation had several sensors. The result was a high cost system that limited air bag use to high-end cars. In principle, a highly reliable MEMS crash sensor could be developed that would allow system cost to be reduced to the point where it would become standard in every car.

Auto manufacturers were understandably nervous – if an air bag sensor fires at the wrong time, it ***causes*** an accident! To address that concern, the ADXL50 incorporated a self-test feature that simulated an acceleration event by applying an electrostatic potential. Proper sensor response showed that both the MEMS and the supporting electronics were functioning properly.

The fears went beyond self-test. Would polysilicon fail due to fatigue caused by automobile vibration after a few years? In the early '90s, there was insufficient fatigue life data on polysilicon. On-chip integration allowed ADI to use closed loop electronics that generated an electrostatic force to oppose motion caused by an acceleration force. With this design, there was no motion, thus eliminating the issue of fatigue.

There's a Lesson here: Listen to your Customer and respond to his or her concerns.

This example illustrates how Material Properties can drive MEMS product design. Polysilicon actually has excellent fatigue properties. ADI switched to open loop designs after the fatigue data was generated.

Even with these conservative design features, the ADXL50 failed the initial qualification tests. Something was causing the output to very slowly drift. The effect could only be observed in high temperature accelerated tests. The cause was linked to the unpassivated MEMS surface. However, the solution was not obvious because standard IC passivation cannot be applied to suspended microstructures.

The solution took advantage of the fact that the MEMS die was mounted in the package with a thermoplastic polyimide-siloxane block copolymer. A short furnace bake (Figure 3) was inserted into the assembly flow in order to use the die attach as a source of organic vapors. The high temperature vapors generated in this process reacted with the polysilicon MEMS surface to create a monolayer passivation. The now-passivated ADXL50 readily passed qualification tests. As a historical point, the ADXL50 was the first integrated surface micromachined product of any type to be produced in high volume.

What's the Lesson here? MEMS products are extremely surface sensitive. This example described an electrical effect. As discussed below, mechanical stiction is more common.

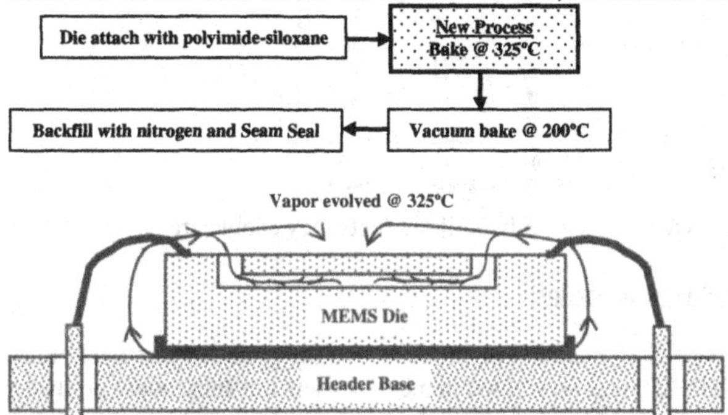

Figure 3: Organic vapors from die attach passivate MEMS.

Reusable engineering

Things are easier after the first product. ADI built design and manufacturing capabilities to produce air bag sensors. They then used these capabilities to build MEMS gyros and low-g multi-axis accelerometers. In essence, they followed the IC industry model: simply change the mask set to build new products in an existing fab.

It was not that simple. Beyond the difference in sensitivity, there were substantial packaging issues [7]. This is not a surprise because MEMS products often face packaging challenges. By building multi-axis sensors in an existing capability, it was financially possible for ADI to wait until the market learned how to use these products. This is a critical point because MEMS devices are not stand-alone products. Rather, they are ***disruptive technology components*** that enable ***higher-level products*** to do things that previously couldn't be done. If those higher level products don't yet exist, it takes several years to develop them. That was the case with these multi-axis sensors.

Sales grew slowly in first 5~6 years, before taking off like a rocket (Figure 4). The range of applications for multi-axis sensors is truly astonishing because virtually every new hand-held device has one. Yole estimates that the multi-axis sensor market is growing at 30% per year [8], so there are now about a dozen suppliers fighting for market share.

The Lesson: Competition follows the money. Once a large market is established, other suppliers will enter and compete for market share.

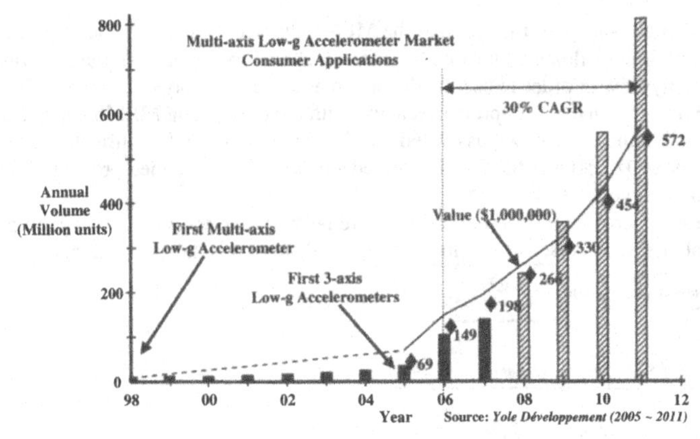

Figure 4: Breakthrough MEMS devices grow slowly until products they enable are developed.

Product Architecture and Level of Integration

ADI established the air bag sensor market with products that integrate support electronics and MEMS on one chip. The ADXL50 was assembled in metal headers but subsequent products were in ceramic cavity packages. Several years later, Bosch and Motorola (now Freescale) introduced competitive products. Their product architecture and packaging was quite different. Both suppliers produced the MEMS sensor and the support electronics on different wafers. MEMS die were capped with silicon, sealed with screen printed glass frit, assembled with a circuit die and molded in plastic. The Motorola designs sensed acceleration in the vertical axis while ADI and Bosch offered in-plane sensing products. Why these differences?

ADI utilized its strengths in fab process development and leading-edge signal processing when it integrated support electronics and MEMS on one chip. The output from MEMS inertial sensors is affected by mechanical stress. By using cavity packages, they avoided these stresses. Bosch and Motorola also capitalized on their respective strengths. Bosch had the world's largest hybrid manufacturing plant, so they were well-equipped to develop products sealed with screen printed glass frit. Motorola was in production of glass-mounted pressure sensors that were packaged in plastic. Thus, all three companies built on the expertise they already had in order to minimize risk, cost and time-to-market.

The Lesson: Do what you do best.

The Price-Cost Spiral

As noted above, many companies now offer multi-axis motion sensors for games, cell phones and a wide range of consumer devices. Most of them make MEMS wafers in one plant, purchase CMOS wafers from a foundry and mold them in one package. This product architecture raises some interesting possibilities. Competitive pricing is absolutely critical in high volume products, so fab costs must be low. Many MEMS suppliers are shifting from 150 mm to 200 mm wafers in order to reduce unit cost. However, there is another path: feature size. A quick analysis shows that the glass seal controls the die size of multi-axis motion sensors.

Figure 5 schematically shows a MEMS element hermetically sealed with glass under a silicon cap. Depending on the supplier, seal width can vary from 150 microns to over 300 microns. Reducing seal width to 30 microns or less doubles the number of die per wafer. This reduction in unit die cost can be a huge competitive advantage. Thirty microns is too narrow for glass frit and the seal must be hermetic. This suggests that metal seals might be the best answer.

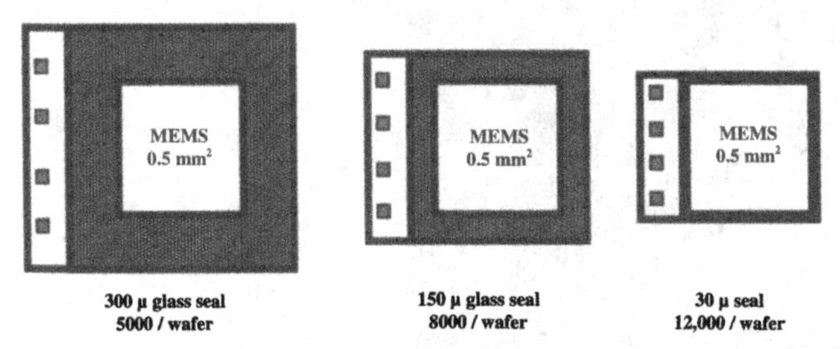

Figure 5: Die shrink from reduction in width of cap seal. Die count assumes 150 mm wafer.

But what metal? Gold is one possibility. Many groups are pursuing 3-D wafer-wafer interconnects based on copper so that is also a possibility.

Screen printed solder paste can't produce seal rings that are only a few 10's of microns wide so eutectic and thermocompression wafer bonding based on plating or sputter deposition are the obvious options. These bond processes require that metal be deposited on both wafers. This is a critical issue because it means that a patterned metal process has to be integrated with the MEMS process.

Gold and copper are deep trap contaminants so they are rigorously excluded from Front End wafer fabs. Back End wafer bonding with copper is feasible for standard IC wafers. However, MEMS release is a Front End process, so it would have to be integrated with deposition and etching of copper seal rings. This is not impossible, but it does require an unusual facility commitment.

Aluminum is not an obvious alternative because the oxide forms instantly. However, aluminum deposition and patterning are standard in IC fabs, so aluminum thermocompression bonding eliminates the process integration barrier. Aluminum wafer-bonded seal rings as narrow as 3 microns are hermetic [9,10].

Thermocompression bonding eliminates high energy interfaces by using high force to bring surfaces into intimate contact. Success requires high-quality alignment, temperature and uniformly-distributed high force. Aluminum is a reactive metal, so once the surface films are dispersed, the underlying metals fuse together with no trace of the interface (Figure 6). Aluminum thermocompression bonds illustrate some interesting Materials Science. For example, stress-induced crystallites nucleate and grow along the edges of the bond interface.

What's the Lesson? Cost reduction efforts will cause glass-sealed MEMS die to be replaced by smaller metal-sealed die. The example used aluminum thermocompression bonding. However, as high die counts are implemented in 200 mm wafers, the high, uniform forces required by thermocompression bonding will challenge the capabilities of commercial wafer bonders. This practical fact gives eutectic bonding a long term advantage.

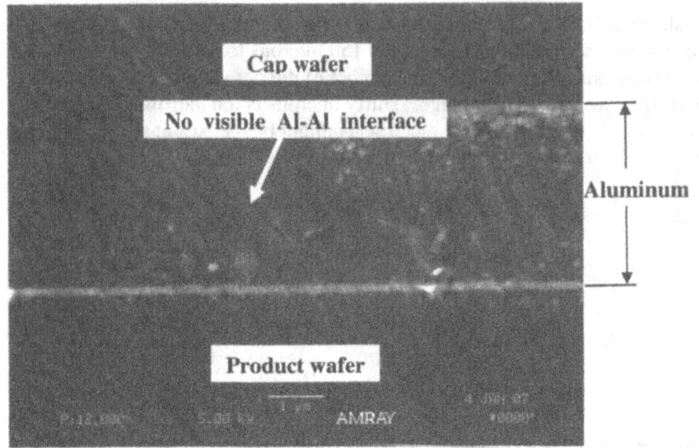

Figure 6: Cross-section of aluminum-aluminum thermocompression bond that had been PAN etched in an attempt to highlight the interface. (Courtesy of Analog Devices, Inc.)

The Tyranny of MEMS Surfaces

The high surface/volume ratio of MEMS devices causes both performance and reliability to be dominated by surface characteristics. A previous example described how an organic passivation suppressed high temperature electrical drift in the ADXL50. However, mechanical stiction is a more widely recognized surface effect. There is no one solution to stiction. Each company has its own design rules that specify pull-off force, surface roughness and a variety of other factors. Package technology also has a major effect.

Some suppliers use anti-stiction treatments. Most of these treatments cover high energy inorganic surfaces with a low energy organic film. ADI encountered stiction when it moved from the ADXL50 metal headers to ceramic cavity packages [11]. The solution was an organic treatment applied during the package seal process. A batch wafer process [12,13] replaced the package-level processes. It forms a one-nanometer thick, organic-rich surface (thicker dielectric films may support surface charges). Repeatability and uniformity are outstanding because the treatment is self-limiting and standard IC equipment is used. ADI's MEMS surface treatments were the first nanoscale processes used in high volume production.

The Lesson: Never rely on development of new equipment. The IC industry has invested countless billions of dollars to develop robust and reliable equipment. Use it. ADI's antistiction processes would have been more difficult to introduce, and more difficult to maintain, if they had required custom equipment.

CONCLUSIONS

The MEMS industry is a major supplier of disruptive technology components that enable higher-level products to deliver new functions at affordable prices. Success requires innovation, manufacturing expertise, market awareness and financial patience. With credibility now established, MEMS product growth will accelerate as new opportunities are recognized.

REFERENCES

1. J. B. Angell, S. C. Terry, and P. W. Barth, *Scientific American*, **248** (4), 44 (1983).
2. S. Majumder, J. Lampen, R. Morrison and J. Maciel, *Microwave Symp. Digest*, **3**, 1935-1938 (2003).
3. W. S. Lindenberger, T. L. Poteat, and J. E. West, U.S. Patent No. 4 524 247 (18 June 1985).
4. J. C. Baumhauer, H. J. Hershey, and T. L. Poteat, U. S. Patent No. 4 533 795 (6 August 1985).
5. I. J. Busch-Vishniac, W. S. Lindenberger, W. T. Lynch, and T. O. Poteat, U. S. Patent No. 4 558 184 (10 December 1985).
6. Anon, *Micronews*, **53**, (Yole Développement SARL, Lyon, December 2006) p. 9.
7. J. R. Martin, U.S. Patent No. 6 358 771 (19 March 2002).
8. M. Potin, *Micronews*, **68**, (Yole Développement SARL, Lyon, April 2008) pp. 6-7.
9. J. Martin, *Proc. SPIE, **6463**, Reliability, Packaging, Testing and Characterization of MEMS/MOEMS VI*, edited by A. Hartzell and R. Ramesham (SPIE, Bellingham, 2007) p. 64630M.
10. C. H. Yun, J. R. Martin, L. Chen and T. J. Frey, *ECS Transactions*, **16** (8), 117-124 (2008).
11. J. R. Martin in *Nanotribology: Critical Assessment and Research Needs*, edited by S. M. Hsu and Z. C. Ying (Kluwer, Dordrecht, 2002) ch. 14.
12. J. R. Martin in *Handbook of Nanotechnology (2nd ed.)*, edited by B. Bhushan (Springer-Verlag, Berlin, 2007) ch. 55.
13. J. R. Martin, U.S. Patent No. 7 364 942 (29 April 2008).

Mater. Res. Soc. Symp. Proc. Vol. 1139 © 2009 Materials Research Society 1139-GG01-05

A novel gap narrowing process for creating high aspect ratio transduction gaps for MEM HF Resonators

S. Stoffels[1,2], G. Bryce[1], R. Van Hoof[1], B. Du Bois[1], R.P. Mertens[1,2], R.Puers[2], H.A.C. Tilmans[1] and A. Witvrouw[1]

[1]IMEC , Kapeldreef 75, 3001 Leuven, Belgium
[2]KULeuven, Kasteelpark Arenberg 10, 3001 Leuven

ABSTRACT

In this work a novel technique to create nanometer sized air gaps for high frequency (HF) mechanical resonators will be presented. The technique is based on the narrowing of initially wide gaps with a conformal 'narrowing' layer. The novelty of this technique is that it enables the creation of narrow high-aspect ratio gaps (e.g. 100nm gaps in 10μm thick layers) without the need for complex lithography or high aspect ratio etching. Furthermore, the electrodes and the resonator itself can be patterned in a single processing step. The process methodology will be explained and validation experiments in a silicon-germanium (SiGe) based technology will be shown. This technology uses low temperature (~450°C) poly silicon-germanium (SiGe) as the structural layer, which can be processed above CMOS, and therefore allows the fabrication of MEM devices above CMOS.

INTRODUCTION

In the field of wireless communication, micro electro-mechanical (MEM) HF resonators are promising replacement components for use in filters and reference oscillators due to their tiny size, high quality factors (5000-100,000) and achievable frequencies ranging from tens of MHz till several GHz [1]. Electrical signals can be filtered with these mechanical resonators, however the signals need to be transduced between the electrical and mechanical domain. This transduction can be achieved by e.g. electrostatic, electrodynamical, piezoelectric, etc... means [2]. Electrostatic transduction with air gaps is the prevalent method used to drive MEM resonators due to the simplicity of the process. However, for these electrostatically transduced air gap MEM resonators, a small air gap in the range of 50-100nm is needed to achieve an efficient conversion between electrical and mechanical energy. In this work we will present a novel process to create nanometer sized air gaps for high frequency (HF) mechanical resonators and evaluate the process in a poly-SiGe based technology.

In prior work, narrow gaps were created by the use of a spacer layer in between the separately deposited resonator and electrode layers [3]. In other research, the narrow gaps were directly defined using expensive advanced lithography in combination with high aspect ratio etching [4]. Our technology differs from these methods as the electrode and resonator are patterned in a single processing step with relatively wide air gaps. The desired width of the gaps is achieved afterwards by narrowing the wide air gaps in a controllable manner. The novelty of this technique is that it enables the creation of narrow high-aspect ratio air gaps (e.g. 100nm gaps in 10μm thick layers) without the need for advanced lithography combined with high aspect ratio etching as reported in literature [3,4]. Moreover poly-SiGe, which was chosen as the structural layer, can be deposited at low temperature (≤450°C). This temperature is compatible with CMOS processing [5,6] and therefore allows the fabrication of MEM resonators above CMOS.

In the next section the process concepts will be explained. The following section will show the implementation of the process in a poly-SiGe based technology and the experiments performed to evaluate the feasibility. Finally, conclusions and recommendations will be presented in the last section.

PROCESS CONCEPTS

The trench narrowing process starts from relatively wide (~ 0.5μm) transduction gaps between the resonator and electrode structures. These structures are formed in the structural layer by normal i-line or DUV lithography and etch. A good control of litho and etch uniformity is needed to have an initial gap, which is uniform over the entire wafer. The actual gap-narrowing is achieved by depositing a second conformal layer. Important process parameters for the conformal layer are roughness, conformality and uniformity. The layer roughness will determine the minimal achievable gap, conformality is needed to have a uniform gap over the height of the sidewall, and a uniform layer is needed to reach a good wafer level uniformity.

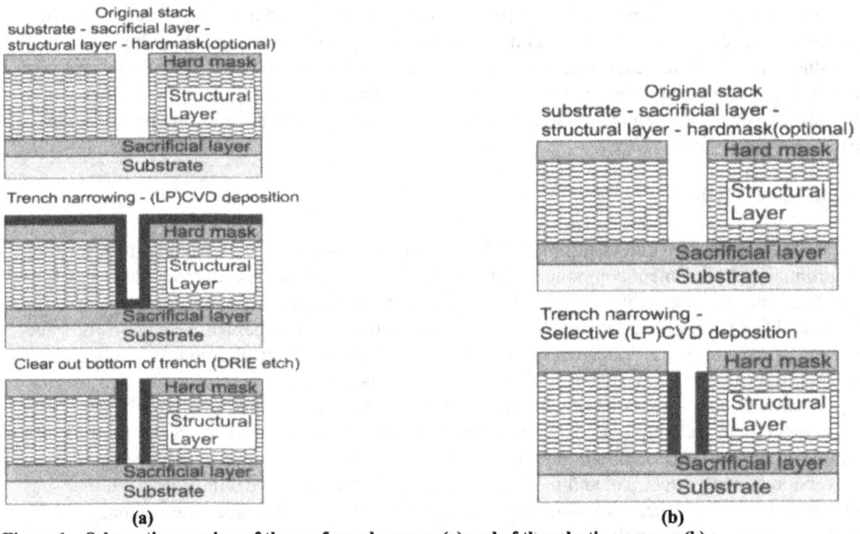

Figure 1 – Schematic overview of the conformal process (a) and of the selective process (b).

Two different processes are possible in case of a SiGe MEMS technology. The first process uses a boron doped (B-doped) CVD SiGe layer, which deposits conformally over the entire wafer (Figure 1a). This layer is thus also deposited on the bottom of the gap, where it should be removed to avoid short-circuiting the electrode and resonator. The removal can be done using a DRIE etch with sidewall passivation, which needs to be specially tuned to leave the sidewalls intact.

A second process uses a CVD SiGe layer (Figure 1b), which has an incubation time [7] for growing on oxide but not on the structural layer (SiGe). This layer will preferentially deposit on the gap sidewalls (structural layer) and not on the bottom (sacrificial oxide) or top (oxide hard mask). The need for performing an etch after deposition is therefore relaxed or totally eliminated. In the case of an undoped selective layer, we might need to add an extra, preferentially CMOS-compatible, annealing step to diffuse boron atoms from the structural layer into the CVD deposited SiGe layer and to activate them (e.g. 30' at 450°C for 89% of Ge [8]).

EXPERIMENTS

These process concepts were experimentally verified using a SiGe based MEMS technology. Using this technology, the structures were patterned in a SiGe structural layer, while silicon oxide (SiO_2) was used as sacrifical layer. Two sets of experiments were performed. We first validated the conformal process (Figure 1a), by performing a first set of gap narrowing experiments on etched SiGe/SiO_2 structures. A second set of experiments was performed to evaluate the feasibility of the selective process (Figure 1b). These experiments were performed on blanket wafers of SiGe and SiO_2. A SiGe deposition recipe was developed on these wafers with the aim of achieving a good selectivity between SiGe and SiO_2. The thickness of the deposited layer was monitored on the two different substrates and compared to determine the selectivity for the different deposition recipes.

Conformal process

The conformal process was verified on unreleased MEMS structures, processed in a SiGe technology. The MEMS structures were fabricated on top of a silicon (Si) substrate by first depositing a 1 µm SiO_2 (sacrificial layer), followed by a PECVD deposition of 4µm SiGe (structural layer) and finally a deposition of a 1.4µm SiO_2 hard mask. The hard mask and structural layer were sequentially patterned using a reactive ion etch (RIE). This RIE structural etch, reduced the thickness of the oxide hard mask to 1µm.

The conformal deposition was evaluated on the fingers of an interdigitated comb drive. This comb drive consists of a set of blades, which are 0.5µm wide and spaced 0.5µm apart. Gap narrowing experiments were done at a heater temperature of 480 °C (± 465°C wafer temperature) and 4Torr, 420sec deposition, using SiH_4, GeH_4, and optionally B_2H_6 as precursors. Figure 2 shows 4µm deep SiGe trenches after partial refill with a B-doped CVD SiGe layer having ~75% Ge. It is clear that a good sidewall coverage was obtained and that the SiGe layer was also deposited on the bottom and top SiO_2 layers. The thickness of the CVD SiGe-layer deposited on the top, bottom and sidewall surface, was approximately 170nm, 130-150nm and 150-160nm, respectively. The material on the bottom of the trench needed to be etched away to avoid shorts, while leaving the sidewalls intact. A Bosch process [9] (SF_6 etch gas, C_4F_8 passivation gas) was used to perform this etch. The sidewalls needed to be protected during the SF_6 etch, therefore the Bosch process was started with a passivation step. This passivation step ensures that a thin polymer layer protects the sidewalls during the etch step. For the case of 6 seconds passivation and 7 seconds etch almost no material was removed from the sidewalls after a single cycle, while ~60nm of SiGe was removed from the top and bottom layer. Repeating this DRIE etch cycle 4 times effectively removed the short-circuiting bottom layer with minimal impact on the CVD SiGe thickness, which was still 140-160nm after the etch. Figure 4 shows a 4μm thick SiGe disk resonator, with transduction gaps narrowed from 600nm to 270nm using the conformal narrowing technique.

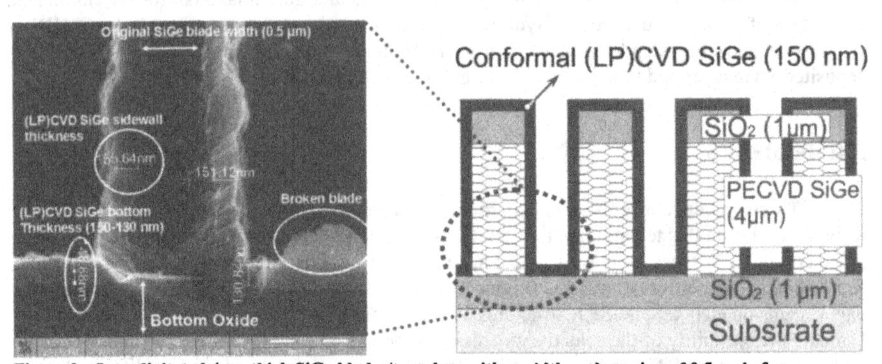

Figure 2 – Interdigitated 4μm thick SiGe blades/trenches, with a width and spacing of 0.5μm before narrowing, conformally deposited with boron doped SiGe, 150nm thick. The SiGe blades are standing on a SiO_2 bottom layer. The second blade broke off while dicing. (Left SEM picture; right schematical overview of the cross-section)

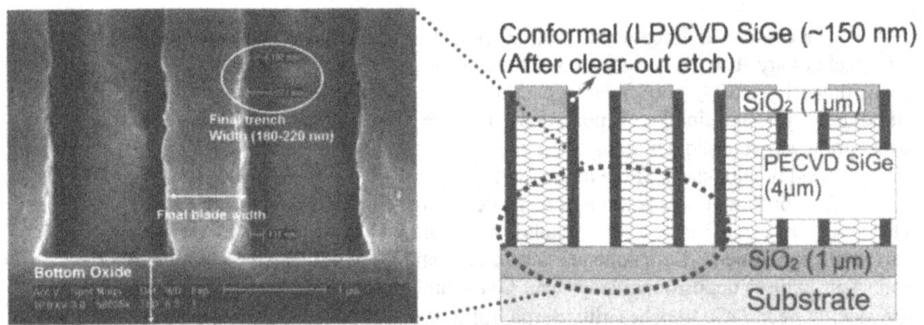

Figure 3 - Trench bottom after four times applying a 6s passivation / 7s etch cycle for a Bosch type DRIE etch. The SiGe layer has clearly been removed from the bottom of the gap. The final gap width is equal to 180-220nm, meaning that the CVD deposited SiGe layer was still 140-160nm. (Left SEM picture; right schematical overview of the cross-section)

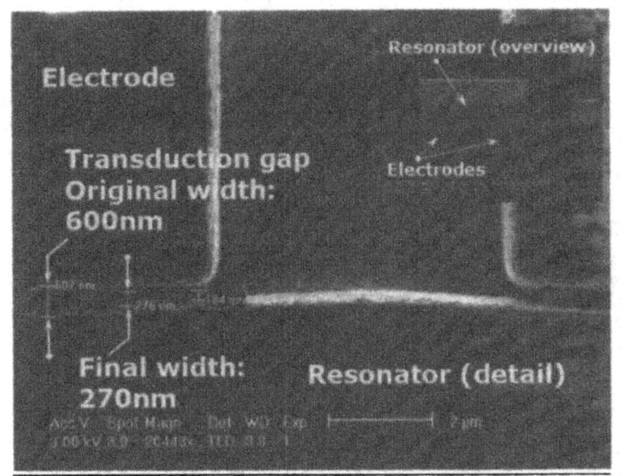

Figure 4 - 4µm thick SiGe disk resonator from which the transduction gaps have been narrowed from 600nm to 180-270nm with the process of Figure 1a.

Selective process

Experiments for the selective process, shown in Figure 1b, were performed using a silicon substrate with either a blanket 1µm PECVD SiGe or a 50nm thermal SiO_2 layer deposited on top. The different blanket layers were chosen to simulate the different possible substrates for the CVD deposition. A CVD SiGe layer was then deposited on top using different recipes to study the growth behavior on the different materials. This SiGe layer was grown at 480°C heater temperature (± 465°C wafer temperature) and 4Torr, using SiH_4, GeH_4, and B_2H_6 as precursors. The hydrogen (H_2) flow was varied and the effect on the incubation of the CVD SiGe layer was studied, both on thermal oxide and PECVD SiGe. Datapoints were taken for different deposition times and H_2 flows (Figure 5a). The reference process had a H_2 flow of around 200sccm, while the maximum had an added H_2 flow of 1000sccm. The deposition time was varied from 150 seconds to 1200 seconds. From the results it can clearly be seen that the incubation time is increased when the H_2 flow is increased; i.e. the formation of the seed layer is inhibited by adding more H_2. The incubation time for CVD SiGe growth on thermal oxide was for all cases longer compared to the incubation time on PECVD SiGe. It is this fact that allows for a selective deposition of the CVD SiGe layer on SiGe with respect to SiO_2. After the seed layer is formed the deposition rates on thermal oxide and PECVD SiGe were the same, as the curves for growth on both substrates are parallel for each different H_2 flow. A second important observation is that, when adding more H_2, the incubation time increases more on the thermal oxide substrate than on the PECVD SiGe substrate. The substrate selectivity thus increases for increasing H_2 flow, reaching a maximum at 800sccm H_2 flow. At this point the difference in deposition thickness of the CVD SiGe on both substrates is 172nm . This means that with the selective recipe we can reach a minimal gap dimension of 150nm when starting from an initial gap of 0.5µm, before forming an incubation layer on the bottom oxide. However, a scanning electro microscopy inspection of the thermal oxide substrate during the incubation time showed a localized 3D growth of the deposited SiGe layer at incubation sites (Figure 5b). Thus, the selective SiGe process does not form a shorting layer on the bottom of the trench, but the localized growth at the incubation sites of the selective SiGe layer could form particles during the release step.

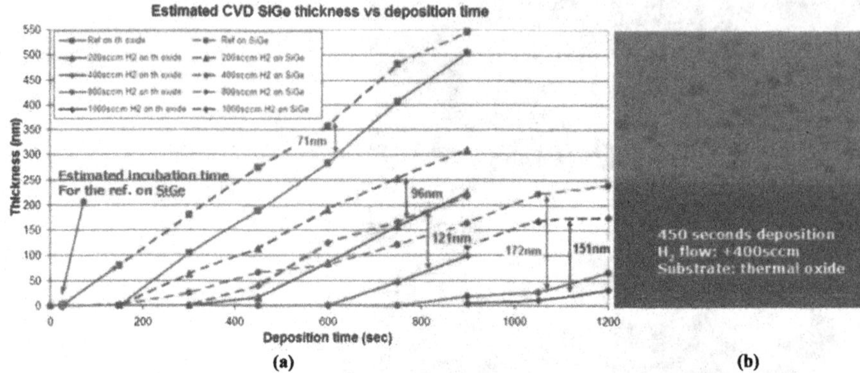

(a) (b)

Figure 5 – a.) Selectivity experiments for a CVD SiGe layer deposited on a thermal oxide and on a PECVD SiGe layer. The H_2 flow was the parameter varied for the experiments. Maximum selectivity is reached for a H_2 flow of +800sccm. For the reference process on PECVD SiGe we estimated the incubation time by taking the intercept of the growth curve with the time axis. b.) SEM picture of the deposited SiGe film on thermal oxide, during the incubation time. It is obvious that a 3d growth occurs at nucleation sites.

CONCLUSIONS

We have presented a novel process for creating narrow air gaps in MEM HF resonators. The process is based on the narrowing of initially wide gaps to the desired width. The process was successfully applied using a SiGe based technology for CMOS-integrated HF-MEM resonators. Results showed the possibility of trench narrowing with a boron doped CVD SiGe deposition, followed by a 'bottom clearing' etch step. Current results show an improvement of the transduction gap aspect ratio (AR) from 8:1 to 20:1. We also showed the feasibility of using a selective deposition with a boron doped CVD SiGe layer by increasing the H_2 flow during the deposition. The selective recipe was shown to have a maximum thickness difference of 172nm between thermal oxide and PECVD SiGe substrates. This selectivity was reached for a H_2 flow of 800sccm. Future work includes using a thicker conformal CVD layer and further optimizing the DRIE etch, which could further improve the AR. The selective recipe will be deposited on a patterned structure to validate the results achieved on blanket wafers. Finally the process will be integrated in a full process flow for HF-MEM resonators, yielding measurable devices.

REFERENCES

[1] C.T.C. Nguyen, *IEEE trans. Ultrason., Ferroelect., Freq. Contr.*, vol. 54, no. 2, pp. 251-270, Feb. 2007
[2] H.A.C. Tilmans, *J. of Micromechanics and Microengineering*, Vol.7, pp. 285-309, 1997
[3] J. R. Clark, W.-T. Hsu, and C. T.-C. Nguyen, *Technical Digest, IEEE IEDM*, Dec. 11-13, 2000, pp. 399-402.
[4] S. Pourkamali, F.Ayazi, *Technical Digest, MEMS 2004*, pp. 813-816
[5] A. E. Franke, et.al. , *IEEE/ASME Journal of MEMS*, **12**, 160-171 (2003).
[6] A. Schreurle, et. al., Proc. *IEEE MEMS 07*, pp. 39-42.
[7] H.C. Lin, et. al, *J. Electrochem. Soc.*, vol. 141 No. 9 (09/1994), pp. 2559-2563.
[8] S. Sedky, Proc. MRS, Vol 729 (2002), pp. 205-213
[9] F. Laermer, A. Schilp, K. Funk, M.Offenberg, *Technical Digest MEMS'99*, pp. 211-216, Florida, USA, 1999.

Mater. Res. Soc. Symp. Proc. Vol. 1139 © 2009 Materials Research Society 1139-GG01-09

Dicing of Fragile MEMS Structures

Peter Lange[1], Norman Marenco[1], Sven Gruenzig[1], Stephan Warnat[1], and Thilo Semperowitsch[2]
[1]Fraunhofer Institute for Silicon Technology, Itzehoe, Germany
[2] Accretech GmbH, Munich, Germany

ABSTRACT

Stealth dicing appears to be the appropriate method to singulate surface sensitive microelectromechanical devices, without producing dicing debris or using water jet. The operating mode of the stealth dicing will be explained. A three dimensional integration project with open micromachined structures is taken as a reference to evaluate the advantages and limitations of this new dicing method. The preconditions for successful singulations will be discussed. Optical and SEM inspection results after successful separation will be discussed.

INTRODUCTION

Conventional singulation of MEMS (Micro Electro Mechanical System) devices is carried out after protection of the fragile structure. Particles generated during dicing can block mechanical movements, mechanical load or vibration can deteriorate hyperfragile structures and, for example for bio-MEMS, cutting fluid can interact with sensitive layers. Protection is mostly done either by a capping process or by a resist, both on a wafer level. Membrane sensors for flow and pressure are usually protected by a temporarily layer, which is removed afterwards. However, this procedure can cause yield loss. Inertial sensors, micromirrors and rf-MEMS are permanently protected by a capping layer initially for functional reasons. Devices which have open structures on both sides, generally afford new techniques because sensitive structures cannot be enclosed completely since it could hinder functionality. Maho IR laser technology for (stealth) dicing is claiming to be the ultimate separation technology for open and fragile structures. Separation takes place by generating a polycrystalline area with a high dislocation density within the wafer along the dicing line and subsequent tape expansion thereby avoiding any particle generation. Typical examples are microphones as well as ink jet printheads, which are already fabricated in mass-production.

Recently a chip to wafer assembly has been proposed in the European project DAVID (Downscaled Assembly of Vertically Interconnected Devices), in which a sensor chip is bonded face down on top of an ASIC wafer (3-D integration of MEMS and ASIC) [1]. Therefore it is needed to singulate before the MEMS is placed on the ASIC structure. This MEMS is surface micromachined with features in the submicron range and thus very sensitive against particle contamination. For a characterization of the dicing process with respect to particle generation, chipping occurrence and environmental conditions, the complete process flow has to be taken in mind: the dicing and the separation of chips by expansion has to carried out in environments with at least a cleanroom class 100. The further transport to the pick and place tool and the die placement has to be done also under controlled conditions.

EXPERIMENT

In this section we will give a description of the test structure for the evaluation of the stealth dicing method. The mechanism of heat generation inside the silicon wafer while leaving the surface unattached will be explained in a phenomenological way. The subsequent method of separation and wafer conditions therefore will also be commented.

Description of the test structure

The test structure is given in Fig. 1 in a schematical view (left) and a close up of the processes structure (right). This MEMS is a surface micromachined resonator structure with minimum linewidth of 0.8 µm in a comb structure which features the ISIT thick epipoly-Si process with a height of 11 µm [2]. This device with an aspect ratio of 1:14 is extraordinarily sensitive to any particle contamination and thus appropriate for evaluation purpose. In the scope of the DAVID project we separate the MEMS dies before they could be bonded on an ASIC wafer. We assumed that only stealth dicing showed the potential to solve this ambitious task.The stealth dicing method, however, is a relative new separation method and mostly not well understood. We would like to explain the mechanism of separation in view of the published literature and give a phenomenological view of the optical processes which generate the heat for the conversion of the crystalline to a highly distorted polycrystalline phase of the material, which acts as a precursor for the separation.

Figure 1: Schematical view of the test structure in surface micromachining technology (left side). On the right side a close up of the real structure taken with an SEM is shown.

Stealth dicing

Stealth dicing is a singulation method, which is separating dies from a wafer without generating particles, damage, or heat in the device area. Local laser absorption is induced in the vicinity of a focal point inside the wafer bulk making use of the temperature dependence of absorption coefficient. Only the neighborhood of a focal point is heated and reaches higher temperature than the melting point. By rapid cooling cracks are generated which act as precursor for subsequent singulation by expansion of the wafer. Many articles have been published up to now on the phenomenon of stealth dicing [3,4,5], mostly concerned with the explanation of the crack development as a initial state for singulation. We would like to add some remarks concerning the interaction of light (photons) with matter dealing with the question how the melting process is started. Absorption of laser light within the silicon is the matter of question. Fig. 2 shows the transmission for silicon with 400 μm thickness and 1 Ωcm resistivity (transmission is the inverse of absorption, neglecting reflectance). Laser ablation processes mostly are done with wavelength below 1 μm. The cross-section for absorption is high in this area, respectively the transmission goes down to zero. In the range above 1 μm the transmission increases strongly. For the stealth dicing a short pulsed IR laser with a wavelength of 1.064 μm (E=1.16 eV) is applied, In the transition range the absorption is low but not zero since its located in the fundamental edge of the Si semiconductor. This energy is slightly above the bandgap (E_{gap} = 1.14 eV) of Si at room temperature.

The absorption of photon energy and its conversion to lattice heat is dominated by the the creation of electron – hole pairs:

- direct optical absorption of photons with energies higher than the bandgap energy Egap (T). For silicon this is an indirect transition from valence- to conduction band (under the assistance of phonons). The population of the conduction band with electrons makes free carrier absorption possible.

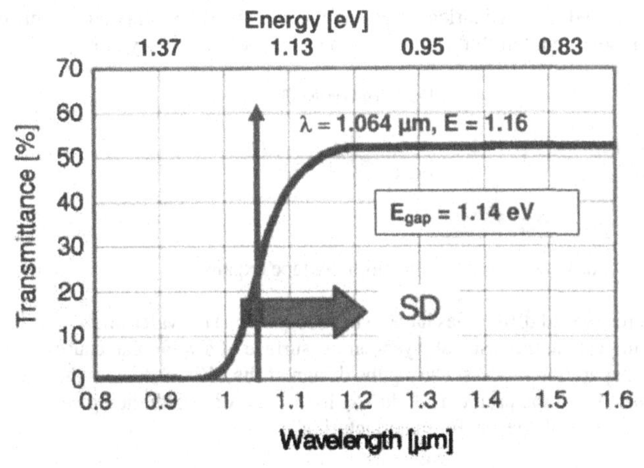

Figure 2: Spectral transmittance of a wafer with 400 μm thickness and 1 Ωcm resistivity (after [3]). In addition the laser wavelength of 1.064 μm is indicated, also the bandgap energy E_{gap} and a scaling in eV.

- optical induced free carrier absorption is a transition from conduction band to higher states. For Si and ns laser pulses free carrier generation is more effective as recombination of carriers [6]. The heating of the lattice by thermalization of excited electrons through collisions with phonons and other carriers is dominant as compared to carrier recombination.

For the onset of heating up it is therefore necessary that the generation of electron-hole pairs is higher than recombination and thermalization effects. Therefore the right pulse length of the laser, the appropriate power and a focus point at the diffraction limit (smallest possible extension of laser beam) has to be chosen. Power and focus point are determining the photon density; the pulse length is important for the limited diffusion of the thermal wave, which extends to a few µm [7] for pulses in the ns range. Once the heat is generated locally, the absorption probability increases exponentially with temperature thus creating a nonlinear heating effect, which in result provides enough energy for local melting and evaporation processes. Once this highly distorted area is created, then after rapid cooling the distribution of compressive and tensile stress enables the development of vertical cracks as described by Monodane et al [5]. However, the surface remains unattached since the photon density aside the focus area is too low to create an excitation rate higher than the recombination rate. This is the extraordinary difference to laser ablation processes, in which material is removed from the wafer surface down the bottom, thus generating debris along the dicing line.

Separation of chips

After the laser treatment the unseparated wafer has to be fixed on a dicing tape. This tape is going to be expanded by a cylindrical stage and the dies should be singulated. This process was described in more detail in Ref. [2] and is shown here schematically in Fig.3.

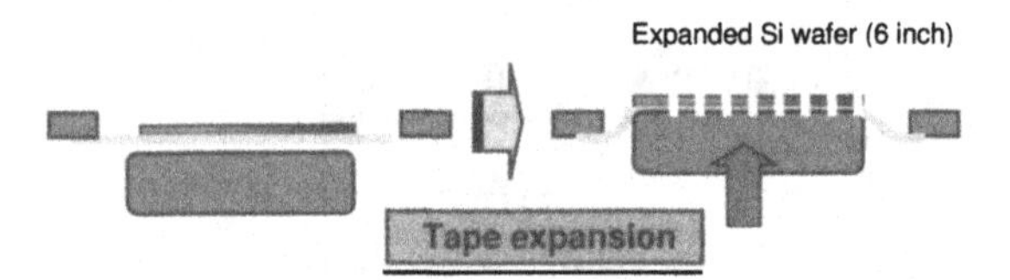

Figure 3: Schematical view of the die separation by tape expansion

However, in the real world difficulties have to be recognized. The wafer thickness, the doping level and, very important, the residual layers on the surface of a wafer can cause severe problems for the separation process. During the design of the MEMS devices the designer have not always been aware of the problems of dicing. In the case of stealth dicing we need to consider some elements of "design for manufacturing".

Wafer conditioning for stealth dicing

The DAVID project is a development project in which the process flow was improved from run to run. In first loops there appeared areas on the wafer, not used for active devices, which were covered with aluminum (see Fig. 4 a). A singulation by tape expansion was not possible. After a mask redesign and subsequent processing this error was corrected. Afterwards we recognized that the whole wafer surface was covered with the thick epi-poly Si with a thickness of 11 μm. As an outcome the crack propagation in mono-crystalline Silicon (the substrate) is interrupted when it approaches the poly-Si region. Strong deviations from the cutting line are the results, as shown in Fig. 4 b. The mask for the appliance of the DRIE process for structuring the thick epi-poly Si film had to be rewritten considering these experiences. Last but not least, we had some residuals from the bond frame for the chip to wafer bonding, a gold-tin alloy, which prevented separation. The residuals were only located in a small edge ring to a resist stripping process. This last problem was overcome by cutting a maximum square out of the wafer. This step and the epi-poly removal enabled a successful separation of the dies. From this experience we can give the recommendation, that in the dicing lines:

- no aluminum, in general no metals, are allowed. Strong absorption effects prevent a successful focusing of the laser beam inside the wafer.
- no active layers, in particular, thick layers for surface micromachining, are allowed, because linear crack propagation is hindered.

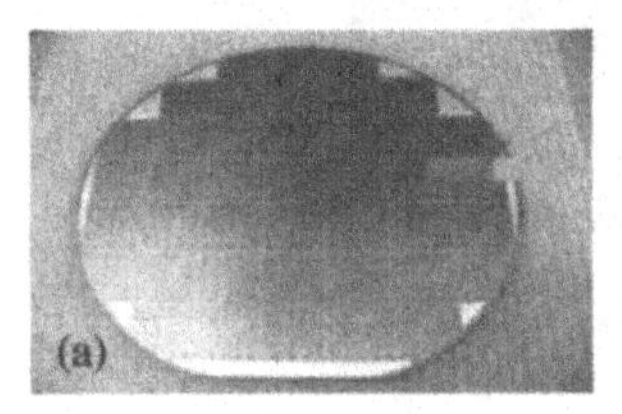

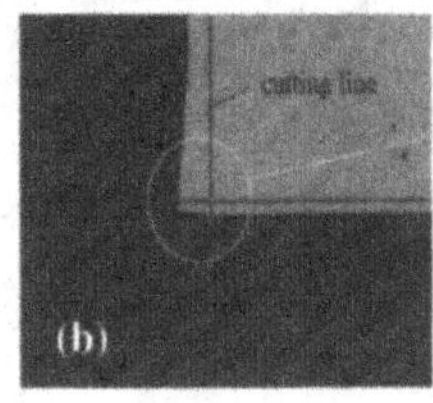

residuals

Figure 4: MEMS wafer with aluminum residuals (a) and with displaced cutting lines due to thick (11 μm) epi-poly Si on the surface (b).

RESULTS AND DISCUSSION

Considering all the aforementioned details we achieved the first results for the stealth dicing of open surface micromechanical structures. First we performed an optical inspection of the separated dies on the tape and we found no significance of particle/debris on the wafer front side. For the inspection in a scanning electron microscope (SEM) we used a commercial pick and place tool, which lifts the dies with a tiny vacuum nozzle and transferred the dies under controlled atmosphere. For these inspection we performed a statistical probing from different places on the tape. All of them have been found to be in an excellent condition. Two SEM photographs are shown in Figure 5 as representatives for the different investigated samples. There was no chipping, peeling and the cutting quality appeared perfect, although some thin layers remained on the backside of the wafer. Due to the process flow these are a sacrificial layer (silicon oxide) and a buried poly (Silicon), all together in a range of about 2 μm thickness.

CONCLUSIONS

The demonstrator, the DAVID device, is fabricated in an R&D project and reveals some technical deficiencies: design for manufacturing has not been considered. Nevertheless the results from stealth dicing of these devices look very promising. The results from optical and SEM control show no particles and chipping effects. Beyond that, stealth dicing appears to be the ultimate solution for separation of open MEMS dies. In addition a short explanation of the heating process during laser processing has been given on the basis of the interaction of light with matter.

Figure 5: SEM micrograph of a separated MEMS die after successful stealth dicing (left), a close up of this die is shown also (right).

ACKNOWLEDGMENTS

We would like to acknowledge very gratefully the support by Accretech, Tokyo which performed the Mahoh (stealth)-dicing. The presented work is part of the DAVID project [1], which is funded from beginning of 2006 until end 2008 by the European Commission within the Sixth Framework Programme (ref. IST-027240).

REFERENCES

[1] N. Marenco, H. Kostner, W. Reinert, and G. Hillmann, Smart System Integration, Barcelona (2008), Proceedings, p. 351

[2] P. Merz, W. Reinert, K. Reimer, B. Wagner, Konferenzband Mikrosystemtechnik-Kongress 2005, D/Freiburg, VDE Verlag, p.467-470

[3] M. Kumagai, N. Uchiyama, E. Ohmure, R. Sugiura, K. Atsumi, and K. Fukumitsu, IEEE Transaction on Semiconductor Manufactoring, **20 (3)**, 2007, p.259

[4] E. Ohmura, F. Fukuyo, K. Fukumitsu, and H. Morita, JAMME, **Vol 17 (1-2)**, 2006, 381

[5] T. Monodane, E. Ohmura, F. Fukuyo, K. Fukumitsu, H. Morita, and Y. Hirata, JLMN-Journal of Laser Micro/Nanoengineering, **Vol. 1 (3)** 2006, 231

[6] M. F. von Allmen, Coupling of laser radiation to metals and semiconductors, in M. Bertolotti (Ed), Physical Processes in Laser-Material Interactions, 1983 Plenum Press, New York, page 49

[7] Private communication with H. Bernt

Microdevices and Micro/Nanofluidics

Mater. Res. Soc. Symp. Proc. Vol. 1139 © 2009 Materials Research Society 1139-GG02-01

BioMEMS Technologies for Regenerative Medicine

Jeffrey T. Borenstein

Biomedical Engineering Center, Charles Stark Draper Laboratory, Cambridge MA 02139 U.S.A.

ABSTRACT

The emergence of BioMEMS fabrication technologies such as soft lithography, micromolding and assembly of 3D structures, and biodegradable microfluidics, are already making significant contributions to the field of regenerative medicine. Over the past decade, BioMEMS have evolved from early silicon laboratory devices to polymer-based structures and even biodegradable constructs suitable for a range of *ex vivo* and *in vivo* applications. These systems are still in the early stages of development, but the long-term potential of the technology promises to enable breakthroughs in health care challenges ranging from the systemic toxicity of drugs to the organ shortage. *Ex vivo* systems for organ assist applications are emerging for the liver, kidney and lung, and the precision and scalability of BioMEMS fabrication techniques offer the promise of dramatic improvements in device performance and patient outcomes.

Ultimately, the greatest benefit from BioMEMS technologies will be realized in applications for implantable devices and systems. Principal advantages include the extreme levels of achievable miniaturization, integration of multiple functions such as delivery, sensing and closed loop control, and the ability of precision microscale and nanoscale features to reproduce the cellular microenvironment to sustain long-term functionality of engineered tissues. Drug delivery systems based on BioMEMS technologies are enabling local, programmable control over drug concentrations and pharmacokinetics for a broad spectrum of conditions and target organs. BioMEMS fabrication methods are also being applied to the development of engineered tissues for applications such as wound healing, microvascular networks and bioartificial organs. Here we review recent progress in BioMEMS-based drug delivery systems, engineered tissue constructs and organ assist devices for a range of *ex vivo* and *in vivo* applications in regenerative medicine.

INTRODUCTION

The field of regenerative medicine has emerged in response to the urgent and increasing need for new therapies for diseases and injuries resulting in tissue and organ loss and failure. According to the United Network for Organ Sharing (UNOS) [1], there are more than 100,000 Americans on waiting lists for a vital organ transplant, and the number grows rapidly each year. Current therapeutic avenues center on cadaveric transplanted organs; in some cases donors with a genetic match may provide one of their two kidneys, or part of their liver in a split liver donor transplant. The use of animal organs, or xenotransplantation, is under active investigation, but numerous challenges remain. In all of these instances, the recipient's immune system represents a significant challenge in terms of cost, quality of life and safety. Ultimately the goal of the field of regenerative medicine is to provide an unlimited source of laboratory-grown tissues and organs based on a patient's own cells, and ready on demand at the time when it is required.

Regenerative medicine is often considered as the use of cell-based therapies to treat diseases by regenerating or restoring lost structure and function in tissues and organs. In this overview the definition is broadened to include innovative technologies for the repair or restoration of tissue and organ function; the specific role of cells depends upon whether the technology incorporates cells into the regenerative therapy directly, or recruits or treats cells in the body through the introduction of specific materials or compounds into the target tissue or organ. By this definition, regenerative medicine extends beyond the development of cell-based therapies to include drug delivery systems that address sites of disease, organ assist devices that augment or restore function, as well as tools for regenerative medicine such as in vitro models to test the safety and efficacy of drugs and biologics for restoring function. In the following, four broad areas of regenerative medicine are covered: implantable drug delivery systems, in vitro models for developing new therapeutic compounds, tissue-engineered organ replacement systems, and organ assist devices.

BIOMEMS FOR DRUG DELIVERY

Over the past few decades remarkable advances in the development of new therapeutic compounds for treating diseases have been made. These advances provide a tremendous opportunity for improving patient outcomes but also present new challenges for safe and efficacious delivery systems. Conventional drug delivery methods utilize ointments, oral or injectable medications that are introduced systemically and travel to the site of disease [2]. Three of the principal challenges of these conventional methods include the ability of drugs to maintain their bioactivity when delivered systemically, the occurrence of adverse reactions resulting from interactions with other tissues and organs, and patient compliance. The emergence of polymer-based systems based on controlled release, novel carriers such as nanoparticles and liposomes, and wearable patches for extended administration have dramatically expanded the field of drug delivery and have addressed many of these challenges. The emergence of BioMEMS technology has spurred advances in drug delivery systems ranging from the use of microneedles to avoid pain and enable skin-patch-based administration of compounds to implantable microsystems based on microelectronic or microfluidic release mechanisms. BioMEMS fabrication techniques can be used to construct polymeric particulates with precise control over the shape and size [3]; this technology enables targeted oral drug delivery for numerous applications in which bioavailability is a challenge. Here several advances in BioMEMS-based drug delivery will be reviewed.

Microneedle Technology

Silicon micromachining technology has been used to develop microneedle arrays that essentially replace injections with conventional large needles, avoiding pain and providing opportunities for wearable injection systems. Early work in this field focused on the development of silicon microneedle arrays using reactive ion etching techniques, but this work has now been extended into polymer and metal microneedle array fabrication using silicon structures as masters for replication techniques [4]. Microneedles are small enough that they do not stimulate pain receptors, and have other advantages over conventional injections such as administration by untrained personnel. These structures have been shown to be capable of delivering drugs transdermally in human subjects without pain, using molecules that would otherwise not be able to pass through the skin [5]. Current microneedles are as small as 5 microns in diameter at the tip, with a base diameter of roughly 250 microns and a height less than one millimeter. Biodegradable microneedles offer the potential for temporary application with no residue or adverse effects after application of a microneedle patch.

Microneedles with on-chip positive displacement MEMS-based pumps enable continuous drug delivery systems that are completely miniaturized and integrated using BioMEMS technologies. A team at UC Berkeley has reported a silicon micromachined drug delivery system capable of delivering quantities as small as 1 nanoliter/sec for thousands of cycles without evidence of wear or failure [6]. This system enables precise control over drug infusion rates for drug delivery systems, and can also be applied towards sample collection for bioanalysis.

Implantable Microfluidic Delivery Systems

Many diseases require repeated administration of drugs over long periods of time, ranging from months to years to a patient's lifetime, with precise control over dose for reasons of safety and efficacy. Further, many conditions represent particular challenges for delivery due to the inaccessibility of the target tissue or organ. For these reasons, fully implantable, self-controlled drug delivery systems are being developed to treat diseases ranging from diabetes and cancer to neurological disorders, and vision and hearing loss. Ideally, these fully implantable systems would be run in a closed-loop fashion, utilizing data from an indwelling sensor to identify changing conditions and calibrate the dosage accordingly. Even without on-board sensing, these systems represent an extraordinary challenge in terms of overall size and integration of multiple functions, including power, electronic control, communications, drug storage, and the drug release mechanism itself.

One emerging approach in this arena is a class of microfluidic drug delivery systems, in which a liquid drug formulation is introduced at the site of disease by a micropump according to a previously specified delivery profile. Chung *et al.* [7] have developed an electrokinetic pump for low-power drug delivery, utilizing the electrokinetic effect to eject drugs from an array of microwells. Responsive hydrogels have also been employed as microfluidic drug delivery systems [8]; these hydrogels can be designed to respond to changes in the environment such as glucose levels, pH or other stimuli to undergo volume expansion. A disposable MEMS-based micropump known as the Insulin Nanopump™ [9] has been developed for diabetes treatment; a permanent electronics module is integrated with a disposable reservoir and pump containing a week's supply of insulin.

Microfluidic drug delivery devices for regeneration of lost function represent another avenue for implantable systems, and one of the principal opportunities in this arena is the treatment of sensorineural hearing loss. The condition affects 28 million Americans and is the most prevalent birth defect, occurring in 3 of every 1000 births in the United States. Recent advances in molecular biology have identified new targets and compounds potentially capable of restoring function in the hair cells lining the cochlear tubes, but these therapies cannot reach their intended targets through systemic delivery or using currently available delivery approaches.

New research is exploring the use of a microfluidic delivery system that is partially implantable but will ultimately be fully implanted within the mastoid cavity, directly introducing drug to the cochlea through a cannula 100 microns in diameter [10-11]. A next generation system currently in development utilizes microfabricated components comprising layers of laser- or micromachined polymers functioning as fluidic capacitors, resistors, valves and microactuators. Figure 1 laser-machined polyimide vias that form layers in the microfabricated valves and actuators. Initial in vivo results have been obtained using a guinea pig model, demonstrating a window of delivery parameters resulting in safe administration of test compounds with effects on hearing extending beyond 1 cm depth into the scala tympani using a cochleostomy to gain access to the cochlear fluid. A flow sensor has been integrated into the system to provide real-time feedback and potentially closed-loop control.

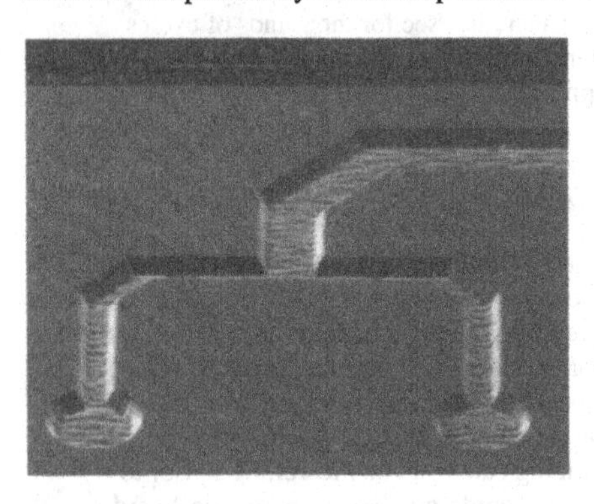

Figure 1. SEM image of laser-machined vias in polyimide films for use in layered microvalve and micropump structures for intracochlear drug delivery system.

Microelectronic Controlled Delivery

Microelectronically controlled reservoirs form the basis of a drug delivery system whose inception was inspired by fabrication technologies used in the semiconductor industry [12]. Individual reservoirs arranged in a planar two-dimensional array are addressed with a pre-programmed microprocessor and can be integrated with wireless telemetry and closed-loop control with on-board sensors. One of the principal challenges involves the actual mechanism used to burst open each individual well; drugs stored within the well must be hermetically sealed and protected from the in vivo environment until they are released. Numerous techniques for opening wells are possible; the goal is to minimize power consumption and maximize the reliability of the well-opening technique for specific implantable drug delivery and biosensing applications.

ENGINEERED TISSUE CONSTRUCTS: IN VITRO MODELS

The field of tissue engineering and regenerative medicine has always been motivated by the long-term goal of developing on-demand replacement tissues and organs in the laboratory. While great progress has been made towards this goal for specific tissue types such as skin, bone and cartilage, full replacement organs grown in the laboratory remain an extremely challenging goal and therefore many nearer-term applications of engineered tissue constructs have emerged. Microfabricated or BioMEMS-based engineered tissue constructs for in vitro models possess many advantages over traditional cell culture models, primarily relating to the ability of BioMEMS technologies to more accurately replicate the microenvironment within tissues and organs. Microfabrication techniques provide tools for controlling the microenvironment of cells and tissues through the use of chemical gradients [13], topographic features at the microscale and nanoscale [14-15], spatial control for co-culture [16-17], as well as replication of the fluid mechanical forces [18]. Applications of these technologies towards in vitro models include the establishment of platforms for evaluating and predicting drug toxicity using human cells, and drug development and discovery platforms.

Hepatotoxicity Models

Current *in vitro* models for hepatotoxicity rest upon approaches such as organotypic liver slices and conventional cell culture techniques, but a large number of microfabricated systems are emerging. One of the principal challenges in this area has been the difficulty in sustaining the phenotype of primary hepatocytes in an artificial microenvironment. Pioneering work by Bhatia *et al.* [19] has demonstrated that micropatterned co-cultures of hepatocytes with other cell types in specified configurations can lead to maintenance of hepatocyte phenotype in culture for extended periods. Recently Khetani et al have demonstrated a platform for hepatotoxicity testing using primary hepatocytes co-cultured in well plates using micropatterned collagen spots [16]. Micromechanical devices in which silicon micromachined combs derived from inertial MEMS structures have been cultured with hepatocytes and fibroblasts and used to control cell-cell interactions have also been reported [20]. Microstructured perfusion devices [21] and microfluidic bioreactors have also been developed as platforms for hepatotoxicity testing. Microfabrication techniques have been used to construct devices capable of perfusing liver cells with oxygenated media without subjecting the hepatocytes to direct hydrodynamic forces, which can cause severe damage to the cells [22]. Microfluidic devices housing rat hepatocytes in a chamber separated from a microvascular network with a membrane bilayer have been used to demonstrate protein synthesis and metabolic function, showing greatly increased function vs. conventional static culture over extended periods of time [17].

Systems are now being developed that integrate multiple organoid structures in a single device. Human cell culture-based systems may not accurately mimic in vivo toxicity without replicating the interactions between different organs and tissues, and therefore chips containing microscale culture systems replicating different organs are being developed [23].

ENGINEERED TISSUE CONSTRUCTS: TISSUE AND ORGAN REPLACEMENT

Microfabricated tissue constructs can provide in vitro models for discovery, screening and research applications, as described earlier, but can also form the basis of direct therapies for tissue and organ loss or failure. Tissue engineering approaches for therapy are often categorized as either cell-based therapy without scaffolds, cell-free scaffolds, or the use of scaffolds that are pre-populated with cells [24]. In either of the latter two cases, microfabrication technology is a powerful tool for replicating the microarchitecture of tissues and organs, emulating the biomechanical properties of scaffolds and resulting tissues through structural modifications, and controlling the transport of soluble signaling factors and therefore the interactions between the various cell types populating the tissue construct.

Microvascular Networks

Progress in the field of tissue engineering has been gated by the lack of an intrinsic microvasculature in growing tissues, thereby depriving thick and complex tissues of oxygen and nutrients. Several approaches have been taken towards the development of an engineered microvascular network as the backbone of tissue constructs, typically using microfabrication techniques to fashion bifurcating microfluidic channels in a range of scaffolding materials, populating these networks with endothelial cells and demonstrating functional behavior. Early work utilized PDMS microfluidic networks modeled after physiological vasculature, populated with endothelial cells and cultured over periods of several weeks [25]. In Figure 2, a PDMS microfluidic network seeded with endothelial cells is shown after 4 weeks in culture. Numerous techniques have been used to generate BioMEMS-based master molds for artificial capillary network formation, often with the goal of modifying the geometry of the channel cross-section [26]. These early studies have been augmented by efforts to construct biodegradable scaffolds populated with endothelial cells for implantable tissue applications [27-28]. Endothelial cells have been shown to form confluent layers in bioresorbable materials such as Poly(lactic-co-Glycolic Acid) and PolyGlycerol Sebacate. More recently, calcium alginate hydrogels have been microfabricated into fluidic networks to mimic the transport properties of the microcirculation [29]; nutrients can travel either through the microchannels or traverse the bulk of the scaffold. Endothelial and perivascular cells have been introduced into microfluidic channel networks formed in a collagen gel, demonstrating physiological barrier function and cellular organization [30].

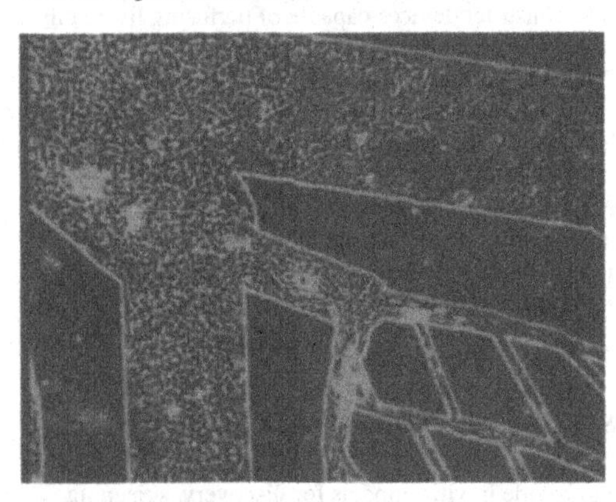

Figure 2. Image of endothelial cells cultured for 4 weeks in a PDMS microfluidic scaffold, where the smallest channels are roughly 20 microns in diameter.

Neural Tissue Engineering

Development of neural tissue and brain tissue represent a huge challenge but an enormous opportunity as researchers look for new approaches towards the treatment of neurodegenerative diseases and traumatic injuries affecting neural function. Critical elements of neural tissue engineering include appropriate choice of scaffolding materials, matrix materials that enable the formation of three-dimensional tissues, and environmental cues such as nanotopographic features on the substrate and the application of electric fields during cell culture. In a pioneering study, Dertinger *et al.* [13] demonstrated the use of a microfluidic mixer to generate chemical gradients needed for neuronal development. Neurons from the hippocampus were cultured using this mixer, leading to polarization of the neurons and elongation of one end, leading to axon formation. Chemical and mechanical cues such as ligand patterning, chemical gradients and microtopography can be investigated in a competitive fashion, to determine the relative strength of each effect [31]. One such study on hippocampal neurons indicated that contact guidance cues from the surface topography are predominant [32]. The use of nanotopography to explore contact guidance cues and their effects on engineered tissue constructs is a burgeoning field but is beyond the scope of this review.

Cardiac Tissue Engineering

Microfabricated scaffolds provide precise arrangements of multiple cell types for complex organs such as the liver and the kidney. Another aspect of microfabricated scaffolding that enables advances in regenerative medicine is the ability of the microfabrication technique to provide more realistic mechanical properties of the scaffolding. Nowhere is this more important than in cardiac tissue engineering, where the mechanical behavior of the replacement tissue is critical in replicating the properties of healthy cardiac tissue at multiple length scales [33]. Clinically, the fact that myocardium cannot regenerate after an injury has spurred intense efforts to reproduce cardiac tissue in vitro, and microfabrication technology has played an important role in this process [34]. Cell patterning arrays and microfluidic bioreactors have been used to generate cardiac organoids, forming the basis for eventual generation of cardiac replacement tissue in the laboratory.

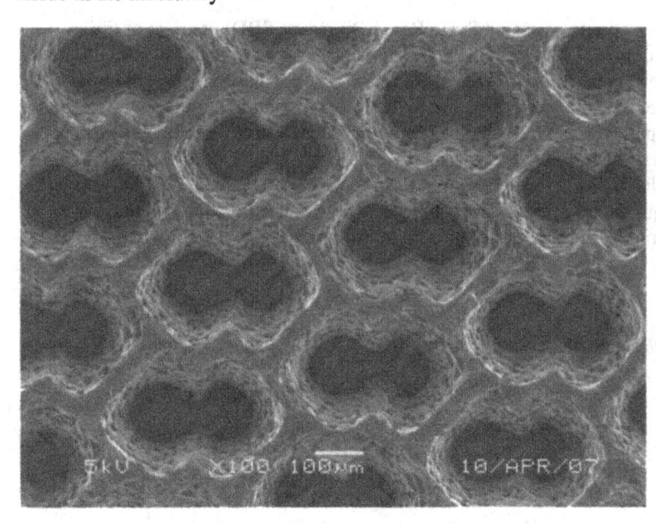

Figure 3. SEM image of PolyGlycerol Sebacate scaffold machined using a laser to form pores for cardiomyocyte seeding, with anisotropic biomechanical features that mimic cardiac tissue.

Recent work by Engelmayr *et al.* [35] has demonstrated the ability of microfabrication technology to produce cardiac patches with properties similar to native tissue. Using a laser to machine accordion-like geometries into a bioresorbable scaffold, in concert with tuning of the polymer synthesis and preparation protocol, precise mechanical properties of the scaffold could be obtained. Anisotropic electrophysiological properties of engineered tissue, cultured from neonatal rat heart cells, were obtained using these microfabricated honeycomb scaffolds. In Figure 3, a section of an anisotropic polyglycerol sebacate scaffold machined with a laser to form pores for seeding cardiomyocytes is shown.

BIOMEMS FOR ORGAN ASSIST DEVICES

Engineered tissue constructs for replacement of organ function remain the ultimate goal of tissue engineering and regenerative medicine. As a bridge to this long-term goal, many demonstrations of BioMEMS-based organ assist devices have emerged. As in the cases of in vitro models and organ replacement, the principal advantage of these systems is the ability to replicate the microenvironment of tissues and organs. Exciting advances in renal, hepatic and pulmonary assist devices have been demonstrated, principally through the use of microfluidics and BioMEMS fabrication techniques that are capable of being scaled up to provide the capacity required for human therapy. Early generations of these technologies may replace existing devices in intensive care units and in the clinic, but ultimately BioMEMS-based approaches will lead to home care and wearable systems that improve outcomes and greatly enhance the quality of life for patients.

BioMEMS-based Renal Assist Devices

Current practice for patients with End Stage Renal Disease (ESRD) consists of thrice-weekly visits to a dialysis clinic for 4-hour sessions using various types of dialysis procedures, including hemodialysis, peritoneal dialysis, or hemofiltration. Home dialysis with nocturnal treatments each night has emerged as an alternative to clinic-based therapy, but the fundamental paradigm of tethering the patient to a large piece of equipment for extended periods of time while the blood is purified of undesired solutes has remained unchanged for the past several decades. The advent of BioMEMS microfabrication technology has presented an opportunity to miniaturize the dialysis equipment, increase the efficiency of clearance of small solutes as well as middle molecules, and reduce inflammatory responses by improving blood flow in the dialyzer cartridge.

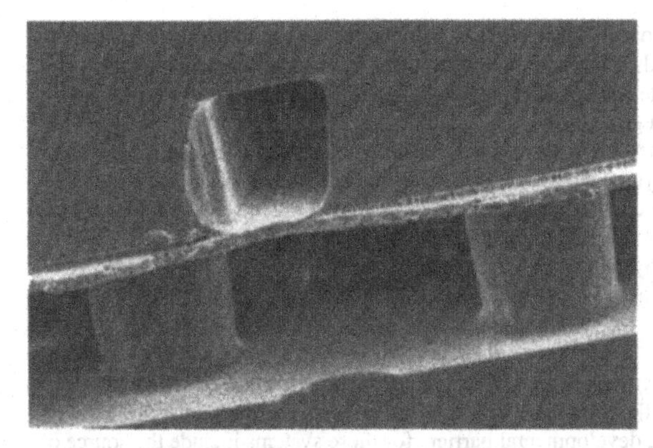

Figure 4. SEM image of membrane bilayer device for renal assist device, showing side view of microvascular channel (top) separated from dialysate chamber (bottom, supported by cylindrical posts), with a thin polyethersulfone membrane in between.

Kaazempur-Mofrad *et al.* [36] have reported a MEMS-based dialysis cartridge in which the hollow fibers in conventional cartridges are replaced by stacks of microvascular networks sandwiched between shallow flat-plate dialysis chambers. Rather than forcing the blood through the long path length of the hollow fibers, blood in this system is run through a network of microfluidic channels modeled after the physiology of the microcirculation of the kidney. In Figure 4, a cross-sectional SEM image illustrates the semi-permeable membrane sandwiched between two PDMS channel layers, one for the circulation and the other for the dialysate. Lee *et al.* have developed a microfluidic dialysis cartridge incorporating a polymer nanofiber membrane integrated into a PDMS microfabricated module [37]. The Nissenson group has proposed a BioMEMS-based system [38] containing filtration and resorption modules to accomplish full renal replacement; this approach is acellular but utilizes a specialized membrane structure to reclaim the excess water and needed solutes from the waste stream of the glomerular filtration module.

While opportunities for a cell-free renal replacement system such as that of Nissenson et al are very promising, the challenge of replacement the complex array of functions performed by the epithelial cells lining the nephron is immense. Several groups are exploring BioMEMS-based approaches for constructing renal replacement systems populated by kidney stem cells or other cell types designed to replicate the full function of the nephron. A renal microchip constructed using PDMS microfluidics has been cultured with Madin Darby Canine Kidney (MDCK) cells, for applications in nephrotoxicity testing and renal replacement systems [39]. Humes *et al.* have investigated the use of nanofabricated porous membranes [40] integrated into a tissue engineered bioartificial kidney for treatment of acute kidney injury. These silicon nanopores much more closely mimic the geometry of the glomerulus because they are in a slit-like configuration, thereby leading to more highly optimized separation of solutes from the blood for hemodialysis applications.

Liver Assist Devices

Current liver assist systems are used as a bridge to transplantation, and progress in this field has been difficult. Conventional hollow fiber liver assist devices have been commercialized and used to treat patients with liver failure, but most patients wait for a cadaveric transplant or split liver graft from a donor. Bioartificial liver (BAL) devices represent an opportunity to treat these patients for extended periods using cultured hepatocytes in a large bioreactor capable of augmenting liver function. In one configuration, a parallel plate BAL device was constructed using plates micromachined with grooves to house hepatocytes while protecting them from excessive shear stresses [41]. The flow rates required for delivery of oxygen to the BAL reactor are high enough to cause hydrodynamic damage to the cultured hepatocytes in a conventional configuration, so hepatocytes are cultured in concentric grooves in a radial flow structure. In this embodiment, the system did not require an oxygen permeable membrane to separate the hepatocytes from the flow, thereby simplifying the system.

Artificial Lung Devices

Most BioMEMS-based artificial organ devices require the integration of cells with microfabricated scaffolding, and the principal challenges are not engineering-related but rather biological questions. Principally, developmental barriers for these systems include the source of the cells, which may not be available from patients with end stage organ disease, potential immune reactions with the patient, and challenges associated with maintaining cellular phenotype in an artificial microenvironment. Therefore organ assist devices that do not require a cellular component are of interest, and microfabricated ECMO (ExtraCorporeal Membrane Oxygenator) systems represent a particularly significant opportunity. The surface area for gas exchange in the human lung is roughly 100 m^2, two orders of magnitude larger than conventional hollow fiber oxygenators, and the surface area to blood volume ratio is at least an order of magnitude larger in the lung than in current ECMO devices. Burgess *et al.* [42] are developing an artificial lung based on MEMS fabrication technology using stacked plates which provide far greater gas exchange area than conventional ECMO devices, using bifurcated microchannel networks carrying blood underneath a thin oxygen-permeable PDMS membrane. Stacks of 6 layers were constructed and endothelialized to reduce coagulation in the microvascular networks, and very high gas permeance values were obtained. In another study, Hoganson *et al.* [43] demonstrated efficient transport of oxygen and carbon dioxide in a microfabricated PDMS artificial lung device. Thin PDMS membranes and thin PDMS-coated porous membranes demonstrated high efficiency transfer of both oxygen and CO_2. In all of these cases, the ability to generate high surface areas for blood contact with oxygen sources moves the ECMO device much closer to the human lung in terms of performance and efficacy.

SUMMARY

The field of regenerative medicine and related developments in tissue engineering, organ assist devices, and drug delivery systems, is expanding rapidly due to the pressing need for solutions for patients suffering from tissue loss and organ failure. Drug delivery systems and tissue engineering technologies are emerging in response to the growing need for new therapeutic approaches. Many of the challenges facing the field are related to fundamental biological problems associated with sourcing cells and maintaining their function in engineered tissue constructs produced in the laboratory. Microfabrication and BioMEMS-based tools and technologies offer the opportunity to replicate the microenvironment in ways that will enable researchers to overcome many of the fundamental barriers in regenerative medicine. Over the next several years, these opportunities are likely to result in new medical technologies for implantable and smart drug delivery systems, improved *in vitro* assays for drug development and testing, engineered tissue constructs and organ assist and replacement devices to meet these needs.

ACKNOWLEDGMENTS

The author would like to gratefully acknowledge the support of Draper Laboratory and the Center for Integration of Medicine and Innovative Technology (CIMIT), and technical contributions of Sarah Tao, Joseph Charest, George Engelmayr, Lisa Freed and Chris Bettinger in the preparation of this manuscript.

REFERENCES

1. http://www.unos.org
2. W.M. Saltzman, *Drug Delivery* (Oxford Univ. Press, Oxford UK, 2001.)
3. S.L. Tao and T.A. Desai, *Adv. Mater.* **17** 1625 (2005).
4. D.V. McAllister, P.M. Wang, S.P. Davis, J-H Park, P.J. Canatella, M.G. Allen and M.R. Prausnitz, *Proc. Nat. Acad. Soc.* **100** 13755 (2003).
5. S. Kaushik, A.H. Hord, D.D. Denson, D.V. McAllister, S. Smitra, M.G. Allen and M.R. Prausnitz, *Anest. Analg.* **92** 502 (2001).
6. J.D. Zahn, A.A. Desmukh, A.P. Pisano and D. Liepmann, *Biomed. Microdev.* **6** 183 (2004).
7. A.J. Chung, D. Kim and D. Erickson, *Lab. Chip.* **8** 330 (2008).
8. D.T. Eddington and D.J. Beebe, *IEEE JMEMS* **13** 586 (2004)
9. http://www.debiotech.com/products/msys/insulinpump.html
10. Z. Chen, S.G. Kujawa, M.J. McKenna, J.O. Fiering, M.J. Mescher, J.T. Borenstein, E.E. Swan and W.F. Sewell, *J. Controlled Release* **110** 1 (2005).
11. J. Fiering, M.J. Mescher, E.E.L. Swan, M.E. Holmboe, B.A. Murphy, Z. Chen, M. Peppi, W.F. Sewell, M.J. McKenna, S.G. Kujawa and J.T. Borenstein, *Biomedical Microdevices*, in press.
12. J. Santini, M. Cima and R. Langer, *Nature* **397** 335 (1999).
13. S.K.W. Dertinger, X. Jiang, Z. Li, V.N. Murthy and G.M. Whitesides, *Proc. Nat. Acad. Sci.* **99,** 12542 (2002).
14. R.G. Flemming, C.J. Murphy, G.A. Abrams, S.L. Goodman and P.F. Nealey, *Biomaterials* **20,** 573 (1999).
15. C.J. Bettinger, B. Orrick, A. Misra, R. Langer and J.T. Borenstein, *Biomaterials,* **27,** 2558 (2006).
16. S.R. Khetani and S.N. Bhatia, *Nature Biotech.* 26 120 (2008).
17. A. Cararro, W.M. Hsu, K.M. Kulig, W.S. Cheung, M.L. Miller, E.J. Weinberg, M.R. Kaazempur-Mofrad, J.T. Borenstein, J.P. Vacanti and C. Neville. *Biomed. Microdevices* **10** 795 (2008).
18. H. Lu, L.Y. Koo, W.M. Wang, D.A. Lauffenburger, L.G. Griffith and K.F. Jensen, *Anal. Chem.* 76 5257 (2004).

19. S.N. Bhatia, U.J. Balis, M.L. Yarmush and M. Toner, *FASEB J* **13** 1883 (1999).
20. E.E. Hui and S.N. Bhatia, *Proc. Nat. Acad. Sci.* 104 5722 (2007).
21. M.J. Powers, D.M. Janigian, K.E. Wack, C.S. Baker, D. Beer Stolz and L.G. Griffith, *Tissue Engineering* **8** 499 (2002).
22. M. Zhang, P.J. Lee, P.J. Hung, T. Johnson, L.P. Lee and M.R. Kaazempur-Mofrad, *Biomedical Microdevices* 10 117 (2008).
23. V. Viravaidya, A. Sin and M.L. Shuler, *Biotechnol. Progress* **20** 316 (2004).
24. A. Khademhosseini, J.T. Borenstein, R. Langer and J.P. Vacanti, *Proc. Nat. Acad. Sci.* **103** 2480 (2006).
25. J.T. Borenstein, H. Terai, K.R. King, E.J. Weinberg, M.R. Kaazempur-Mofrad and J.P. Vacanti, *Biomedical Microdevices* **4** 167 (2002)
26. G.J. Wang, C.L. Chen, S.H. Hsu and Y.L. Chiang, *Microsyst. Technol.* **12** 120 (2005).
27. K.R. King, C.J. Wang, M.R. Kaazempur-Mofrad, J.P. Vacanti and J.T. Borenstein,*Adv. Mater.* **16** 2007 (2004).
28. C. Fidkowski, M.R. Kaazempur-Mofrad, J.T. Borenstein, J.P. Vacanti, R. Langer and Y. Wang *Tissue Engineering* **11** 30 (2005).
29. M. Cabodi, N.W. Choi, J.P. Gleghorn, C.S. Lee, L.J. Bonassar and A.D. Stroock, *J. Amer. Chem. Soc.* **127** 13788 (2005).
30. K.M. Chrobak, D.R. Potter and J. Tien, *Microvasc. Res.* **71** 185 (2006).
31. J.L. Charest, M.T. Eliason, A.J. Garcia and W.P. King, *Biomaterials* **27** 2487 (2006).
32. N. Gomez, S. Chen and C.E. Schmidt, *J.R. Soc. Interface* **13** 223 (2007).
33. M.R. Kaazempur-Mofrad, E.J. Weinberg, J.T. Borenstein and J.P. Vacanti, "Tissue Engineering: Multi-Scaled Representation of Tissue Architecture and Function," in Complex Systems Science in Biomedicine, (Kluwer Academic - Plenum Publishers, New York, 2003.)
34. R. Iyer, B. Plouffe, S.K. Murthy and M. Radisic, "Microreactors for Cardiac Tissue Engineering," in Micro and Nanoengineering of the Cell Microenvironment, eds. A. Khademhosseini, J.T. Borenstein, M. Toner and S. Takayama (Artech House, Boston, 2008.)
35. G.C. Engelmayr Jr., M. Cheng, C. J. Bettinger, J. T. Borenstein, R. Langer, and L. E. Freed, *Nature Materials* Nov 2 2008 (DOI:10.1038/nmat2316).
36. M.R. Kaazempur-Mofrad, N.J. Krebs, J.P. Vacanti and J.T. Borenstein, *Proc. 2004 Sensors and Actuators Conf.*, (Transducers Research Foundation, Cleveland OH, 2004.)
37. K.H. Lee, D.J. Kim, B.G. Min and S.H. Lee, Biomedical Microdevices **9** 435 (2007).
38. A.R. Nissenson, C. Ronco, G. Pergamit, M. Edelstein and R. Watts, *Hemodialysis Int'l.* **9** 210 (2005).
39. R. Baudoin, L. Griscom, M. Monge, C. Legallais and E. Leclerc, *Biotechnol. Prog.* **23** 1245 (2007).
40. H.D. Humes, W.H. Fissell and K. Tiranathanagul, *Kidney Intl.* **69** 1115 (2006).
41. J. Park, Y. Li, F. Berthiaume, M. Toner, M.L. Yarmush and A.W. Tilles, *Biotechnol. Bioeng.* **90** 632 (2005).
42. K.A. Burgess, H.H. Hu, W.R. Wagner and W.J. Federspiel, *Biomedical Microdevices* Epub, PMID 18696229, 2008.
43. D.M. Hoganson, J. Anderson, B. Orrick and J.P. Vacanti, Amer. Assoc. Thoracic Surgery Conf., 2008.

Poster Session

Mater. Res. Soc. Symp. Proc. Vol. 1139 © 2009 Materials Research Society 1139-GG03-01

Design and Fabrication of MEMS Piezoelectric Rotational Actuators

Danny Gee, Wayne A. Churaman, Luke J. Currano, and Eugene Zakar
U.S. Army Research Laboratory, 2800 Powder Mill Road,
Adelphi, MD 20783, U.S.A.

ABSTRACT

As Microelectromechanical Systems (MEMS) continue to mature and increase in design complexity, the need to exploit rotation in MEMS devices has become more apparent. An in-plane piezoelectric rotational actuator is proposed that provides free deflections on the order of 1.5° with applied biases of less than 25V and nanoampere currents. Moments up to 6×10^{-8} N·m, corresponding to forces of 125 μN, were measured using MEMS cantilever springs. The actuator utilizes the low-power, high-force characteristics of lead zirconate titanate and a coupled, dual offset-beam design to provide effective rotational displacement. The resulting power consumption is three orders of magnitude less than current electrothermal rotational designs.

INTRODUCTION

The vast majority of MEMS actuators generate translational motion. A few rotational actuators have been documented in the development of micro-robotics, locomotives, and other biologically-inspired devices [1-3]. While piezoelectric materials are scrutinized for limited strain and fabrication complexities, they exhibit excellent characteristics that are attractive for MEMS actuators. Piezoelectric actuators have demonstrated wide bandwidths, high sensitivity, and large stroke forces [4]. Also with high power densities, piezoelectric actuators generally exhibit the most efficient transduction mode [5]. Lead zirconate titanate (PZT) is a well-studied piezoelectric material that is used in MEMS for its attractive thin film properties.

Previous research in piezoelectric rotational actuation has been largely limited to optical microstages that operate with non-planar behavior [6]. As the longitudinal piezoelectric coefficient for PZT is nearly double the value of the transverse coefficient, it is convenient to develop out-of-plane actuators. This presents a unique challenge for applications that require in-plane actuation. In this report, an in-plane, piezoelectric MEMS rotational actuator is proposed. The underlying operation of the rotational actuator is presented and the process sequence of the device fabrication is detailed. Preliminary results of the actuator are given and the defects associated with device fabrication are examined within this paper.

THEORY

Operational Concept

The premise behind the rotational actuator is similar to a previously designed electrothermal actuator [7]. The actuator is composed of two parallel, yet offset beams that are connected to a free perpendicular beam, which is shown in Figure 1. A released cantilever is also integrated into the device to serve as a resisting spring to the actuator movement.

Figure 1. Labeled image of a fabricated rotational actuator.

The two offset beams are the actuation beams where deformation occurs. A piezoelectric stack, consisting of a PZT film sandwiched between a top and bottom metal electrode, serves as the transducer. When a bias is applied to the two electrodes, the PZT polarizes and transforms along its transverse expansion (d_{31}) mode. Since the electric field is applied across the film thickness, the actuation beams compress axially and pull on the yoke that connects the two beams. This generates a torque around the central point of the yoke, which causes the amplification beam to rotate. The angle of rotation is dependent on the stiffness of the two actuation beams and the magnitude of the torque.

Fabrication

The rotational actuator was fabricated on a silicon-on-insulator (SOI) wafer with a 2μm silicon device layer over 1000Å of buried silicon dioxide. The device silicon is used as a structural support for the actuating piezoelectric layer, while the buried oxide protected the base of the device silicon from etching during the release of the actuator. The piezoelectric stack was fabricated on top of the device layer. Initially, a thin layer of PECVD silicon dioxide was deposited, followed by the sputtering of a seed layer of titanium and then platinum for the bottom electrode metal. The piezoelectric layer was deposited with sol-gel PZT on the wafer. Metallization for the top electrode was done by sputtering platinum onto the wafer, illustrated in Figure 2A.

The top electrode was formed by patterning and ion milling the top metal layer (Figure 2B). The PZT and bottom metal were subsequently ion milled with a separate mask to shape the actuator (Figure 2C). A PZT wet etch (DI water, HCl acid, and HF acid) was used to expose the bottom metal for electrical contact (Figure 2D). In areas where the metals and PZT were removed, the underlying PECVD-deposited oxide was etched away by a reactive ion etch to expose the device silicon.

The device silicon layer was patterned by deep reactive ion etching (DRIE) to delineate the beams and contact pads (Figure 2E). The exposed buried oxide was removed by RIE to allow for etching and undercutting of the silicon handle wafer (Figure 2F). The etch was briefly continued into the handle wafer by DRIE. A thick photoresist was spun on the wafer to fill the trenches around the devices and was patterned to protect the sidewalls of the devices. The silicon handle wafer was etched with XeF_2, undercutting the beams and releasing the device

(Figure 2G). Lastly, a buffered hydrofluoric (BHF) oxide etch was performed to remove the buried oxide from the bottom of the beams. The wafer was ashed in oxygen plasma to remove any remaining residues and particulates (Figure 2H).

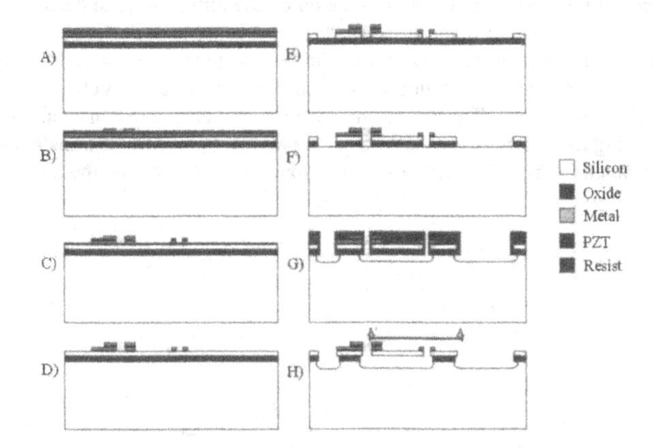

Figure 2. Cross-sectional view of the fabrication process sequence for the rotational actuator.

RESULTS

Actuator Data

Test data was recorded for devices with an actuation beam length of 500μm and a thickness of the 2μm device silicon layer and an additional PZT thickness of 1μm. The resisting springs were patterned only in the device silicon to be 500μm long and 2μm thick. The dimensions of the actuation beam were 485μm long (from the center of the yoke to the tip), 25μm wide, and 2μm thick. Various actuator beam and resisting spring widths were tested to determine deflection trends, generated forces, and the stiffness of the actuators. The actuator beam widths tested were 5, 7, and 10μm, while the resisting spring widths for force testing were 10, 15, and 20μm. Devices without resisting springs were also included to test free deflections. Measurements of the deflections were taken from the vernier scale, as shown in Figure 3.

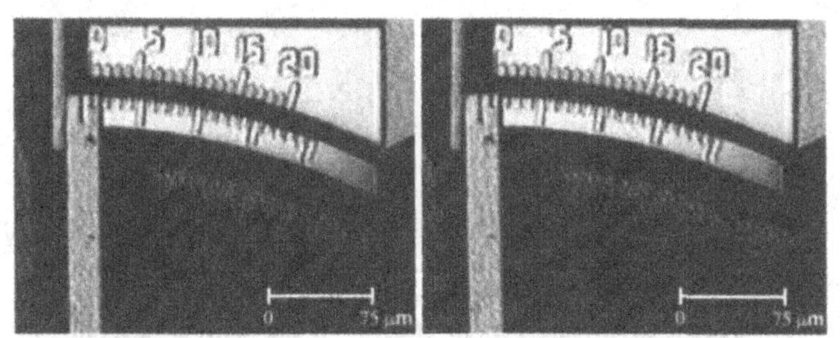

Figure 3. Images of actuator with no applied bias (left) and 21V bias (right).

Devices were tested by varying the applied voltage until a pair of vernier marks aligned. The deflection angle gradations of the angular vernier scale were 0.15°; therefore, the error associated with this on-chip measurement was estimated as +/- 0.075°. To determine the change in deflection, the measured angle was subtracted by the initial angle of deflection under no bias, θ_0, which is inherent from intrinsic stresses. Figure 4 consists of deflection graphs for four 5μm-wide actuators with various resisting spring widths, including one with no resisting spring. The data confirmed the expected operation of the actuators, as larger deflections occurred for thinner springs under the same bias. Thinner springs have a smaller spring constant, which represents a smaller opposing force against the deflecting actuators. Approximate translational displacements can be determined by trigonometry, given the angle of deflection and the length of the amplification beam.

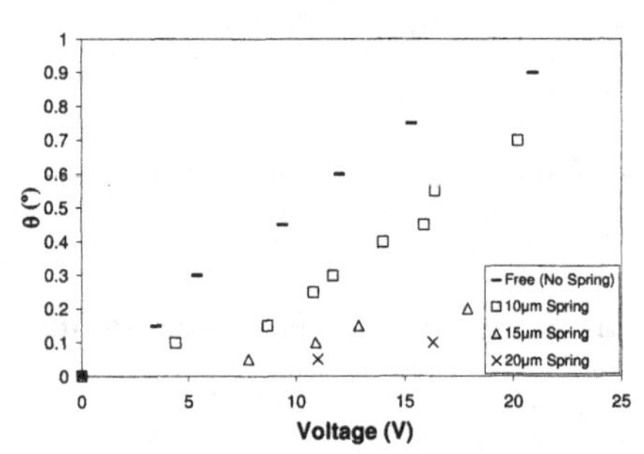

Figure 4. Deflection angle over the biasing range for various springs for a 5μm-wide actuator.

The effectiveness (deflection per unit power) for the maximum free deflection of the piezoelectric versus electrothermal actuators is shown in Figure 5. As electrothermal actuators are characterized for their ability to generate large forces (v-beam designs) or large displacements (u-beam designs), they serve as a good standard for comparison. Although limited to smaller deflection angles and forces, the piezoelectric actuators require orders of magnitude of less power per angle of deflection due to a much smaller current draw.

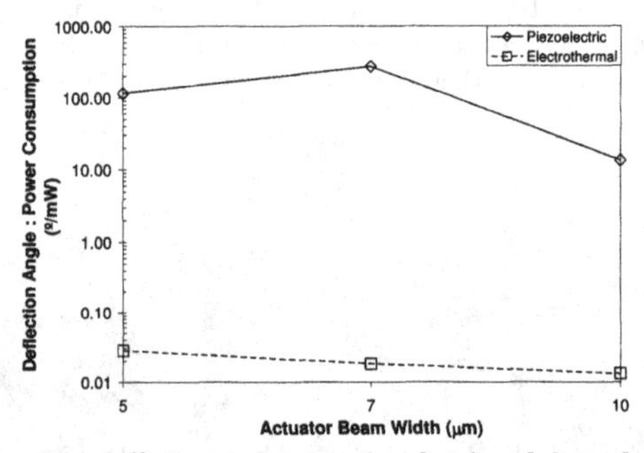

Figure 5. Comparison of effectiveness between piezoelectric and electrothermal rotational actuators.

To normalize the data from various voltages, the values were extrapolated from nearest neighbors to determine deflections at a common voltage. Assembling all of the data for varying spring stiffnesses and actuators, a torque-rotation graph is constructed for various actuator widths, which is displayed in Figure 6. As torque and rotation are analogous to force and displacement, a corresponding force-displacement graph is also shown. The area under a curve is the feasible region for the actuator output, with the maximum force equal to the y-intercept and the maximum displacement equal to the x-intercept. Given a required force, the maximum displacement can be read off of the curve, and the same is true for a given required force. In addition, by comparing the measured force-displacement relationship for a given set of actuators of the same dimensions at the common bias, the slope of a line represents the stiffness of the actuator. The generated actuator spring constants are credible, as wider actuators consist of a larger volume, which have a greater stiffness.

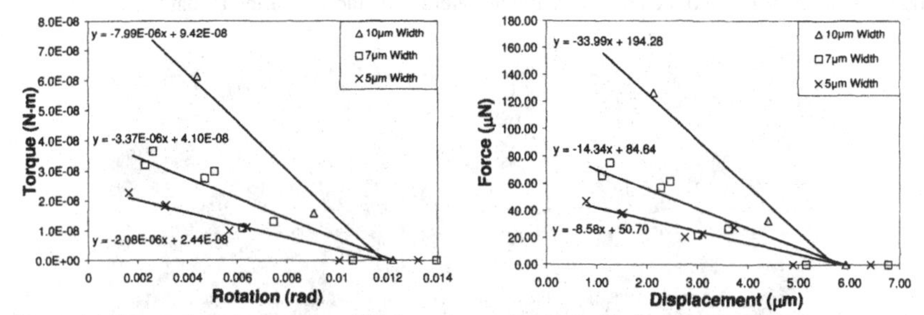

Figure 6. (a) Torque-rotation graph and (b) its corresponding force-displacement graph for actuators of various widths at a constant voltage.

Actuator Defects

Originally, devices were released with the buried oxide intact. These actuators suffered from significant upward bending of the spring and amplification beams and were deemed defective as the amplification beam would actuate past the bent resisting springs without any contact. In Figure 7a, an equivalent bending angle of 6.5° was approximated from the length of the amplification beam and the height difference between the vernier scales. When the buried oxide layer was removed with a wet etch in BHF for a second set of wafers, the beams were planar, as shown in Figure 7b. Therefore, the bending was attributed to the compressive residual stress of the buried oxide.

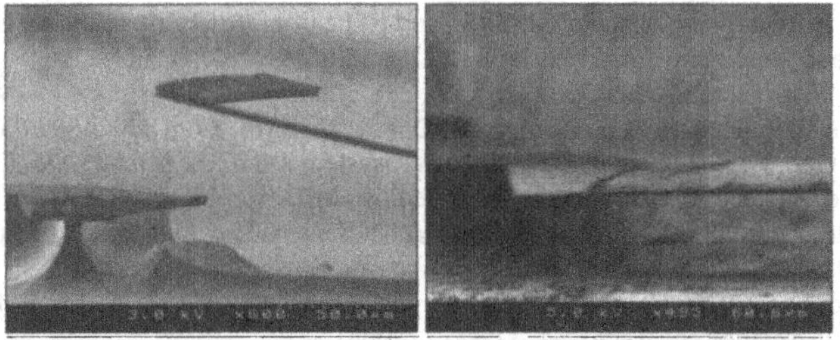

Figure 7. SEM images of (a) beam bending and (b) planar beams after oxide removal.

As the rotational actuators were tested, the device failure modes were documented. An applied bias above ~25V typically resulted in dielectric breakdown. The remnants of a beam after dielectric breakdown can be seen in Figure 8a. In some cases, the current would creep towards the power supply limit, but the device would be turned off prior to physical breakdown. When the electric field was reapplied, the current would continue to creep upwards, signifying irreparable damage to the actuator.

Delamination of the piezoelectric stack from the device silicon was observed for some devices that underwent the BHF etch to remove the buried oxide layer (Figure 8b). The authors speculate that imperfections in the photoresist sidewall coverage allowed the BHF etch to attack the PECVD-deposited oxide and the titanium adhesion layers in some areas. Furthermore, the BHF etched areas in the PZT, forming pinholes and fissures, thereby degrading the quality of the film. By controlling the BHF etch time and the quality of the sidewall resist, the piezoelectric material can be protected, hence increasing the yield and stability of the actuators.

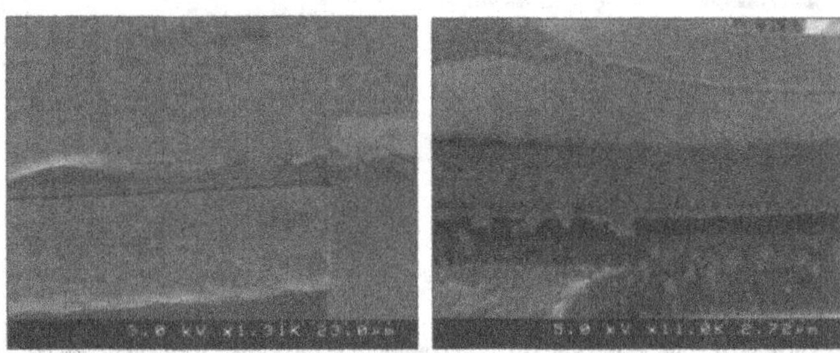

Figure 8. SEM images of (a) dielectric breakdown of beams and (b) stack delamination.

CONCLUSIONS

We have presented an in-plane, piezoelectric rotational actuator along with the experimental data for actuators of various dimensions. A maximum free deflection of 1.5° was observed, while maximum moments of $6x10^{-8}$ N·m, or forces up to 125 μN, were calculated. Addressing the buried oxide within the actuator was critical to the device fabrication, as residual stresses caused significant beam bending, while oxide etching also created material inconsistencies. Future work includes the design optimization for maximum deflection and the development of a finite element model and an analytical model.

REFERENCES

1. J. Yan, R. Wood, S. Avadhanula, M. Sitti., and R. Fearing, IEEE International Conference on Robotics and Automation, **4**, 3901-3908 (2001).
2. T. Morita, Sensors and Actuators A, **103** (3), 291-300 (2003).
3. K. Oldham, J. Pulskamp, R. Polcawich, M. Dubey, JMEMS, **17** (4), 890-899 (2008).
4. H. Xu, T. Ono, D. Zhang, and M. Esashi, Microsystem Tech., **12** (9), 883-890 (2006).
5. T. Abe and M. Reed, IEEE Workshop Proceedings MEMS, 164-169 (1994).
6. A. Schroth, C. Lee, S. Matsumoto, and R. Maeda, Sensors and Actuators A, **73** (1-2), 144-152 (1999).
7. L. Currano, D. Gee, W. Churaman, M. Dubey, P. Amirtharaj, M. Yu, and B. Balachandran, Advances in Science and Tech., **54**, 372-377 (2008).

Mater. Res. Soc. Symp. Proc. Vol. 1139 © 2009 Materials Research Society 1139-GG03-04

Measurement and Analysis of Structural Damping in Silicon Carbide Microresonators

Sairam Prabhakar, Frederic Nabki, Mourad El-Gamal, and Srikar Vengallatore
McGill University, Montreal, Quebec, H3A 2K6, Canada.

ABSTRACT

The design of microresonators with high natural frequencies (1 MHz to 1 GHz) and low structural damping is essential for devices used for applications in communications. Here, we report experimental measurements of damping at low pressure and ambient temperature in electrostatically-actuated, metallized silicon carbide microresonators. Comparison of the measured values with the predictions of a model for thermoelastic damping indicates that the contribution of this mechanism to the measured damping ranges from 10% to 50% over a broad frequency range (3 MHz to 30 MHz).

INTRODUCTION

Microresonators hold promise for miniaturized, low power signal-processing and wireless radio-frequency communications. For a number of these applications, it is desirable for the microresonators to attain natural frequencies of vibration in the 1 MHz – 1 GHz range to match the frequencies of the signals of interest, and to exhibit low damping ($Q^{-1} < 10^{-4}$) to improve the frequency selectivity of devices used for filtering applications. These requirements provide the motivation for developing surface-micromachining technologies using materials that exhibit high acoustic velocities (such as silicon carbide and diamond-like carbon), and for fundamental studies of the mechanisms of structural damping in micromachined resonators.

Recently, a low-temperature (<300 °C), integrated circuits (IC)-compatible, surface-micromachining process was developed at McGill University [1]. The structural material used in this process is a stress-controlled, amorphous, Hexoloy silicon carbide thin film that is deposited using sputtering, and polyimide is used as the sacrificial material. The details of the surface-micromachining steps and microstructural characterization of the SiC thin films have been described elsewhere [1,2]. Here, we report the first studies of structural damping at low pressure in Hexoloy SiC microresonators.

DEVICE GEOMETRY AND DAMPING MEASUREMENTS

The devices tested in this study were doubly-clamped, electrostatically-actuated, metallized SiC micromechanical beam resonators (Figure 1). Each beam is a trilayered laminated composite (56 nm Al/ 78 nm Cr/ 2 μm SiC). The Young's modulus (E), density (ρ), thermal conductivity (k), coefficient of linear expansion (α), and specific heat (C) of these materials are listed in Table I. The amorphous SiC thin film is electrically non-conducting, and the aluminum layer is used to enable electrostatic actuation [1]. The chromium layer is used as an etch-stop during reactive-ion etching of the structural layer using NF_3. The dynamic characteristics of 36 micromechanical beam resonator devices were measured at low pressures (~1 mTorr) and ambient temperature using a network analyzer (HP 8753D VNA). The quality factor was measured from the 3 dB bandwidth (Figure 2). These beams had lengths ranging from 32 μm to 64 μm, width of 8 μm, and natural frequencies of vibration ranging from 3 MHz to 30 MHz.

Table I. Structural and material properties of the trilayered composite microresonator.

Material	Thickness	E (GPa)	ρ (Mg/m^3)	k (W/m/K)	α (1/K)	C (J/m^3/K)
Hexoloy-SiC	2 μm	260	3.1	126	4×10^{-6}	2.1×10^{6}
Cr	78 nm	200	7.9	94	6×10^{-6}	3.6×10^{6}
Al	56 nm	70	2.7	220	24×10^{-6}	2.4×10^{6}

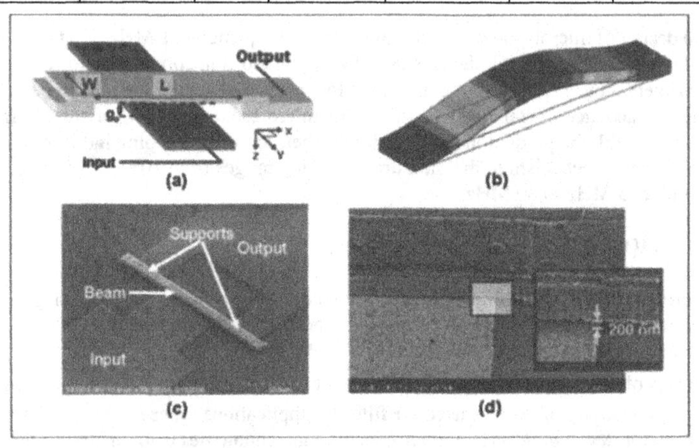

Figure 1. (a) 3D schematic of a doubly-clamped resonator. (b) Simulated fundamental mode shape of the beam. (c) Scanning electron micrograph of a doubly-clamped SiC microresonator. (d) Magnified image showing the 200 nm gap between the beam and the substrate [1].

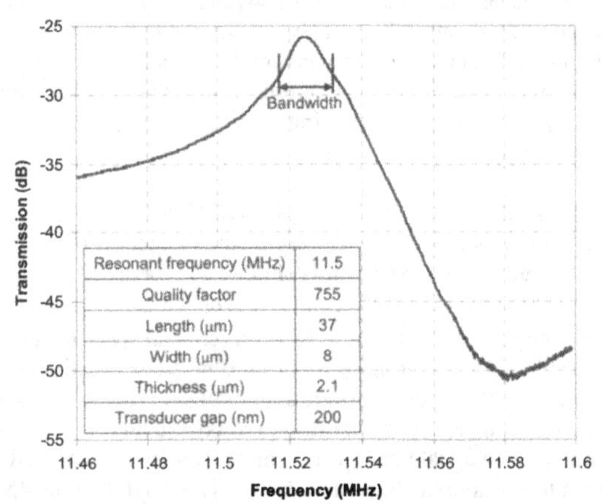

Figure 2. Frequency response of a SiC microresonator device measured at a pressure of 1 mTorr using a 20 dB gain amplifier. The quality factor was measured from the 3 dB bandwidth.

The inverse quality factor of the 36 devices ranged from 9.6×10^{-4} to 2.7×10^{-3}. For many applications, it is desirable to reduce these values by about an order of magnitude. As the first step towards that goal, we aim to identify the dominant mechanisms of energy dissipation in these microresonators. Damping in vacuum-operated flexural-mode microresonators is primarily due to thermoelastic damping, internal friction, and anchor losses due to radiation of elastic energy into the substrates. A detailed analysis of thermoelastic damping (TED) is a natural starting point for our analysis because this mechanism establishes the absolute lower bound on structural damping in flexural-mode microresonators. Therefore, a comparison of the measured damping with the predicted thermoelastic damping can establish (i) whether TED is dominant in these structures, and (ii) accurately quantify the total contributions from anchor losses and internal friction to the measured structural damping. The next section presents a detailed model for TED in layered composite micromechanical beams.

ANALYSIS OF THERMOELASTIC DAMPING

Thermoelastic damping in flexural resonators results from the coupling between inhomogeneous strain and temperature fields. The vibration of the beam generates an oscillating strain gradient which, in turn, generates an oscillating temperature gradient due to thermoelastic coupling. This finite temperature gradient leads to irreversible heat conduction, entropy generation, and thermoelastic damping. The first exact theory for TED in laminated composites was developed by Bishop and Kinra for symmetric, tri-layer, plate resonators [3]. This model was specialized to symmetric, three-layered, composite microbeams by Vengallatore [4]. Subsequently, Prabhakar and Vengallatore [5] developed a model to calculate TED in asymmetric bilayered beam resonators. Here, we build upon these studies to develop a generalized framework to calculate TED in asymmetric laminated composite microbeams with an arbitrary number of layers.

Consider a laminated composite beam with n layers. The length L, width w and thickness $y_n - y_0$ are defined by an *x-y-z* coordinate system as $0\le x\le L$, $0\le z\le w$ and $y_0\le y\le y_n$, as shown in Figure 3.

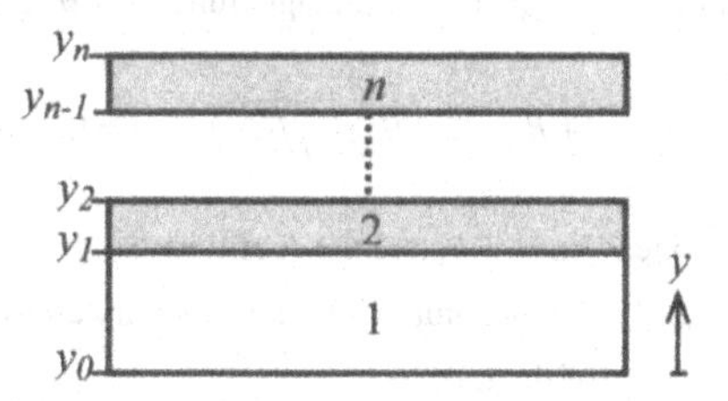

Figure 3. Schematic cross-section of a n-layered micromechanical beam.

The beam undergoes pure flexural vibrations with angular frequency ω. The time harmonic transverse displacement of the beam is given by:

$$Y(x,t)=Y_0(x)exp(i\omega t), \tag{1}$$

where the mode shape $Y_0(x)$ is determined from the boundary conditions at the ends of the beam span. The longitudinal strain within the beam is given by:

$$\varepsilon = \bar{\varepsilon}(x,y)exp(i\omega t) = (y - y')\, d^2Y_0/dx^2\, exp(i\omega t), \tag{2}$$

where y' is the location of the neutral axis. Assuming every material layer to be linear thermoelastic, the longitudinal stress within the beam is given by:

$$\sigma_j = E_j\varepsilon = E_j\bar{\varepsilon}\, exp(i\omega t) = \bar{\sigma}_j\, exp(i\omega t) \quad j = 1,\ldots n. \tag{3}$$

For pure flexural vibration, the hydrostatic stress within the beam is equal to the longitudinal stress, σ_j.

The linearized, thermoelastically-coupled equation for heat conduction across the thickness of the beam is [3-5]:

$$k_j\, \partial^2\theta_j/\partial y^2 = C_j\, \partial\theta_j/\partial t + \alpha_j T_0\, \partial\sigma_j/\partial t \tag{4}$$

Here, $\theta = T - T_0$ is the temperature change relative to the equilibrium temperature of the beam. This equation has to be solved for $\theta_j(y,t) = \bar{\theta}_j(y)exp(i\omega t)$, subject to boundary conditions along the beam thickness, as well as appropriate interface conditions within the laminated composite. For the beam vibrating in vacuum, the outer faces are adiabatic. Therefore, we have

$$\partial\theta_1/\partial y\big|_{y=y_0} = 0; \quad \partial\theta_n/\partial y\big|_{y=y_n} = 0 \tag{5}$$

Assuming perfect thermal contact between the layers, the first set of interface conditions are:

$$\theta_j\big|_{y=y_j} = \theta_{j+1}\big|_{y=y_j} \quad \text{for } j = 1,\ldots,(n-1) \tag{6}$$

From continuity of heat flux across the layers, the second set of interface conditions are:

$$k_j\, \partial\theta_j/\partial y\big|_{y=y_j} = k_{j+1}\, \partial\theta_{j+1}/\partial y\big|_{y=y_j} \quad \text{for } j = 1,\ldots,(n-1) \tag{7}$$

The boundary-value problem represented by Eq. (4) with boundary conditions and interface conditions of Eqs (5-7) can be solved by the method of eigenfunction expansions, or by employing Green's functions [6]. These solution techniques use the eigenfunctions of the homogeneous version of Eq. (4) as building blocks. In the absence of the source term, Eq. (4) reduces to the classical diffusion equation, with eigenfunctions $\phi_{j,m}$ having the form:

$$\phi_{j,m} = A_{j,m}\, cos\left(\beta_m\sqrt{k_j/C_j}\; y\right) + B_{j,m}\, sin\left(\beta_m\sqrt{k_j/C_j}\; y\right) \quad j = 1,\ldots,n; \quad m = 1,\ldots,\infty \tag{8}$$

It remains to determine the eigenvalues β_m, and the coefficients, $A_{j,m}$ and $B_{j,m}$. To this end, the eigenfunctions, $\phi_{j,m}$ from Eq. (8) are substituted into the boundary and interface conditions to yield a system of linear equations of the form:

$$\left[\mathbf{M}(\beta_m)\right]_{2n\times 2n} \begin{Bmatrix} A_{j,m} \\ B_{j,m} \end{Bmatrix}_{2n\times 1} = \{0\}_{2n\times 1} \tag{9}$$

For non-trivial solutions of this system of equations, the determinant of the matrix **M** must be zero:

$$|\mathbf{M}(\beta_m)| = 0 \tag{10}$$

This condition yields a transcendental equation for the eigenvalues, β_m. Due to the homogeneity of the system of equations in Eq. (9), the coefficients $A_{j,m}$ and $B_{j,m}$ are determined in terms of any one of them. Having obtained the eigenvalues and eigenfunctions, the solution to the one-way coupled equation of heat conduction can be assembled as [6]

$$\bar{\theta}_j = -T_0 \sum_{m=1}^{\infty} \frac{i\omega}{\beta_m^2 + i\omega} \left(\sum_{j=1}^{n} \int_{y_{j-1}}^{y_j} C_j \phi_{j,m}^2 \, dy \Big/ \sum_{j=1}^{n} \int_{y_{j-1}}^{y_j} k_j \sigma_j \, \phi_{j,m} \, dy \right) \phi_{j,m} \tag{11}$$

The thermoelastic temperature change, θ_j, is a complex valued quantity, reflecting the fact that the temperature change is out of phase with the applied stress. The thermal strain produced by the thermoelastic temperature change within the beam is [7]:

$$\bar{\varepsilon}_j^{th} = \alpha_j \bar{\theta}_j \tag{12}$$

The work lost per cycle of vibration in each layer is given by [3]:

$$\Delta W_j = -\pi \, wL \int_{y_{j-1}}^{y_j} \sigma_j \, Im\left(\bar{\varepsilon}_j^{th}\right) dy \tag{13}$$

The peak strain energy within each layer during a cycle of vibration is [3]:

$$W_j^{max} = \frac{w}{2} \int_0^L \int_{y_{j-1}}^{y_j} \bar{\sigma}_j \bar{\varepsilon}_j \, dx \, dy \tag{14}$$

The total work lost due to thermoelastic coupling and the total strain energy within the beam may be obtained by summing over the individual layers. Hence, the magnitude of TED is obtained as:

$$Q_{TED}^{-1} = \frac{1}{2\pi} \sum_{j=1}^{n} \Delta W_j \Big/ \sum_{j=1}^{n} W_j^{max} \tag{15}$$

Equations (1) to (15) represent a theoretical framework to compute thermoelastic damping in laminated composites with an arbitrary number of layers. This framework was implemented numerically within an object-oriented C++ computational environment to compute the frequency dependence of TED. The object-oriented paradigm lends itself naturally to the computation of TED in laminated composites. Each laminate of the composite structure represents an object within the C++ implementation. Every laminate object has its own thermal and mechanical field variables and material constants. Dynamic memory allocation enables the creation of a user-defined number of laminates during runtime. The transcendental equation for the eigenvalues, Eq. (10), is solved numerically using the method of bracketing and bisection, while the eigenfunction coefficients are evaluated using LU decomposition and backsubstitution for matrix inversion in a reduced form of Eq. (9). Efficient numerical algorithms derived from Press *et al.* [8] were used within the computational routine.

COMPARISON OF MEASURED DAMPING AND PREDICTED TED

Figure 4 shows the measured damping values for the 36 Al/Cr/SiC devices tested in this study. The damping ranges from 9.6×10^{-4} to 2.7×10^{-3} over a frequency range of 3 MHz to 30 MHz. Also shown on Figure 4 is a solid line corresponding to the computed value of thermoelastic damping. The material property values listed in Table I were used as inputs in this calculation. The comparison of the measured damping and predicted TED is consistent with the expectation that this mechanism establishes an absolute lower bound on structural damping. Depending on the frequency, the contribution of TED to the structural damping ranges from 10% to 50% in these metallized SiC microresonators. Our future efforts will focus on understanding the role of other damping mechanisms (anchor losses and internal friction) in these devices.

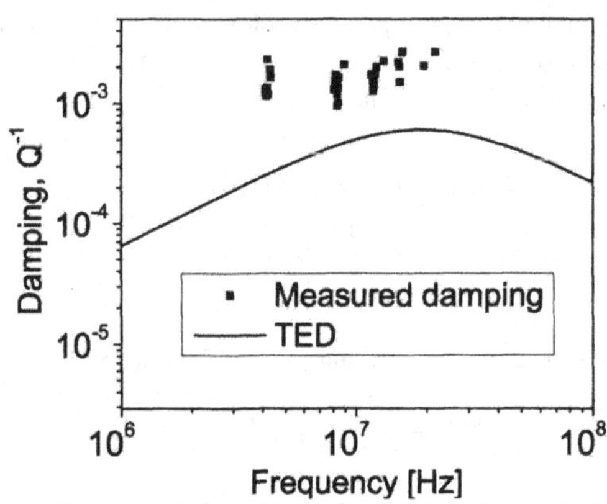

Figure 4. Comparison of measured damping and predicted thermoelastic damping (TED) in doubly-clamped metallized SiC microresonators.

ACKNOWLEDGEMENTS

The microresonators were fabricated at the NanoTools Microfabrication Facility at McGill University. Financial support from NSERC and the CRC program is gratefully acknowledged.

REFERENCES

[1] F. Nabki, T.A. Dusatko, S. Vengallatore, and M.N. El-Gamal, *Tech. Dig. Solid-State Sensors, Actuators, and Microsystems Workshop,* Hilton Head, SC, pp. 216-219, 2008.

[2] J. Crocker, G. Sosale, and S. Vengallatore, *Proceedings of the 21st Canadian Congress of Applied Mechanics* (CANCAM), June 3-7, 2007, Toronto, Canada.

[3] J.E. Bishop and V.K. Kinra, *Int. J. Solids Structures*, **34**, 1075 (1997).

[4] S. Vengallatore *J. Micromech. Microeng.*, **15**, 2398 (2005).

[5] S. Prabhakar, and S. Vengallatore, *J. Micromech. Microeng.*, **17**, 532 (2007).

[6] M.N. Ozisik, *Heat Conduction*, (John Wiley and Sons, New York, 1993)

[7] B.A. Boley and J.H. Weiner, *Theory of Thermal Stresses*, (Wiley, New York, 1960).

[8] W.H. Press, S.A. Teukolsky, W.T. Vetterling and B.P. Flannery, *Numerical Recipes in C++: The Art of Scientific Computing*, (Cambridge University Press, Cambridge, 2002).

Mater. Res. Soc. Symp. Proc. Vol. 1139 © 2009 Materials Research Society 1139-GG03-05

Fabrication and Hot Switching Behavior of Electroplated Gallium Spheres for MEMS

Yoonkap Kim and David F. Bahr
Mechanical and Materials Engineering, Washington State University, Pullman WA USA

ABSTRACT

Liquid metal microscale switches, often using mercury, are sometimes used in place of solid-solid contact switches because of the ability to minimize damage from switching and the ability to make good contacts for electrical and thermal conductivity. However, mercury has potential health and safety problems, and is difficult to use at high frequency (kHz) operation due to poor adhesion between the liquid-solid contacts. One alternative to the mechanical and chemical problems of a liquid mercury switch is using soft metals, such as gallium or tin, as a solid metal sphere for switches that can melt at moderate temperatures. Ga micro-spheres for switching operations were deposited on a substrate consisting of photolithographically patterned W films on SiO_2 and Si substrates by electroplating, and the applicability for use in a microscale switch was investigated by characterizing the macro structure, hardness, and electrical performance during switching. The resistivity of the electroplated Ga droplets was similar to the theoretical value for pure Ga, and suggests that the electrodeposited Ga will be suitable for a solid MEMS switch. The hardness of the Ga sphere was 5.7 MPa. This suggests a maximum of ~40 μN can be applied to each 50μm radius Ga contact in the current configuration for switching applications. When the Ga spheres were investigated for electrical performance during hot switching, the resistance increased over six switching cycles, but the original lower resistivity was recovered after a 393 K thermal reflow process.

INTRODUCTION

Micro electro mechanical systems (MEMS) basically consist of using processing developed for microelectronics processing to fabricate devices with both electrical and mechanical functionality [1,2]. Various types of micromechanical switches have been introduced since MEMS were developed [3-7]. These micromechanical switches are useful in applications for wide operating temperature ranges, radiation insensitivity, and usually have a high on-off impedance ratio. There is particular interest in having both a good thermal and electrical contact when the switch is closed. Almost all micromechanical switches have been designed with solid-to-solid contacts [8]. Micromechanical switches have various problems such as contact bounce, high contact resistance, noise, slow rise times, and a short operational lifetime due to mechanical wear and tear. These problems can be solved by using liquid metals because they show a fast signal rise time, high contact area, and low contact resistance, and can reduce wear and tear since it is a liquid phase [9]. There are several pure low melting point liquid metals that could be used for MEMS switches such as gallium, mercury, sodium, potassium, phosphorous, and tin. The representative liquid metal is mercury for micro electrical switches because mercury can be used in the liquid state at room temperature. However, liquid mercury has potential health and safety problems [10]. Also, the mercury droplet can be difficult to use at high frequency (kHz) operation because it does not always have good adhesion between liquid-solid contact [10]. For maintaining the liquid-solid contact, high mechanical force should be applied to mercury droplets, and moreover the external force can be one factor to detach the droplets from the original positions. Therefore, there is interest in finding alternative metals for thermal and electrical switches. In the present study, we investigated gallium (Ga) as a contact metal for use in MEMS switches because Ga has similar properties to mercury, such as low melting temperature and good thermal conductivity, but Ga can also be used as a solid and then re-melted to "heal" a damaged switch. To understand the structure of Ga micro-droplets for switching operations, the Ga droplets were deposited on the fabricated samples by electroplating, and the behavior was investigated by characterizing the macro structure, hardness, thermal reflow and electrical performance in a simulated MEMS structure, consisting of tungsten (W) films on silicon (Si) and SiO_2 substrates.

EXPERIMENT

The samples were fabricated from single side polished 3 inch diameter (100) silicon (Si) wafers. To increase the conductivity for electroplating and the electrical experiments, Boron (B) was used to dope the Si. Silicon dioxide (SiO_2) was grown as a thin (120-180μm) oxide layer on both sides of the wafer using wet oxidation after boron doping. A patterned mask of thirty dies with various size dots was used for the photolithography process. After patterning the photoresist, the oxidation layer was chemically etched in Buffered Oxide Etch (BOE). A 10nm thick titanium / tungsten (Ti / W) layer for adhesion was DC sputtered on the polished side of the wafer, and a 100nm thick W was then sputtered on the surface of Ti / W. After a photolithography process, the W and Ti / W layers were chemically etched in H_2O_2. The wafer was cleaned with 5 rinsing steps with acetone, isopropyl alcohol, and deionized (DI) water, and then diced into thirty test samples.

The electrolyte consisted of gallium chloride and 140 ml solution of deionized (DI) water and hydrochloric acid (HCl). The pH of the solution was kept between 1.5 and 2.5, adjusted by adding HCl, and the concentration of gallium was 2.5*M*. The conditions for the most uniform deposition used a current density, a cathode current, and a cathode potential of 1A/cm^2, between 0.622A and 0.942A, and -0.6 to -0.7 V respectively, and the temperature was kept between 40°-50°C. A platinum wire and the sample were placed in a glass beaker as a H_2 evolution anode and cathode respectively [11]. Ga selectively plated on the W surfaces with only small amounts depositing on the B doped Si.

The thermal reflow test was carried out in a furnace and ambient atmosphere. Reflow test samples were put in the hot furnace. After the desired time was reached, the samples were cooled by switching off the furnace, the samples were taken out from the hot furnace.

The test sample with micro-droplets was fixed in place with a vacuum chuck. To apply a voltage to the test sample, two tungsten (W) tips were used to contact the micro-droplet for the negative electrode and the B doped Si surface. The W tips were mounted on micro actuators which can be controlled in the X, Y, and Z axes, and were connected with vacuum pumps and the power supply. After a 3 V DC potential was applied with power supply, the W tip lowered into contact for all hot switching experiments. A Fluke 189 digital multimeter was used to monitor current to produce steady state current-voltage curves, while for hot switching the voltage across a fixed resistor was monitored with an oscilloscope to determine the transient current performance. After the electrical performance testing the droplets were subjected to another thermal reflow cycle, when the resistivity and switching behavior was tested again. The morphology of the samples was observed using an FEI SIRON field emission scanning electron microscope.

RESULTS and DISCUSSION

Ga droplets were electroplated on the W spots for 78 seconds, and then thermal reflow was carried out for the droplets at 100°C for 10 min. The result and structures are shown as SEM images in figure 1 Before thermal reflow, the surface shapes of Ga droplets were rough as shown in figure 1 (a), but after thermal reflow, the shapes were relatively smooth, and formed more spherical geometries as seen in figure 1 (b) which improves reproducibility of the contact.

(a) Before thermal reflow

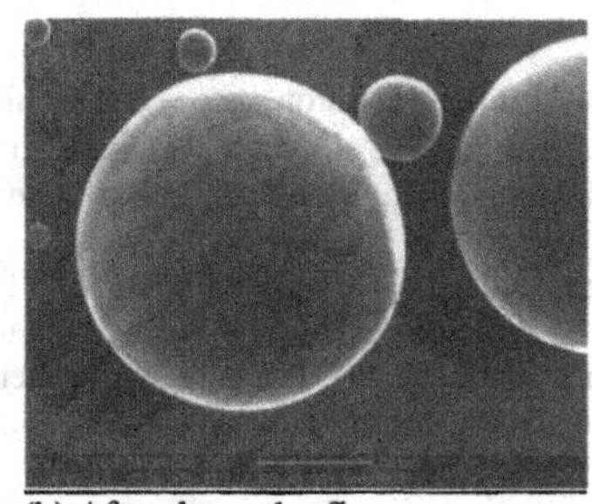

(b) After thermal reflow

Figure 1. SEM pictures of electroplated gallium (Ga) droplets on 45 μm W target in 0.2 mm SiO_2 holes after and before thermal reflow at 100°C for 10 min.

The electrical properties of Ga droplets were determined by current-voltage scans. Figure 2 shows an average of 10 I-V curves of the Ga droplets. In figure 2, the slope represents the total conductance. The non-linear behavior has not been investigated in detail, but may be due to creep and deformation around the contact during the test, which would increase the contact area at the gallium – tip contact and lead to a lower resistance (and hence the higher current at increasing voltages as the experiment was started at low voltages and the voltage increased with time). The total resistance was 31.8Ω.

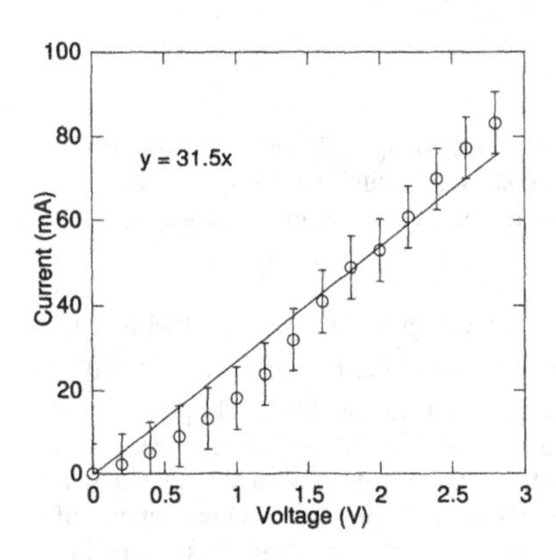

Figure 2. Current-voltage relationship for 10 deposited Ga droplets on W.

The total resistance consists of the resistances of Ga, W, and B doped Si. The resistance is

$$R = \rho \frac{l}{A}, \tag{1}$$

where l is the length of the conductor, A is the cross-sectional area of the conductor, ρ is the electrical resistivity (Ω cm). The total resistance (R_{total}) is

$$R_{total} = \rho_{B\text{-}Si} \frac{l_{B-Si}}{A_{B-Si}} + \rho_W \frac{l_W}{A_W} + \rho_{Ga} \frac{l_{Ga}}{A_{Ga}} \tag{2}$$

The resistivity value of Ga droplets (ρ_{Ga}) was calculated from the equations (2) by using the value of $R_{B\text{-}Si}$ and R_W. The ρ_{Ga} value was equal to $1.1 \pm 0.6 \times 10^{-5}$ Ω cm. It implies that the Ga droplet has high conductivity for MEMS switches, and matches what would be expected for this material [12].

Figure 3 shows electrical performance of the Ga droplet switch at a constant voltage (3V DC). Repeated switching shows the second contact is a lower resistance contact, but the steady state resistance increases with the third and further contacts in figure 3 (a). The contact degrades with switching due to deformation and defect generation.

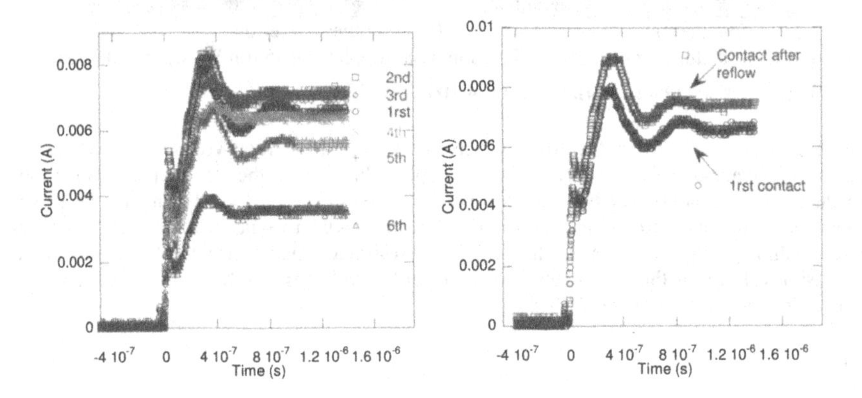

Figure 3. The switch behavior of the Ga droplet at a constant voltage (3V DC). (a) First series of electrical switching tests showing degradation with continued switching. (b) Electrical performance after thermal reflow, where performance now exceeds the initial conditions.

This phenomenon is particularly defined by figure 4. Figure 4 shows the SEM images of the defected Ga droplet after switching and the healed droplet after thermal reflow at 120°C. After electrical switching there were some defects on the surface of the Ga droplet, so the contact area between the Ga droplet and W tip was decreased and the resistance increased as increasing contact cycles. However, after thermal reflow, the Ga droplet in the switch was healed, and then the resistance was restored to the initial value; figure 3 (b). Other reports of switches in metallic contacts during indentation that show changes in resistance as a function of asperity contact [15] note differences in the position and the contact area and a general increase in roughness with increased switching cycles . However, in this study even though the Ga droplets can suffer damage during switching, the defects can be fixed by a moderate thermal reflow due to the properties of Ga, and the reliability and performance of the solid-solid micro scale electrical switch can be restored to the initial condition.

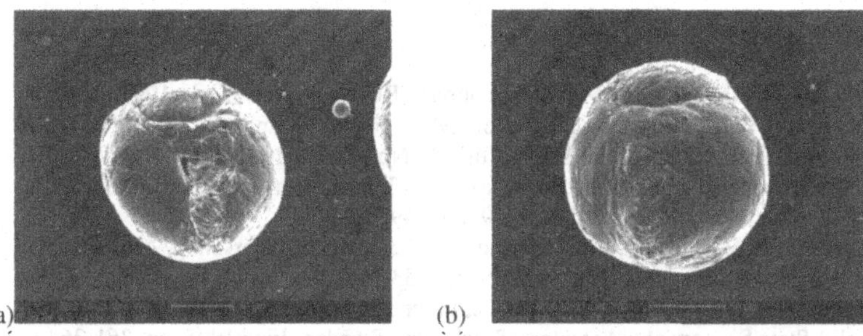

Figure 4. SEM images of the Ga droplet after switching showing damage on the top of the droplet (a) before and (b) after thermal reflow at 120°C, where the damage has be removed.

The hardness of the Ga micro-spheres were measured using the Vickers technique. The indentation of the cross section of the Ga spheres the impression was not a perfect square, and did show pin cushioning around the hardness test impression. The average of six Vickers hardness values of the Ga was 17.2 HV. The maximum applied stress can be calculated with the Vickers hardness value of the Ga droplets and mean pressure [13]. For the electroplated Ga, the estimated flow stress is 60 MPa. The maximum applied load prior to plastic deformation is

$$\mathrm{P} = \frac{R^2(\pi\sigma_y)^3}{0.227E^2} \tag{3}$$

where R is radius and E is elastic modulus [14]. For sphere surface, the radius should be combined the radius of the Ga droplet with the W tip. The radius of the Ga droplet, which was deposited on the 60μmW dot, and the W tip were 50μm and 15 μm respectively. The elastic modulus of the Ga was 9.8GPa, so the maximum load for these switching conditions should be approximately 40 μN.

CONCLUSIONS

The behavior of a Ga micro-droplet was investigated by characterizing the structure, hardness, and electrical performance of the features. The initial surface shape of electroplated Ga was rough, but the surface smoothed and formed a spherical geometric after a thermal reflow. The Ga droplet had low resistivity ($(1.13 \pm 0.6) \times 10^{-5}\ \Omega$ cm), and When the switch cycles were repeated several times, the resistance increased due to the deformation and defects from contact between the W probe and Ga droplet. However, after a 120 °C thermal reflow the resistance returned to the initial value. The hardness of the Ga droplet was 17.2 HV, suggesting that the maximum applied stress and load that can be applied to a Ga droplet in this MEMS device is 60 MPa and 40 μN respectively. Therefore, electroplated Ga micro droplets can be used for electrical switching under substantial mechanical loads, and moderate damage that may occur during switching can be eliminated with a moderate thermal reflow process.

REFERENCES

1. M. Madou, Fundamentals of Microfabrication, CRC Press, New York, 1997.
2. A. Witvrouw, H.A.C. Tilmans, I. De Wolf, Microelectronic Engineering **76** (2004), 245-257.
3. K. Petersen, Proceedings of the IEEE, Vol. **70**, No. 5, May 1982, pp. 420-457.
4. M. Sakata, Proceedings of the Tech Digest IEEE Micro Electro Mechanical Systems, Salt Lake City, UT, USA, February 1989, pp. 149-151.
5. S. Roy, and M. Mehregany, Proceeding of the IEEE MEMS Workshop, Amsterdam, The Netherlands, January-February 1995, pp. 353-357.
6. E. Hashimoto, Y. Uenishi, and A. Watabe, Proceeding of the International Conference on Solid-State Sensors and Actuators, Stockholm, Sweden, June 1995, pp. 361-364.
7. E.A. Sovero, R. Mihailovich, D.S. Deakin, J.A. Higgins, J.J. Yao, J.F. Denatale, and J.H. Hong, Proceedings of the SBMO/IEEE MTTT-S IMOC'99, **1**, Rio de Janeiro, Brazil, August 1999, pp. 257-260.
8. J. Kim, W. Shen, L. Latorre, and C.J. Kim, Sensors and Actuators A, **97-98**, 2002, pp. 672-679.
9. J. Simon, S. Saffer, and C.J. Kim, IEEE MEMS'96 Proceedings (1996), pp 515-520.
10. L.T. Taylor, J. Rancourt, and C.V. Perry, United States Patent Number 5478978, 1995.
11. S. Sundararajan and T.R. Bhat, J. Less-common Metals, **11**, 1966, pp. 360-364.
12. P.D.L. Breteque, Industrial and Engineering Chemistry, **56**, 1964, pp. 54-55.
13. G.E. Dieter, McGraw-Hill Book Company, London, 1988.
14. M.A. Meyers and K.K. Chawla, Prentice-Hall, New Jersey, 1984.
15. C.M. Doelling and T.K. Vanderlick, J.Applied Physics **101**, 2007, p. 124303.

Mater. Res. Soc. Symp. Proc. Vol. 1139 © 2009 Materials Research Society 1139-GG03-06

Fast and Controlled Integration of Carbon Nanotubes into Microstructures

Wenjun Xu[1], Chang-Hyeon Ji [2], Richard Shafer[2], Mark G. Allen[2]
[1]Department of Polymer, Textile and Fiber Engineering,
[2]Department of Electrical Engineering, Georgia Institute of Technology, 791 Atlantic Drive, Atlanta, GA, 30032, U.S.A.

ABSTRACT

In this paper, we report the results of a rapid and room temperature integration approach for the selective and structured deposition of carbon nanotubes (CNTs) into three-dimensional microstructures. The approach exploits electrophoretic deposition (EPD) from an aqueous suspension of CNTs, together with suitably patterned and electrically-energized microstructure-bearing substrates. Uniform 2-D and 3-D micropatterns of CNTs on wafer scale have been achieved in less than 4 minutes with controllable thicknesses ranging from 133nm to several micrometers. Orientation of the deposited CNTs was observed in microstructures with certain dimensions. Surface hydrophobicity of the microstructures was found to be critical in achieving well-defined micropatterning of CNTs. A hydrophobic microstructure surface leads to the selective patterning profiles of CNTs, while a hydrophilic surface induces CNTs assembly over the entire microstructure, with resultant loss of selectivity. This approach can be further extended to fabricate 3-D micropatterns with multilayer materials on flexible substrate through the aid of transfer micromolding techniques.

INTRODUCTION

CNTs have been intensively investigated since 1991 and exhibit promising potential in fields such as molecular electronics, microsensors, solar cells, field emission devices and biocoatings [1-5]. This high aspect ratio nanomaterial possesses interesting piezoresistive, electrical, physiochemical and mechanical properties that can serve as a key enabler in microsystems such as microsensors and microactuators [6-8]. To integrate this versatile material into miniature systems, suitable approaches are required for localizing CNTs in certain desired areas of the microstructure. In general, there are two approaches to achieve this localization: during CNT synthesis, and post-CNT synthesis. The former approach usually involves pre-patterning of catalyst at the desired sites, followed by a high temperature growth reaction. Although this approach deposits the nanotubes in a desired area directly, the high temperature required as well as the inability to refine the CNTs prior to deposition are potential limitations. The latter approach overcomes both of these limitations, at the expense of extra processing steps.

There has been much previous work to pattern CNTs on substrates using a post-synthetic approach. Terranova et. al. reported assembly of CNTs across micron-sized gaps between electrodes via dielectrophoresis under AC field [9]. Cui utilized polymer-assisted self-assembly to obtain CNT layers on microstructures [10]. An AFM tip has also been used to place individual CNTs onto desired locations in nanodevices.

Good control of the thickness and morphology of CNTs is also important for quantitatively studying their properties in micro-systems and optimize the performance of resultant devices. Electrophoretic deposition (EPD), a technique in which electric fields are used to assist the transport of charged species, has been demonstrated to be capable of uniformly depositing CNTs with controlled mass into large substrates at room temperature, and is a facile, fast and cost effective process [11].

In the present study, the use of EPD approaches together with substrates bearing suitably patterned microstructures was adopted to investigate the selective deposition of CNTs onto large areas in a single step. Three different microstructures were investigated and their suitability for selective CNT deposition evaluated. Uniform 2-D and 3-D micropatterns were successfully generated in less than 4min at room temperature. The assembly behavior and morphology of the CNTs were studied and the effect of surface hydrophobicity of the microstructures was found to be a critical factor for achieving selective deposition.

THEORY AND EXPERIMENT

MWCNTs (Nanostructured & Amorphous Materials, Inc. 95+%) were acidified in a mixture of sulfuric acid and nitric acid (3:1 volume ratio) for 4h through sonication at room temperature. Carboxyl acid groups were thereby introduced into the nanotubes; these acid groups were utilized both to facilitate the EPD, as well as to improve the dispersion of the CNTs in aqueous solution. Three geometric types of microstructures were fabricated on silicon substrates. The first microstructure consisted of interdigitated gold microelectrode arrays, in which the electrodes were deposited on a dielectric layer which was formed on top of the silicon. The second microstructure consisted of microchannels, formed in photoresist, atop a conducting gold layer that had been deposited atop a silicon-nitride-bearing silicon wafer. The third microstructure was similar to the microchannel structure, but instead was fabricated from SU-8 epoxy and consisted of circular wells instead of channels. MWCNTs-COOH were then dispersed in water at a concentration of 0.1mg/ml. The microstructure-bearing substrates and a copper counterelectrode were immersed in the CNT dispersion and electrically energized with an external voltage source, with the copper serving as cathode and substrate serving as anode. The negatively-charged CNTs-COO^- were then electrically directed in water into the microstructures under a constant electric field of 20V/cm applied across two electrodes for various times.

The deposited mass, *M*, can be estimated by the equation 1 [12]:

$$M = \int_0^t \alpha A \mu E c dt \qquad (1)$$

where α is the mass fraction of material deposited on the electrode, A is the area of the electrode, μ is the material mobility, E is the applied electric field, and c is the material concentration in the dispersion. To achieve localized deposition of CNTs on the microscale, it is observed that low c and E are preferred. These two parameters also determine the thickness of the CNTs layer since A, μ and c are constants for a given microstructure/dispersion combination.

The morphology of the CNT patterns was characterized by Zeiss Scanning Electron Microscopy (SEM) Ultra60. Atomic force microscopy (AFM, MultiMode and Dimension 3000, Veeco Metrology) was used for topography and surface roughness study of the SU-8 microstructures before and after oxygen plasma treatment. Static contact angle measurements were performed with a sessile drop method for the surface hydrophobicity test. WVASE32 Woollam Ellipsometer was utilized to study the thickness of the CNTs layer on the microelectrodes.

DISCUSSION

Assembled micropatterns of CNTs in microstructures

The micropatterns generated by EPD as deposited in various microstructures are shown in Figure 1. When no substantial topography was present in the microstructures, the resultant CNT micropatterns typically assembled only onto those conductive metal areas to which electric fields had been applied. The negatively charged CNTs were neutralized and aggregated when reaching those surfaces. It was also observed that immediately adjacent conductive metal areas that were not energized exhibited no deposition. Figure 1a and 1d confirm that CNTs did not assemble on

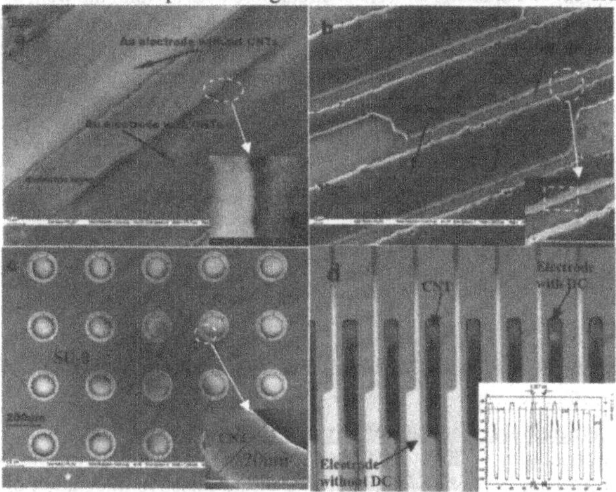

Figure 1. SEM images of CNTs assembly in (a) interdigital microelectrodes, (b) photoresist (PR) microchannels, (c) SU-8 microwell arrays and (d) optical image of the CNTs patterns in microelectrodes, the insert shows an electrode thickness measurement, illustrating the height difference between deposited and non-deposited electrodes.

any micropatterns lacking an applied electric field. Similarly, no deposition was observed on the dielectric materials adjacent to and between the metal areas. The thickness of the deposited CNT layer was approximately 133nm as illustrated in Figure 1d (insert), corresponding to approximately two to three layers of CNTs since the diameter of the CNTs is 50~80 nm. As observed in Figure 1c, CNTs assembled conformally around the microwell inner surface and in addition generated a 20um wide circular belt framing the well opening, thereby creating a continuous 3-D coating in each isolated microwell. Similar assembly behavior was observed in the rectangular microchannels as shown in Figure 1b, where conformal assembly on the channel inner surface and a 5-10um wide rectangular CNTs belt framing the channel was generated. These observations indicate that under these conditions, the deposition of CNTs is in accordance with the conformal coating nature of the EPD process.

Alignment of CNTs

CNTs have been reported to align perpendicular to the electrode substrate (i.e., parallel to the external field lines) under high electric [13] and magnetic fields [14]. In the present study, the CNTs were found to undergo oriented assembly in the rectangular channel and microwell structures when these microstructures were of sufficiently small dimensions (15um and 3um, respectively). For the microchannel structures, the CNTs formed a dense and one dimensional assembly transversely at the bottom of the channel, and also tended to assemble in a parallel fashion on the sidewall of the channel, as observed in Figure 2a. For the microwell structures, the CNTs tend to form a coating around the circular opening in a way that is parallel with the sidewall without clogging the opening, as well as circumferentially within the well, as shown in Figure 2b. However, this orientation effect was only observed in microstructures having openings with relatively small dimensions. For microstructures with larger dimensions, the CNTs seem to assemble in a much less oriented or random manner.

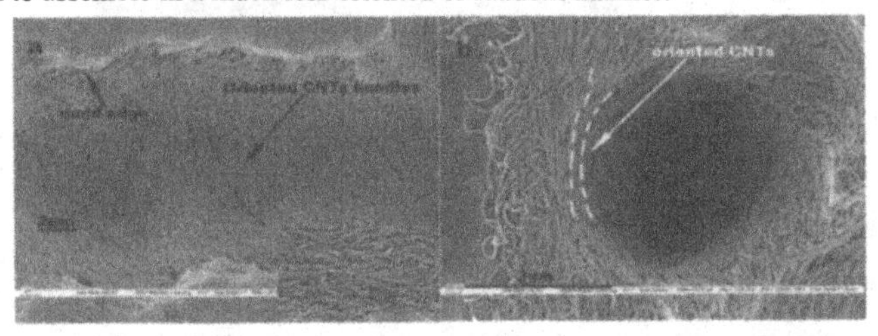

Figure 2. SEM images of aligned CNTs in (a) micro-channel and (b) circular well opening.

Effect of surface hydrophobicity on selectivity of CNT assembly

The surface hydrophobicity of the non-conducting layers, including the top surface exposed to the dispersion, was found to play a significant role in the selective integration of CNTs into microstructures. In the case of interdigitated microelectrodes, the dielectric layer, which is underneath and between the microelectrodes (Figure 1a), was exposed to a brief (i.e., insufficient to completely remove the dielectric layer) hydrofluoric acid (HF) aqueous solution during the electrode patterning process. The contact angle of the dielectric layer after the process decreased significantly as shown in Table I. This rise of the hydrophilicity results in the continuous deposition of the CNTs over the whole microstructure, including the non-conductive dielectric region.

Table I. Contact angle of the dielectric layer

	contact angle without HF etching	contact angle after HF etching
parylene	86.47°	59.54°
SiC	86.05°	43.88°
SiN_x	63.92°	40.41°
SiO_2	38.78°	<10°

This may be due to the fact that CNTs aqueous solution can form a good interface with the hydrophilic surface where the moving CNTs under dc field collide and aggregate with those in the interface layer. This negative effect can be reduced by changing the microelectrode patterning process from etching to lift-off. In this case, the hydrophobicity of the dielectric layers was preserved and 2-D pattern of CNTs on microelectrodes can be achieved as shown in Figure 1a with parylene as the dielectric layer.

A similar hydrophobicity effect was revealed in patterning CNTs into SU-8 microwell arrays. The surface topography of the pristine SU-8 before and after plasma treatment is shown in AFM images in Figure 3. Oxygen plasma increases the surface roughness of the SU-8 from 0.289nm

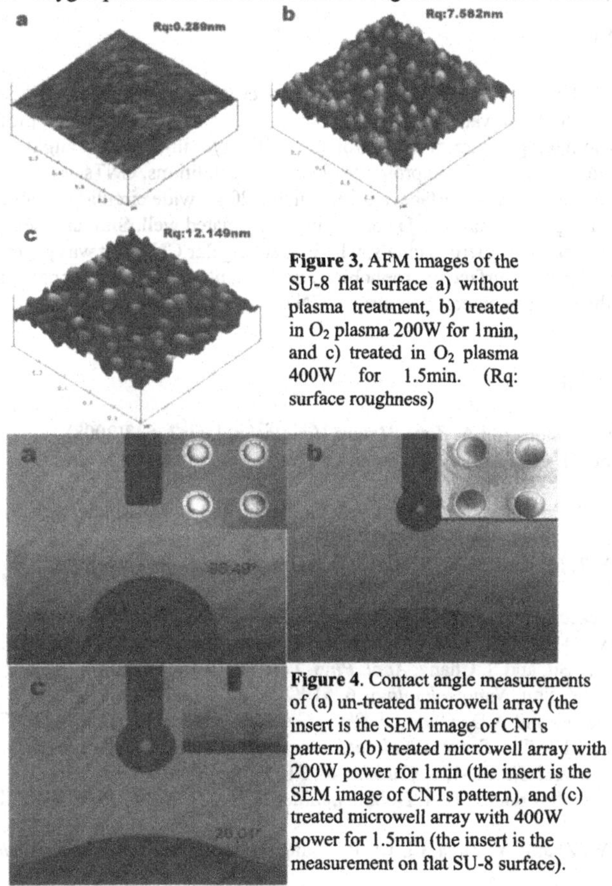

Figure 3. AFM images of the SU-8 flat surface a) without plasma treatment, b) treated in O_2 plasma 200W for 1min, and c) treated in O_2 plasma 400W for 1.5min. (Rq: surface roughness)

Figure 4. Contact angle measurements of (a) un-treated microwell array (the insert is the SEM image of CNTs pattern), (b) treated microwell array with 200W power for 1min (the insert is the SEM image of CNTs pattern), and (c) treated microwell array with 400W power for 1.5min (the insert is the measurement on flat SU-8 surface).

up to 12.149nm. The contact angle of the micropattern dropped from 96.49 ° to 28 °, that is, the surface turned from hydrophobic to hydrophilic. The CNTs then formed a coating over the whole SU-8 microstructure (SEM image insert in Figure 4b) instead of generating a well-defined micropattern in each isolated microwell (SEM image insert in Figure 4a). Another observation is that the microwell patterns exhibit higher hydrophobicity than the flat SU-8 surface in all three stages. As illustrated in Figure 4c, after the oxygen plasma treatment, the contact angle is 28 ° for the microstructure region and is close to 0 ° for a flat (i.e., non-microstructured region) in the field of the wafer. This is believed to be similar to a structure-induced hydrophobicity.

CONCLUSIONS

Uniform 2-D and 3-D micropatterns of CNTs over wafer-scale areas have been successfully achieved via EPD from water-based dispersions. The deposition time was typically less than 4 minutes at room temperature, with controllable CNT layer thicknesses ranging from 133nm to the micrometer range. Under appropriate deposition conditions, CNTs assembled conformally around the microwell inner surface and generated a 20um wide circular belt around the well opening, creating a continuous 3-D coating in each isolated well. Similar assembly behavior was revealed in rectangular microchannels, where a rectangular CNT belt was generated around the channel opening. The surface hydrophobicity of the non-conductive regions in microstructures is critical for the selective micropatterning of CNTs.

REFERENCES

1. K. Jensen, K. Kim, and A. Zettl, *Nature Nanotechnology* **3**, 533(2008).
2. T. Kawano, H. C. Chiamori, M. Suter, Q. Zhou, B.D. Sosnowchik, and L. Lin, *Nano Lett.* **7**, 3686 (2007).
3. S. Berson, R. de Bettignies, S. Bailly, S. Guillerez, and B. Jousselme, *Adv. Funct. Mater.* **17**, 3363(2007).
4. P. Rai, D. R. Mohapatra, K. S. Hazra, D. S. Misra, and S. P. Tiwari, *Appl. Phys. Lett.* **93**, 1921 (2008).
5. A. R. Boccaccini and Q. Chen, *Adv. Funct. Mater.* **17**, 2815 (2007).
6. R. J. Grow, Q. Wang, J. Cao, D.Wang, and H. Dai, *Appl. Phys. Lett.* **86**, 093104 (2005).
7. N.Chang, C. Su, and S. Chang, *Appl. Phys. Lett.* **92**, 063501 (2008).
8. J. Tong andY. Sun, *Nanotechnology*, **6**, 519(2007).
9. M.L Terranova and A.D. Carlo, *J. Phys: Condens. Matter.* **19**, 225004 (2007).
10. W. Xue and T. Cui, *Sensors and Actuators A: Physical* **136**, 510 (2007).
11. R. B. Aldo and S.P.S. Milo, *Carbon* **44**, 3149 (2006).
12. M. Guduru, A. Francis, T. A. Dobbins, *Mater. Res. Soc. Symp. Proc.* **858E**, HH13.29.1 (2005).
13. C. Ma, W. Zhang, Y. Zhua, Li. Jia, R. Zhanga, N. Koratkarb, J. Liang, *Carbon* **46**, 706 (2008).
14. K. Kordas, T. Mustonen, G. Toth, J. Vahakangas, A. Uusimaki, H. Jantunen, A. Gupta, K. V. Rao, R. Vajtai, and P. M. Ajayan, *Chem. Mater.* **19**, 787(2007).

Mater. Res. Soc. Symp. Proc. Vol. 1139 © 2009 Materials Research Society 1139-GG03-07

Through Silicon Vias in Micro-Electro Mechanical Systems

Stephan Warnat[1], Ramona Ecke[2], Norman Marenco[1], Sven Gruenzig[1],
Wolfgang Reinert[1] and Peter Lange[1]
[1]Fraunhofer Institute for Silicon Technology, Fraunhoferstrasse 1, 25524 Itzehoe, Germany.
[2]Chemnitz University of Technology, Center for Microtechnologies,
Reichenhainer Strasse 70, 09126 Chemnitz, Germany.

ABSTRACT

Through silicon vias (TSV) are widely discussed for 3D integration in CMOS devices to pursue the aggressive scaling of the historical Moore's law. Micro-electro mechanical systems (MEMS) can take benefit from this technology in order to combine the MEMS with an integrated circuit (IC) or to enhance the robustness of MEMS. The technological aspects of TSV with regard to the application for MEMS will be introduced. A typical pad structure of such a system is taken to evaluate essential process steps like wafer thinning, TSV hole formation and film deposition.

INTRODUCTION

TSV are used in MEMS for several advantages. They are able to reduce parasitic capacitances and resistances for the signal processing of very small signals, for example in inertial measurement systems, by creating short interconnect. But also they give a chance to overcome measurement problems given by bond wires and organic protection layers (glob top) in harsh environments. For these devices the electrical interconnect area can be fed through the wafer to the backside by vertical interconnects. Mass flow sensors for gases and fluids [1] are a typical example. Corrosion of pads and wire bonds occur if they are not well protected against the media investigated. The glob top limits the device functionality and reliability under various environmental conditions. Figure 1 shows a typical interconnect area of a water flow sensor after a field test. Due to water infiltration of the glob top interface area to the substrate a significant corrosion of the pads can be observed.

Therefore, separation of the active MEMS layers and the interconnections area is preferable, which is possible with through silicon vias (TSV) [2]. TSV connect the front side pad with the wafer back side by a hole etched through the complete wafer thickness, followed by isolation and metallization deposition, structuring of the metal film and a final passivation layer. Solder balls or stud bumps are used to connect these MEMS with a substrate or directly to an application-specific integration circuit.

The requirements/challenges of TSV for MEMS applications are not comparable with TSV for integrated circuit devices. Several TSV approaches for IC require a wafer thinning down to10µm ... 100µm. Such a thickness would change the MEMS performance significantly due to their high sensitivity to stress effect. These effects, however, will even increase with decreasing wafer thickness. This paper therefore shows a TSV integration approach for MEMS wafer with thicknesses > 300µm and an aspect ratio of 4. The TSV formation is performed with standard MEMS and CMOS technologies in order to develop an adaptable process flow for all of our MEMS products. Selected technology-related challenges of the process flow will be presented on a demonstrator and the process robustness will be discussed.

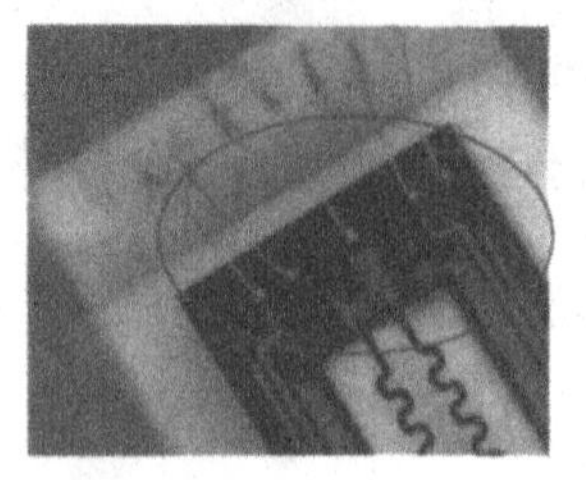

Figure 1: Corroded bond-pads (black areas on the conduction lines) on active MEMS side after field test. A glob top was removed after a field test.

EXPERIMENT

MEMS-Demonstrator

The TSV development for MEMS wafer is performed on demonstrators. The pads for the read out of the MEMS signals may have very different stack compositions. They are strongly dependent of the applied technology whether it is surface or bulk micromachining. In this work we have chosen a pad structure for a typical MEMS application, which is displayed in Figure 2. A TSV integrated in a MEMS wafer is shown, in the close up also the different layers of the pad structure are displayed. The pad structure basically is composed of an isolation -, a metal – and a passivation layer. This film stack for example is in detail a Si_3N_4 from low pressure chemical vapor deposition (LPCVD), a phosphorus silicate glass (PSG) deposited by plasma enhanced chemical vapor deposition (PECVD) and a low temperature oxide (LTO) from a low temperature LPCVD for the isolation layer. A sputtered Ti, TiN and AlCu metallization film and a passivation layer of silicon nitride and silicon oxide deposited by PECVD is completing the pad structure. This composition can affect the stability of the released pad significantly and will be discussed later.

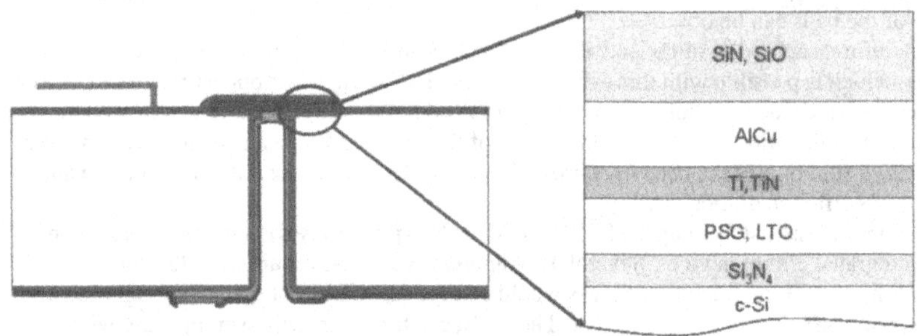

Figure 2: The left sketch shows a TSV integrated in a MEMS wafer. The close up shows a chosen pad structure of a typical MEMS design.

TSV formation

TSV connect a MEMS signal (capacitive, resistive,...) from the wafer front side to the back side and allow a separation of the active MEMS area and the electrical interconnects. Therefore, a hole has to be etched from the wafer back side down to the Ti,TiN layer. Deep reactive ion etching (DRIE) [3] is used for the TSV hole formation. The Si etch rate of the instrument used is too low in order to form the hole in a reasonable time. Therefore a wafer thinning down to 300μm is performed. This thickness allows a handling without any support wafer, which is common in other TSV approaches. DRIE is highly selective against silicon oxides. Therefore, the etchimg stops at the PSG, LTO layer. An isotropic gas phase etching process removes these oxides. SiN is deposited by PECVD inside the hole and opened again at the hole bottom by an anisotropic spacer etch process. Metal organic chemical vapour deposition of TiN and Cu [4] is used as a plating base for the subsequent electro plated Cu film deposition. The Cu film structure is defined by a dry resist film and a wet chemically etching process. SiN is deposited on top of the Cu and structured by a dry etching process in order to act as a hard mask for the subsequent TiN wet etching process. The electrical TSV resistance is measured by a four point measurement probe.

RESULTS AND DISCUSSION

This chapter shows selected results of technological challenges during the TSV formation, which we see as a critical impact for a reliable TSV in MEMS.

Thinning

Wafer thinning is performed in different steps [5]. A coarse grinding removes most of the silicon, before a fine grinding defines the back side surface. A wet chemical stress release etch is required, because mechanical thinning introduces different zones of damages. Handling without this etching will result in low yield due to the introduced large wafer bow by the mechanical stress [5]. The wafers will break during the handling in different process steps. However, also a TSV processing with a stress release etch after the fine grinding has shown a reliability risk. The DRIE etching behavior of these thinned wafers is different compared to blank substrates. A chipping was observed at the entrance of the TSV, like it is shown in the left picture of Figure 3. Arrow (a) indicates the preferential grinding wheel direction. We assume that the observed chipping results from the wheel direction. To give a detailed explanation of the interaction of the damage zones and the wheel direction is out of the scope of this paper. It should be mentioned, that the observed cracks are not comparable with "Mouse-bits" [6] created during DRIE, because "Mouse-bits" are etching time related artifacts and the observed cracks are also observable after short etching times.

Arrow (b) of Figure 3 indicates a rupture line after the DRIE process and a followed silicon nitride deposition formed by PECVD. The scanning electron microscope picture of Figure 3 shows the cross section of this fracture. The SiN film covers the observed cracks homogeneous and KOH pin hole tests have shown that this film is electrically stable. No pin holes were observed, even where the described cracks are formed.

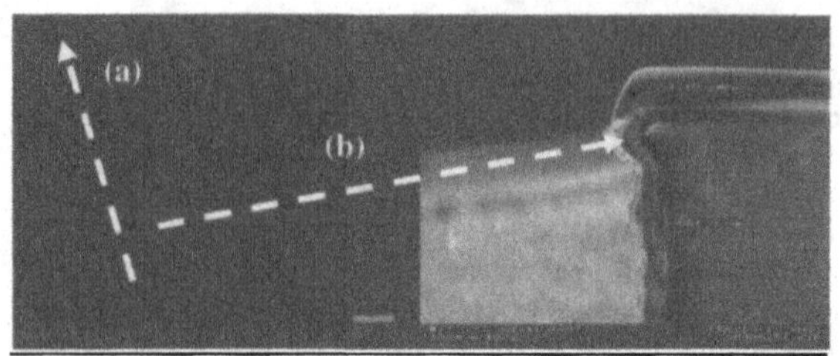

Figure 3: Left:Top view of an etched hole; Right: Cross section of an etched hole which is deposited with an SiN film.

Front side pad stability

The front side layers (metal layers and passivation) must be stable during the complete processing in order to fulfill the requirement of a separated MEMS and interconnect area. The etching of the oxide layer beneath the metal and passivation by an isotropic gas phase etching is the most critical process for the stability. Various oxides show different etch rates. Sacrificial layers like phosphorus-doped silicon glasses etch 19 times faster compared to a thermal oxide [7]. These etch behaviors result in a high vertical under etch. We observed even a higher under etch if PSG is deposited on top of a silicon nitride layer formed by LPCVD. We assume that a local stress between these two layers enhances the vertical PSG etch rate. An under etch over a 100μm distance was observed at our standard working conditions. A temperature increase during gas phase etching reduces the under etch. However, this under etch results in non stable pads during the followed processes. Figure 4 shows a high pad buckling of different pad sizes. A film thickness enhancement of the passivation layer with silicon oxides shows a stable pad. This layer stack in combination with a pad size of 200μm has shown a vacuum stability of $<10^{-4}$mbar, which indicates a stable pad and allows a chance to achieve a hermiticity.

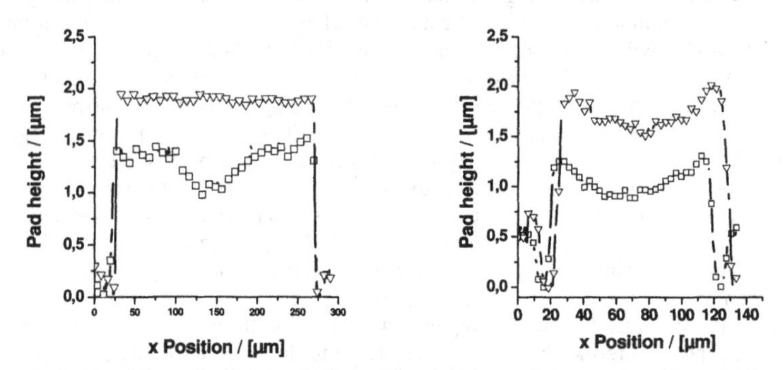

Figure 4: Pad buckling of a 200μm (left graph) and 100μm (right graph) wide pad. The red curves (∇) are with a SiO film on top of the pad structure.

Deposition

Conformal deposition of the isolation and the metallization films is the major challenge in this TSV approach. Common deposition technologies like sputtering or PECVD processes are not developed for the given geometry. Therefore, the deposited films must fulfill the electrical requirements.

The deposited SiN film thickness shows a step coverage of 50% for the technology used {PECVD} [8]. The film must be pin hole free in order to avoid leakage currents. A KOH pin hole test has shown that the film withstands this media for at least 1 hour. The same result was observed when this test was performed before and after the anisotropic spacer etch process. The electrical behavior and structural arrangement of the deposited film was shown earlier by Warnat et al [8].

An adhesion promoter and a diffusion barrier deposition are mandatory for the use of Cu metallization films. A bi-layer stack of Ti and TiN is used often for this purpose. Especially Ti is difficult to deposit conform inside the given geometry. Riedel has shown a process to use only TiN for both requirements, using MOCVD [9]. The deposited TiN film shows a step coverage close to one. It is also observable that the TiN smoothes or even fills surface roughness. Ecke et al have also shown that this TiN film has good barrier characteristics [10].

We are using in our approach a Cu liner as metallization, because thermo-mechanical simulations have shown that the pad reliability decreases by increasing the Cu film thickness [11]. The Cu film deposition behavior is similar to the observed TiN characteristics. The step coverage is close to one and a surface smoothing is observed, too. A specific Cu resistivity of $3.1\mu\Omega cm \pm 0.3\mu\Omega cm$ was determined on top of the grinded surface. We assume that the specific Cu resistivity inside the TSV is comparable to this measured value. Both surfaces, the grinded surface and the sidewalls inside the TSV, show a comparable roughness and the Cu grain structure is similar in the deposited films. This measured value is higher compared to the literature, because the values given are measured on polished wafers. The electrical resistance of two TSV connected in series by Daisy-Chain structures on the wafer front side is higher compared to the theoretical value due to a non optimized oxide layer removal under the metal pad.

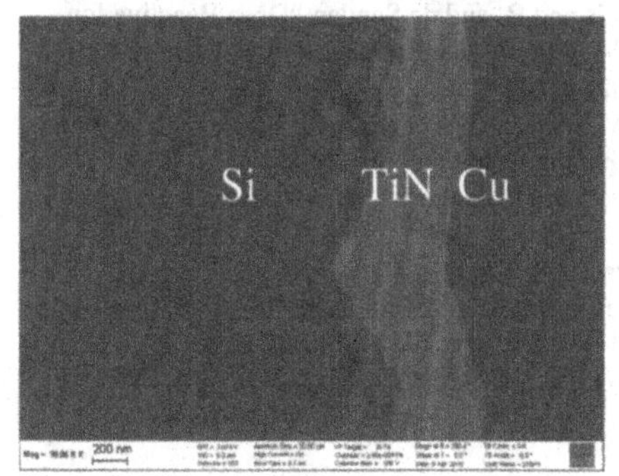

Figure 5: TiN and Cu film inside a 300μm deep TSV deposited by MOCVD. The picture is taken at a TSV depth of 200μm.

CONCLUSIONS

Technological challenges of TSV integration in MEMS devices were presented on a demonstrator. The chosen approach requires a wafer thinning to 300µm. The introduced mechanical stress in combination with the required silicon etching results in a chipping at the entrance of the TSV. The front side pad can be stabilized by a bi-layer passivation; even if an isotropic etching process is used for the oxide removal underneath the pad metal. The step coverage of the isolation and metallization deposited into the TSV shows promising results. Therefore, we conclude that our approach allows the application in harsh environments without corrosion of interconnects in MEMS.

ACKNOWLEDGMENTS

The presented work is part of the DAVID project [12], which is funded from beginning of 2006 until end 2008 by the European Commission within the Sixth Framework Programme (ref. IST-027240).

REFERENCES

1. P. Lange, M. Melani, L. Bertini, M. De Marini, 2nd *Smart Systems Integration Conference*, Barcelona, 2008.
2. S. Warnat, N. Marenco, D. Kaehler and W. Reinert, 8th *Electronic Packaging Technology Conference*, Singapore, 2006.
3. F. Laermer and A. Schilp,, *United States Patent*, **US 5498312**, 1996.
4. A. Klumpp, R. Wieland, R. Ecke and S.E. Schulz "Metallization by Chemical Vapor Deposition of W and Cu", *Handbook of 3D Integration*, ed by P. Garrou, C. Bower and P. Ramm (WILEY-VCH, 2008) pp. 157-174.
5. W. Kröninger, "Fabrication, Processing and Singulation of Thin Wafers", *Handbook of 3D Integration*, ed by P. Garrou, C. Bower and P. Ramm (WILEY-VCH, 2008) pp. 177-208.
6. F. Roozeboom, m.A. Blauw, Y. Lamy, E.v. Grunsven, W. Dekkers, J. F. Verhoeven, E. van den Heuvel, E. van der Drift, E Kessels and R. and de Sanden "Deep Reactive Ion Etching of Through Silicon Vias", *Handbook of 3D Integration*, ed by P. Garrou, C. Bower and P. Ramm (WILEY-VCH, 2008) pp. 47-92.
7. A. Witvrouw, B. Du Bois, P. De Moor, A. Verbist, C. van Hoof, H. Bender and K. Baert, *www.primaxxinc.com/products/cet_papers.html*, last visit Nov. 2008.
8. S. Warnat, M. Hoefer, L. Schaefer, H. Foell and P.Lange, *Mater. Res. Soc. Symp. Proc.* **Vol. 1036**, 2007.
9. S. Riedel, PhD-Thesis, *Technical University of Chemnitz*, 2002.
10. R. Ecke, S.E. Schulz, T. Geßner, S. Riedel, E. Lipp, M. Eizenberg, *Proc. Chemical Vapor Deposition XVI and EUROCVD 14*, 2003-08 (2003) pp 1224-1230
11. T. Falat, K. Friedel, N. Marenco and S. Warnat, *Microsys. Technol.* **Vol. 15,1**, 2009 (in print).
12. www.david-project.eu

Mater. Res. Soc. Symp. Proc. Vol. 1139 © 2009 Materials Research Society 1139-GG03-09

Effects of Aspect Ratio of Micro-sized Photoresist Patterns on Bond Strength between a Si Substrate with AFM Fracture Observation

Chiemi Ishiyama, Akinobu Shibata, Masato Sone and Yakichi Higo
Precision and Intelligence Laboratory, Tokyo Institute of Technology, 4259 Nagatsuta-cho, Midori-ku, Yokohama, Kanagawa, 226-8503, Japan

ABSTRACT

Bond strength between three dimensional micro-sized cylindrical patterns and Si substrate has been evaluated to clarify the effects of the cylinder length vs. diameter ratio, i.e. the aspect ratio, on the bond strength. Cylindrical shape was employed for avoiding ambiguity of loading point under bend conditions. Multiple cylindrical specimens of an epoxy type photoresist, SU-8 with various lengths were fabricated on a silicon substrate under the same photolithographic condition. Bond strength between micro-sized SU-8 and Si substrate under bend loading mode was measured by a mechanical testing machine for micro-sized materials. The maximum bend moment is 9.6×10^{-6} Nm in average and lineally increases with increasing the aspect ratio. On the other hand, the maximum load, i.e. maximum shear load is 106 mN in average and almost constant with increasing aspect ratio. This result suggests that the shear stress near the interface may cause the initiation of delamination. This phenomenon is discussed with three dimensional fracture observation and quantitative analysis of the line profile around the initiation site by AFM.

INTRODUCTION

Recently, chip-making technology has been significantly developed on the basis of two-dimensional photolithography technique, which utilizes fine photoresist patterns on a substrate as masking for etching or molding for sputtering. It has become possible to fabricate more ultra-precise patterns with each passing year. This technique has widened to microfabrication for micro-electro-mechanical system (MEMS) devices in concert with three-dimensional microfabrication technique, micromachining, such as anisotropic etching, reactive ion etching (RIE), LIGA processing and so on. Micro-sized mechanical elements in MEMS devices are made up of the laminating fine pattern films using these techniques, thus, the devices include a large number of interfaces between dissimilar materials. There are many factors of stress concentration and defects at the interfaces, for example, natural defects, residual stress, difference of thermal expansion factor and/or that of Young's moduli and so on. It brings about crack initiation in the vicinity of the interfaces during fabrication or in operation. Although the delamination is one of the most serious problems in MEMS devices, there are no standardized test methods to quantitatively evaluate bond strength for three dimensional microstructures. We have proposed quantitative measurement system for bond strength between three-dimensional micro-sized components and a substrate [1, 2]. A cylindrical shape has been employed as a micro-sized bond test specimen in this system, because multiple specimens can be easily fabricated on a substrate under the same condition. In addition, proper bend loading can be easily applied to the end of the cylindrical specimen.

In this study, bond strength between three dimensional micro-sized components and Si substrate under bend loading condition is quantitatively evaluated using cylindrical shape specimens to clarify the effects of the cylinder length with the constant diameter, in other words, the aspect ratio. When the load is applied to the end of the micro-sized cylinder, maximum bend moment (i.e. maximum tensile stress) at the root edge increases with increasing the cylinder length at the same load. On the other hand, shear stress at the interface between micro-sized cylinder and a Si substrate is constant at the same load in spite of the cylinder length. If the relationship between the maximum bend load and the cylinder length is clarified, it can be classified which stress factor cause the delamination, i.e. maximum tensile strength (maximum bend moment) at the root or shear stress at the interface. After the testing, delamination surfaces were observed using an atomic force micrograph (AFM) to analyze the delamination initiation site. All the results make clear the delamination process and its mechanism.

EXPERIMENT

The materials used in this study are epoxy type photoresist, SU-8 3050 made by Kayaku Microchem and Si substrate of 0.5 mm in thickness. Multiple SU-8 cylinders of 125 μm in diameter were fabricated on a silicon chip about a 15 mm square by photolithography as shown in figure 1 (a). The lithography condition in this study is as follows. A Si substrate was cleaned using alkaline solution, hot pure water, acetone and ethanol just before spin-coating. SU-8 was spin-coated at 2100 rpm for 30 sec on the Si substrate. Film thickness of the SU-8 was naturally uneven in a Si substrate, because the resist has high viscosity, which has been used for fabricating various lengths of the SU-8 cylinders under the same condition. The sheet was pre-baked at 368 K for 60 min, exposed at 300 mJ/cm^2 using UV light, post-baked at 368 K for 6 min, and then, developed using SU-8 developer for 10 min. The diameter and length of each SU-8 cylinder were measured using a confocal scanning laser microscope (SLM). The cylinder length was determined the top of a SU-8 cylinder and Si substrate surface.

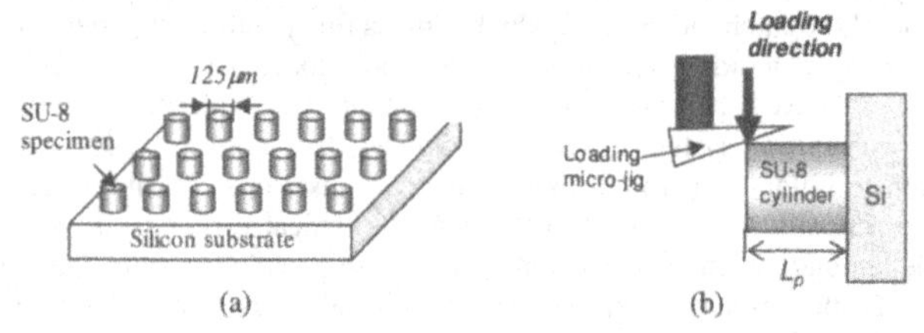

Figure 1. (a) is schematically shown the micro-sized cylindrical specimens on a Si substrate and (b) is micro-sized bond testing system under bend loading condition[1, 2].

Bond tests between SU-8 cylinder and Si substrate were performed under bend loading conditions using a mechanical testing machine, which has been developed in our previous study [3]. Bend load was applied to the end of the SU-8 cylinder at a displacement rate of 0.37 μm/s using a loading jig, which was attached to a stepping motor with precision to 0.1 μm (see figure

1 (b)). The load was measured using a precise strain gauge type load-cell. All the tests were performed at a temperature of 296 K in a clean room.

Before and after the bond testing, the delamination surface on a Si substrate was observed using a polarization micrograph. In addition, the profile of the delamination surface was analyzed using an AFM, XE-100 made by Park Systems. This AFM has a non-contact type of scanning system, thus, the delamination surface can be scanned without crushing the soft delamination surface.

RESULTS AND DISCUSSION

Effects of the aspect ratio on bond bend strength between micro-sized SU-8 cylinders and Si Substrate

Figure 2 shows a typical load-displacement curve obtained from a bending type bond test between a SU-8 cylinder specimen and a Si substrate. All of the load displacement curves are almost linear and the load is suddenly dropped after reaching the maximum load in a brittle manner as shown in figure 2. Therefore, the delamination must abruptly occur at the maximum load; thus, the bond strengths have been determined from the maximum load in this study.

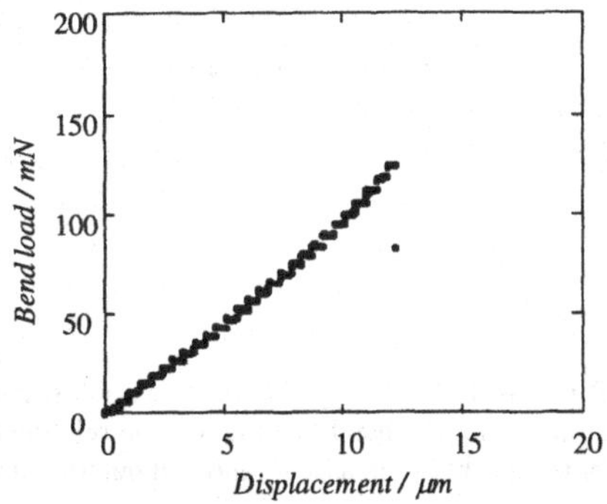

Figure 2. Typical load-displacement curve. The result was obtained from bond testing between micro-sized SU-8 cylinder and Si substrate under bend loading condition.

Table 1. The dimension of the SU-8 cylindrical specimens and their bond strength.

	Diameter / μm	Length / μm	Aspect ratio	Maximim load /mN	Max. bend momemt / 10^{-6} Nm
Meam value	125.1	90.0	0.72	106.1	9.6
Maximum value	125.7	110.0	0.88	124.0	12.6
Minimum value	124.4	71.3	0.57	86.7	6.8
Standard deviation	0.4	11.5	0.00	11.4	1.9

Table 1 concludes the results of bond bend testing between micro-sized SU-8 cylinders and a Si substrate using ten samples. The diameter of the SU-8 cylindrical specimens is 125 μm on average and their variation is less than 1 μm. The cylinder length is 90 μm on average and widely varies within 87 – 106 μm, which is 0.6 - 0.9 on an aspect ratio basis. On the other hand, both the maximum bend load and the maximum bend moment at the delamination vary widely in table 1. These results suggest that the wide variation of bond strength may be caused by cylinder length of the specimens. Figure 3 (a) and (b) show the relationship between the bond strength of micro-sized SU-8 cylindrical specimens and aspect ratio of the SU-8 cylinder with a diameter of 125 μm. The maximum bend moments clearly increase with increasing aspect ratio of the SU-8 cylindrical specimens as shown in figure 3 (a). If the bend moment at the root of the SU-8 cylinder leads to the delamination, it is contradictory to the fact that the maximum bend moment increases with increasing the aspect ratio. However, the maximum load slightly increases but is almost constant without regard to the aspect ratio as shown in figure 3 (b). This result shows that the maximum shear stress at the interface is also almost constant, regardless of the aspect ratio. It is suggested that the shear stress should cause the delamination within the aspect ratio of 0.6-0.9.

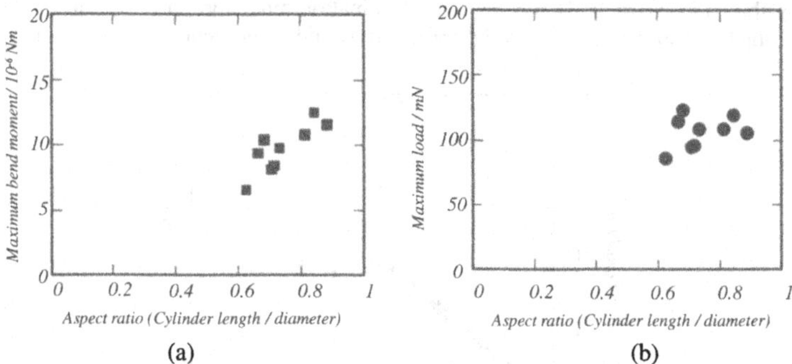

Figure 3. Aspect ratio of micro-sized SU-8 cylinder versus bond strength between micro-sized SU-8 cylinder and Si substrate under the bend loading, (a) the relationship between aspect ratio vs. maximum bend moment and (b) between aspect ratio and maximum bend load.

Fractography of the delamination surface using a polarization micrograph and AFM

Figure 4 (a) shows typical polarization micrographs of a micro-sized SU-8 specimen focused on a Si substrate before testing. There is a visible different area of several microns in width along the just inside of the edge. Figure 4 (b) shows the delamination surface after testing and the arrow shows the loading direction. SU-8 fragments of several microns in width clearly remained at the edge of the cylinder periphery on the delamination surface. The diameter of the fragment circle is several microns larger than 125 μm, which suggests that the SU-8 cylinder has a kind of fillet near the interface. All the observations show that the delamination initiated from the SU-8. In addition, there is bright circle area along the inside of the fragment circle, which seems like bare silicon substrate.

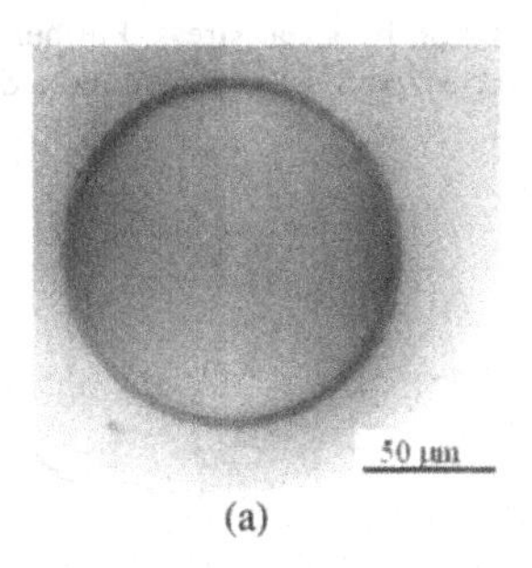

(a)

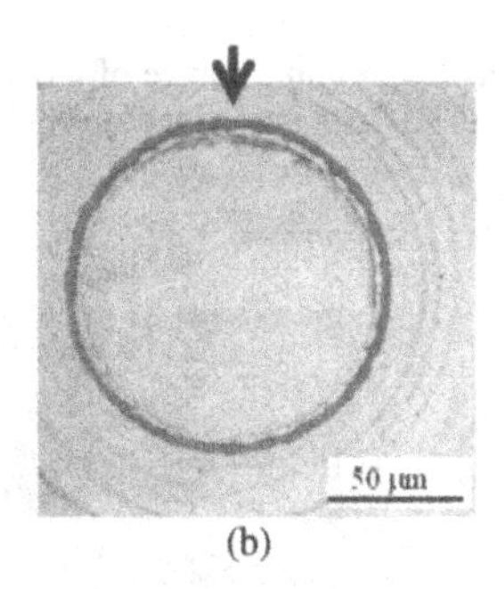

(b)

Figure 4. Microsized SU-8 cylindrical specimens on a Si substrate
(a) Polarization micrographs of a micro-sized SU-8 cylindrical specimen focused on the Si substrate before testing. (b) A delamination surface after testing. The arrow indicates the loading direction.

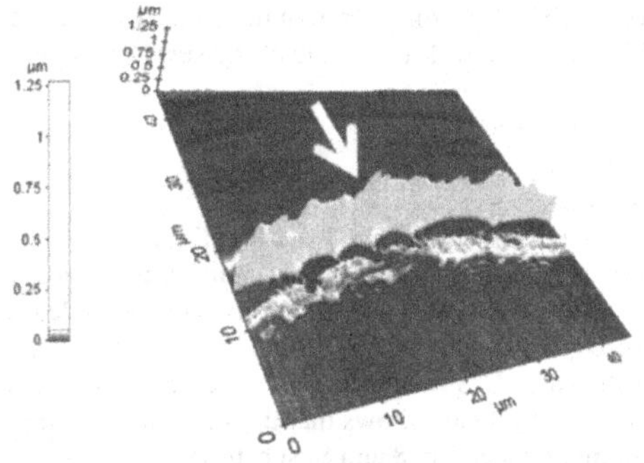

Figure 5. Three-dimensional AFM image of a delamination surface after micro-sized bond testing under bend loading condition.

Figure 5 shows the three dimensional image in the vicinity of the delamination initiation site. The white arrow shows the loading direction. SU-8 fragments were remained on most of the cylinder trace area in figure 5. SU-8 fragments around the edge of the cylinder trace are much thicker than the other area and uneven within 500nm - 1 μm. It is suggested that several micro cracks must coalesce into the main crack at the side surface of the specimen. On the other hand, there is bare-like area inside the edge of the thicker fragment. In addition, the striation-like lines, which are almost perpendicular to the delamination direction, are slightly observed inside the bare-like area on the delamination surface.

Figure 6 shows the line profile of the height on the delamination surface. The height of the bare-like area, point A'of figure 6 (a), is almost the same to that of Si surface, point A of figure 6 (a), as shown in figure 6 (b). The bare area, namely, micro-delamination area from Si substrate is just inside around the SU-8 fragments of the cylinder edge . It is suggested that the micro-delamination area has already existed before testing. In other words, the micro-delamination may have occurred during fabrication. The micro-delamination near the edge of

SU-8 cylinder may become source of the main delamination by shear stress. Further researches for clarifying the delamination mechanism between micro-sized SU-8 cylinder and a Si substrate is needed.

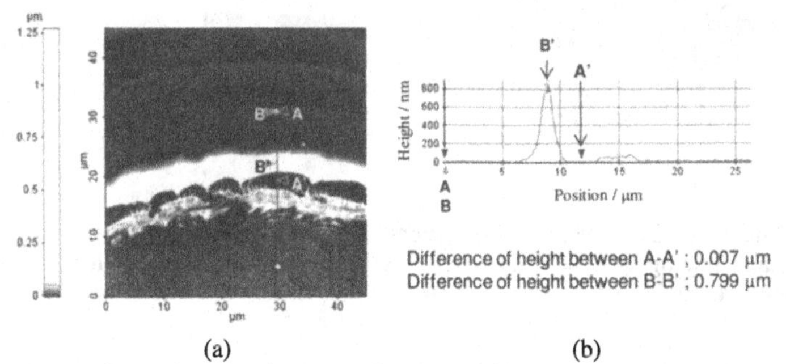

(a) (b)

Figure 6. Line profile of the delamination surface in the vicinity of the crack initiation site. (a) A two dimensional AFM image and (b) Line profile of the surface height.

CONCLUSIONS

Bond bend strength between micro-sized SU-8 cylindrical specimens and a Si substrate has been evaluated to clarify the effects of the aspect ratio of cylindrical shape specimens using an aspect ratio range of 0.6-0.9. The results and discussion are concluded as follows:

1. The maximum bend moment at the delamination occurrence clearly increases with increasing the aspect ratio of SU-8 cylinders, but the maximum load is almost constant regardless of the aspect ratio. This result shows that the shear stress at the interface mainly cause the delamination between micro-sized SU-8 and Si substrate.

2. Micro-delamination areas are observed all along the inside edge of thicker fragments by optical microscope and AFM fractography. It is suggested that the micro-delamination area may be introduced during fabrications, and which may lead to main delamination under the loading.

ACKNOWLEDGMENTS

This work has been supported by Micromachine Center (MMC) Japan.

REFERENCES

1. C. Ishiyama, M. Sone and Y. Higo, *Key Engineering Materials*, **345-346**,1185-1188 (2007)
2. C. Ishiyama, A. Shibata, M. Sone and Y. Higo, *MRS symposium proceedings*, **1052**,217-222 (2007)
3. K. Takashima, Y. Higo, S. Sugiura, M. Shimojo, Mater. Trans. **42**, 68 (2001)

Mater. Res. Soc. Symp. Proc. Vol. 1139 © 2009 Materials Research Society 1139-GG03-17

Resonance Fatigue Testing of Cantilever Specimens Prepared from Thin Films

Kwangsik Kwak, Masaaki Otsu and Kazuki Takashima
Department of Materials Science and Engineering, Kumamoto University, 2-39-1 Kurokami, Kumamoto, Japan

ABSTRACT

Fatigue properties of thin film materials are extremely important to design durable and reliable microelectromechanical systems (MEMS) devices. However, it is rather difficult to apply conventional fatigue testing method of bulk materials to thin films. Therefore, a fatigue testing method fitted to thin film materials is required. In this investigation, we have developed a fatigue testing method that uses a resonance of cantilever type specimen prepared from thin films. Cantilever beam specimens with dimensions of $1(W) \times 3(L) \times 0.01(t)$ mm^3 were prepared from Ni-P amorphous alloy thin films and gold foils. In addition, cantilever beam specimens with dimension of $3(L) \times 0.3(W) \times 0.005(t)$ mm^3 were also prepared from single crystalline silicon thin films. These specimens were fixed to a holder that is connected to an audio speaker used as an actuator, and were resonated in bending mode. In order to check the validity of this testing method, Young's moduli of these specimens were measured from resonant frequencies. The average Young's modulus of Ni-P was 108 GPa and that of gold foil specimen was 63 GPa, and these values were comparable to those measured by other techniques. This indicates that the resonance occurred theoretically-predicted manner and this method is valid for measuring the fatigue properties of thin films. Resonant fatigue tests were carried out for these specimens by changing amplitude range of resonance, and S-N curves were successfully obtained.

INTRODUCTION

The evaluation of mechanical properties including elastic modulus, tensile strength, fracture toughness and fatigue life is necessary to design microelectromechanical systems (MEMS) devices. In particular, fatigue properties of thin film materials are important to design durable and reliable MEMS devices. Fatigue tests of thin film materials have been carried out using proportionally down sized dog-born type specimen or cantilever bending specimen just same as ordinary-sized bulk materials [1]. In these testing methods, cyclic loading frequency is usually up to 10Hz, and it takes much time to obtain fatigue strength after 10^9~10^{10} cycles, which are required for designing MEMS devices. Therefore, On-chip resonating fatigue testing methods at a frequency of several tens of kHz have been developed to reduce testing time at high frequency fatigue region [3-6]. In this testing structure, however, a comb-driven actuator and a specimen are prepared concurrently by photo lithography process. This indicates that this technique can be only applicable for thin films on substrates. For a cantilever specimen, if the length is much larger than the thickness, bending resonance will easily occur by applying vibration. During the resonance vibration, cyclic stress is applied at the fixed end of the cantilever. This indicates that fatigue tests can be performed by bending resonance. In this investigation, fatigue testing method of thin films by bending resonance has been developed and fatigue tests of Ni-P amorphous alloy thin films, gold foils and single crystalline silicon films have been performed.

EXPERIMENT

Materials and Test Samples

The materials used were Ni-P amorphous alloy films, gold foils and single crystalline silicon (SCS) thin films. Ni-11.5mass% P amorphous alloy film was produced by electroless plating onto an Al-4.5mass% Mg substrate. This produced an amorphous layer of 12 μm thickness on a 0.79mm thick substrate [1-2]. The specimen was cut from the Ni-P/Al-Mg in to a rectangular parallelepiped shape, prior to removal of the substrate by dissolution in a NaOH aqueous solution. The length (L), width (W) and thickness (t) of specimens were ≈10mm, 1.5mm and 12μm, respectively as show in Fig. 1(a). Gold foil specimens were prepared from cold-rolled gold tape (99.99% purity) with a width of 1.4mm and a thickness of 10μm. This tape was cut to 10mm in length as shown in Fig. 1(b). The average grain size of the gold foil was 0.73 μm. SCS specimens were fabricated from the top layer of silicon on insulator (SOI) wafers using a photo lithography process. The cantilever beams are 3mm in length, 300μm in width and 5μm in thickness. In addition, a weight is put at the free-end of the specimen as shown in Fig. 2. For SCS specimens, notches were introduced into the center ("A" type specimen) and both sides ("B" type specimen) of the specimen by focused ion beam (FIB) machining as shown in Figs. 3(a) and (b). The notch length of "A" and "B" type specimens were 100μm and 75μm, respectively, which were located 50μm from the fixed end of the specimen.

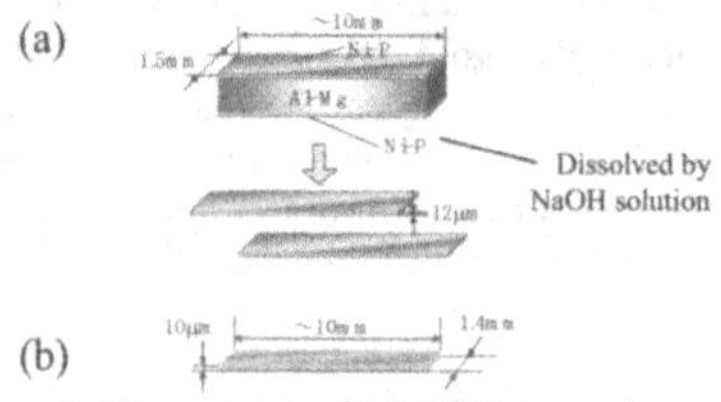

Figure 1. Geometries of (a) Ni-P amorphous alloy thin film specimen and (b) gold foil specimen.

Figure 2. Scanning electron micrograph of single crystalline silicon (SCS) cantilever beam specimen.

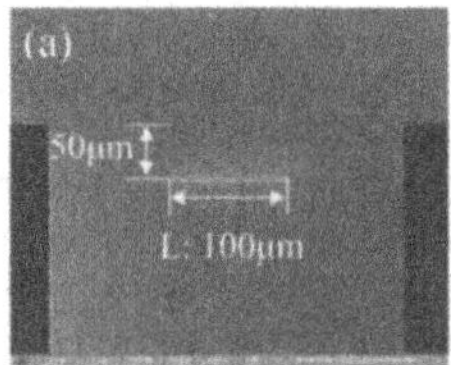

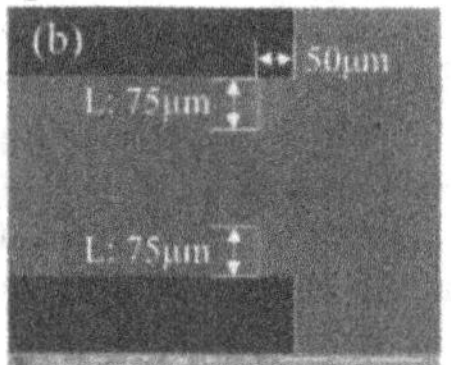

Figure 3. Scanning electron micrographs of SCS specimens. (a) Center notched specimen and (b) double-sided notched specimen.

Fatigue Testing Machine

Figure 4 shows a block diagram of fatigue testing machine. The testing machine consists of an audio speaker (FORSTER FF-77EG 8Ω/5W), a laser displacement meter (KEYENCE LK-G30), a function generator (NF WF1973) and an audio amplifier. An audio speaker is used as an actuator which causes the cantilever type specimen to vibrate. The cantilever specimen is set in a specimen holder and the holder is placed on the top of speaker corn as shown in Fig. 4. A sine wave generated by function generator was used as a driving signal of speaker. The output signal of function generator was amplified by the audio amplifier. Displacement amplitude of specimen was measured by the laser displacement meter. The resonance of specimens was monitored by a CCD camera that was set near the specimen holder. Fatigue tests were carried out at room temperature in air, and the humidity was kept to be 50~60% RH during the tests.

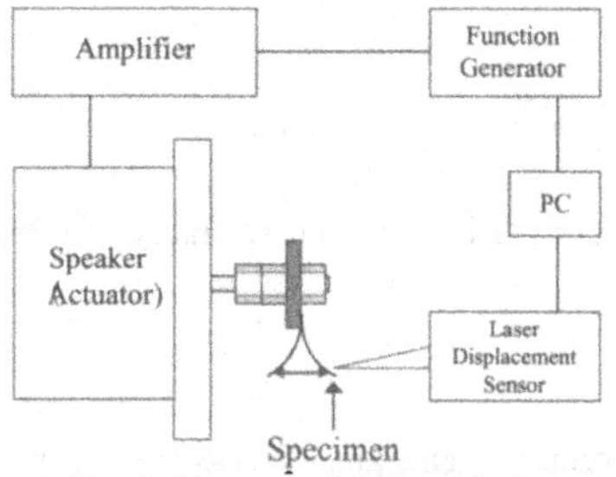

Figure 4. Block diagram of fatigue testing machine

RESULTS AND DISCUSSION

Resonant Behavior

Prior to fatigue testing, the resonant condition of specimen was examined. Figure 5 shows the relation between displacement amplitude range and frequency for Ni-P amorphous alloy thin film specimen. The resonance of specimen is observed at a frequency of 180 Hz. This type of resonance was also confirmed for gold foil and SCS specimen, and the resonant frequency of gold foil specimen was 163 Hz and that of SCS specimen was 175 Hz, respectively. In order to check the validity of this testing method, Young's moduli of the specimens were calculated from the resonant frequencies. The fundamental resonance frequency of a cantilever beam is given by

$$f_c = 0.16154\frac{h}{L^2}\sqrt{\frac{E_e}{\rho}} \tag{1}$$

where, L and h are the length and thickness of the cantilever beam, respectively, E_e is the effective Young's modulus, and ρ is the density of the material. The effective modulus, E_e, is replaced by $E/(1-\nu^2)$, where ν is Poisson's ratio and E is the Young's modulus, if the width of the beam, b, is relatively larger compared to its thickness, h ($b \geq 5h$) [8]. From resonant frequency, the average for Young's modulus for Ni-P amorphous alloy thin films specimen was calculated to be 108.1 GPa, which is consistent with that of the Ni-P amorphous alloy thin film measured by other technique (110 GPa) [9]. The average Young's modulus of gold foil was 62.8 GPa, which is also consistent with that of polycrystalline gold (68.4 GPa) [8]. The Young's moduli obtained were slightly lower than those having been reported values. This is due to the influence of air dumping as the measurements were made in air. These results show that the resonance of cantilever beam occurred theoretically and this technique can be used as fatigue test of thin film specimen.

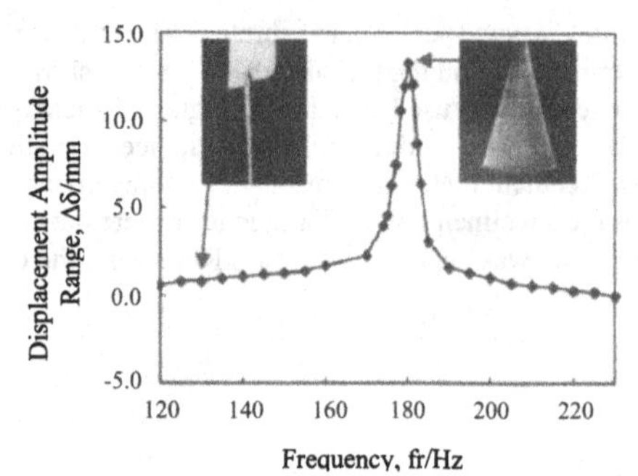

Figure 5. The relation between displacement amplitude range and frequency for Ni-P amorphous alloy thin film specimen.

Stress-Life Curves

Fatigue tests were performed by changing displacement amplitude range. The maximum stress during bending resonance was calculated by equation (2) [7].

$$\sigma_{max} = \frac{2h\delta E}{l^2} \quad (2)$$

where, σ_{max}, l, h, δ and E are maximum bending stress, length, thickness, maximum displacement and Young's modulus, respectively. Figure 6 shows the S-N plots for Ni-P amorphous alloy thin films. As the maximum stress over the fatigue cycle was lower compared to that of static strength of Ni-P amorphous alloy, only few specimens were fatigue fractured and fatigue strength was not able to be determined. Figure 7 shows the S-N plots for gold foils. The fatigue strength was 170 MPa after 10^6 cycles. Figure 8 shows S-N plots for SCS specimens. The vertical axis indicates a ratio of the applied stress to the average static strength in Fig. 8. The fatigue lives are scattered, and similar behavior is often observed for fatigue of silicon films [4-6]. The interesting feature of the plot is that the fatigue does not occur in the range between 10^4 and 10^6 cycles under the stress conditions tested in this investigation.

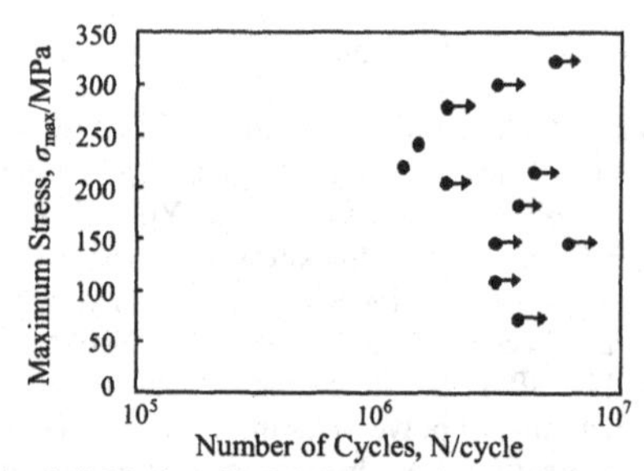

Figure 6. S-N plots for Ni-P amorphous alloy thin films.

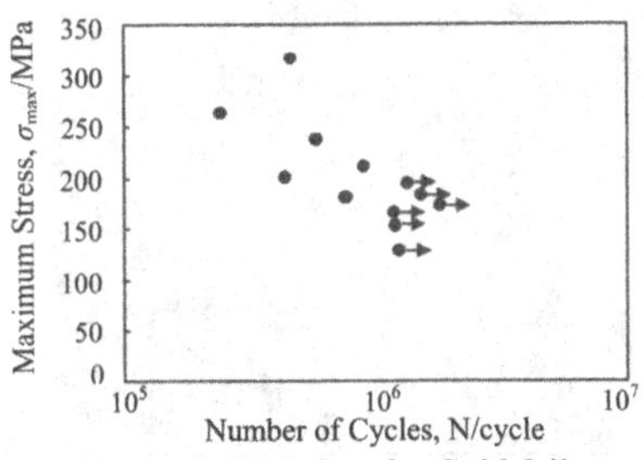

Figure 7. S-N plots for Gold foils.

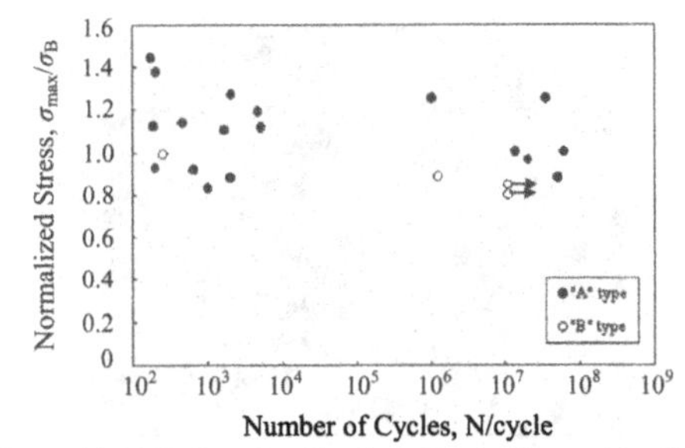

Figure 8. S-N plots for Single crystal silicon thin films.

Fracture surface

Figure 9 shows a scanning electron micrograph of the Ni-P amorphous alloy thin film specimen after a fatigue test. Striations were clearly observed on the fracture surface about 0.5μm intervals. This type of striation is also observed on fatigue surface of Ni-P thin film [2]. This indicates that the specimen was fractured by cyclic loading. In the gold foil specimen, definite asperities are observed near the crack surface, and small grains are also seen in the deformation area (Fig. 10).The size of grain is approximately 0.73μm and this size is comparable to that of grain size of this material. This suggests that the crack has propagated along grain boundary. Figure 12 shows a fracture surface of the SCS specimen. The fracture surface is very flat and shows a cleavage like feature, but some step-like regions are also found at the specimen surface.

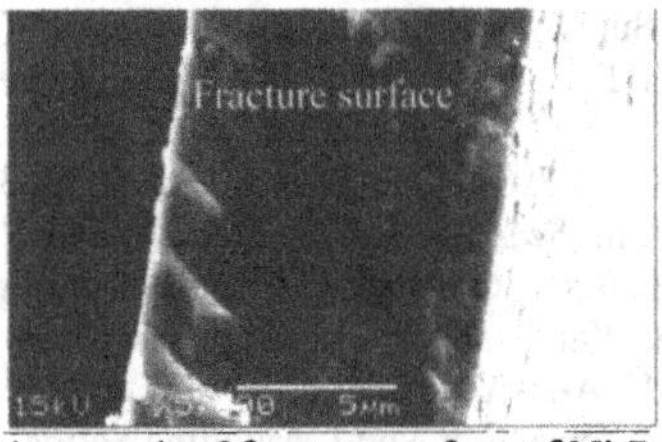

Figure 9. Scanning electron micrograph of fracture surface of Ni-P thin film after fatigue fracture.

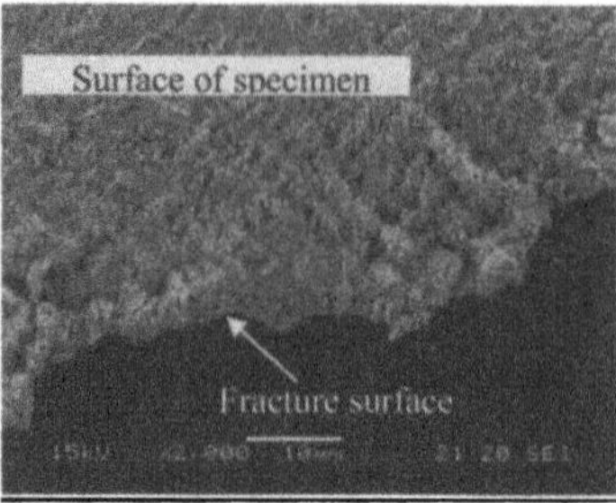

Figure 10. Scanning electron micrograph of fracture surface of Gold foil after fatigue fracture.

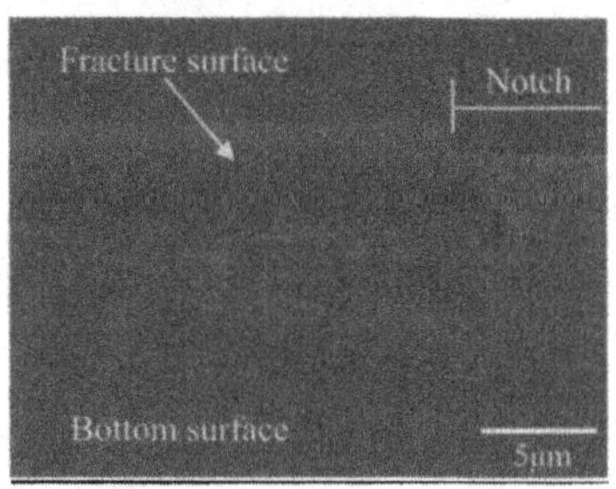

Figure 11. Scanning electron micrograph of fracture surface of SCS after fatigue fracture.

CONCLUSIONS

Bending resonance fatigue tests have been performed for micro sized cantilever beam type specimens prepared from Ni-P amorphous alloy thin films, gold foils and single crystalline silicon thin films. Resonant fatigue tests were carried out successfully for these specimens, and S-N curves were obtained. This testing method is useful for measuring fatigue life of thin film materials.

REFERENCES

1. K. Takashima and Y. Higo, *Fatigue and Fracture of Engineering materials & Structures*, **28**, 703 (2005).
2. K. Takashima, Y. Higo, S. Sugiura and M. Shimojo, Materials Transactions, **42,** 68 (2001).
3. J. –S. Bae, C. –S. Oh, K. –S. Park, S. –K. Kim, H. –J. Lee, *Engineering Fracture Mechanics,* **75**, 4958 (2008).
4. C.L. Muhlstein, E.A. Stach, and R.O. Ritchie, *Acta Mater.*, **50,** 3579 (2002).
5. H.K Liu, B.J. Lee, and P.P Liu, *Sensors and Actuators* A, **140**, 257 (2007).
6. C.L. Muhlstein, S.B. Brown, and R.O. Ritchie, *J. Microelectromech. Syst.*, **10**, 593 (2001).
7. S. Nakano, R. Maeda and K. Yamanaka, Jpn, J. Appl. Phys., **36**, 3265 (1997).
8. C. –W. Baek, Y. –K. Kim, Y. Ahn, Y. –H. Kim, *Sensors and Actuators* A, **117**, 17 (2005).
9. Gong Li, Y.P. Gao, R.P. Liu *Journal of Non-Crystalline Solids,* **353**, 4199 (2007).

Mater. Res. Soc. Symp. Proc. Vol. 1139 © 2009 Materials Research Society 1139-GG03-18

Bending of Pd-based thin film metallic glasses by laser forming process

Yuki Ide[1], Masaaki Otsu[1], Junpei Sakurai[2], Seiichi Hata[2], Kazuki Takashima[1]

[1] Department of Materials Science and Engineering, Kumamoto University, 2-39-1, Kurokami, Kumamoto, 860-8555, Japan
[2] Precision and Intelligence Laboratory, Tokyo Institute of Technology, 4259, Nagatsuta-Cho, Midori-Ku, Yokohama, Kanagawa, 226-8503, Japan

ABSTRACT

Palladium based thin film metallic glasses were plastically bent by laser forming process. Thin films of $Pd_{77}Cu_6Si_{17}$ with a thickness of 0.028 mm and $Pd_{40}Ni_{40}P_{20}$ with a thickness of 0.017 mm were used for specimen. A 50 W YAG laser was employed for forming. Variation of bending angle was investigated by changing working conditions such as laser power, laser operation mode (continuous wave and Q-switch pulsed modes), Q-sw frequency, scanning velocity and scanning number. From the experimental results, both thin films of $Pd_{77}Cu_6Si_{17}$ and $Pd_{40}Ni_{40}P_{20}$ were successfully bent for more than 85 °. The formed thin films did not crystallize but were amorphous. As scanning number increased, bending angle also increased but increment of bending angle per scanning became smaller. When laser power and scanning velocity were changed, bending angle had a peak. When Q-sw frequency was changed, bending angle had a broad peak in $Pd_{77}Cu_6Si_{17}$ case, but that was larger frequency was smaller in $Pd_{40}Ni_{40}P_{20}$ case.

INTRODUCTION

Weight reduction and high performance of electric devices and semiconductor devices are required for appliances and automotive sensors and actuators. MEMS (Micro Electro Mechanical Systems) devices are promising in those applications. In order to enhance reducing the device size and weight, and to improve their performance, it is necessary to fabricate them with not only two-dimensional but also three-dimensional structure in the size of micrometer order. However, since conventional fabrication methods for microdevices are layer integration process like lithography, it is difficult to manufacture three-dimensional structure with large height. In order to make microdevices having three-dimensional structure with large height, new forming methods are required. Plastic working methods, especially bending of thin films, are very efficient ones for those objectives.

Those microdevices are usually made of silicon or silicon compounds because usual micro-manufacturing process such as lithography, LIGA process and so on can be applied to processing those materials. Those materials, however, have crystalline structure, and mechanical and physical properties depend on crystallographic orientation and it is hard to control the distribution of crystallographic orientation in devices to utilize or eliminate anisotropy of the properties. To eliminate dependency of crystallographic orientation on mechanical and physical properties of microdevice material, amorphous materials especially metallic glasses are tried to apply to microdevices. However, metallic glasses are brittle and difficult to deform plastically at room temperature.

Recently, laser forming process is remarkable for forming sheet and foil materials without forming punches and dies. By this method, materials are plastically deformed with thermal stress induced by rapid heating due to laser irradiation. However sheet is formed with stretching by ordinary sheet metal forming processes, sheet is formed with compression at laser irradiated area by laser forming process. Thus, this process may be useful for forming brittle materials that is easy to fracture by tensile stress and very strong against compressive stress. There are a lot of reports about forming sheets of carbon steel, stainless steel [1], aluminum alloy [2-3], titanium alloy [4] and plastics, these are all ductile material, however, a few researches are performed about forming brittle materials by laser forming process [5]. It is also expected that thin film metallic glasses can be bent plastically by laser forming process.

In the present study, working conditions such as laser power (*P*), scanning velocity (*v*), Q-sw frequency (*f*), scanning number (*N*) were changed, and thin film palladium based metallic glass were bent by laser forming.

EXPERIMENTAL METHOD

In the present study, two kinds of thin film palladium based metallic glasses as shown in Table 1 [6-7] were used for specimens. A 50 W YAG laser was used for laser forming. Specimen was fixed to an NC table at only an end like cantilever. For forming, laser beam was irradiated on the specimen and scanned by galvano mirrors perpendicular to the longitudinal direction. After forming, the bending angle was measured using a line scan-type laser displacement sensor.

Working conditions are shown in Table 2. Laser power, scanning velocity, scanning number and Q-sw frequency were changed and bending angle was investigated. Although operation mode of laser device was used both CW and Q-switch pulsed modes, results at only Q-switch pulsed mode are shown in this paper because specimens were melted and fractured at CW mode.

Table 1. Material and size of specimens.

Material	$Pd_{77}Cu_6Si_{17}$	$Pd_{40}Ni_{40}P_{20}$
Glass transition temperature T_g / K	657	578
Crystallization temperature T_x / K	680	651
Length / mm	10	10
Width / mm	1.45	1.45
Thickness / mm	0.028	0.017

Table 2. Laser forming conditions.

Operation mode	Q-sw Pulsed / CW
Wavelength / nm	1064
Frequency, f / kHz	2-30, CW
Laser power, P / W	0.5-7
Scanning velocity, v / mm s^{-1}	15-65
Scanning number, N / times	1-80
Atmosphere	Air

RESULTS AND DISCUSSIONS

Formed Shape and Microstructure

When scanning velocity was v=40 mm/s, Q-sw frequency was f=3.0 kHz and scanning number was N=80 times, bending angle of $Pd_{77}Cu_6Si_{17}$ was θ=89.0 ° at the laser power of P=3.0 W and that of $Pd_{40}Ni_{40}P_{20}$ was θ=86.5 ° at P=1.5 W, respectively.

Figure 1 shows SEM images of surface and cross section of formed specimens. Melting and/or ablation were not observed at the surfaces of both materials. The thickness at the laser irradiated part was slightly increased. This shows bending by temperature gradient mechanism, which is a characteristic bending mechanism of laser forming, happened.

Figure 2 plots micro-XRD patterns at the bending areas of $Pd_{77}Cu_6Si_{17}$ and $Pd_{40}Ni_{40}P_{20}$, respectively. Broad diffraction patterns which are characteristic of amorphous were obtained for both materials. It is confirmed that both materials became amorphous after laser forming.

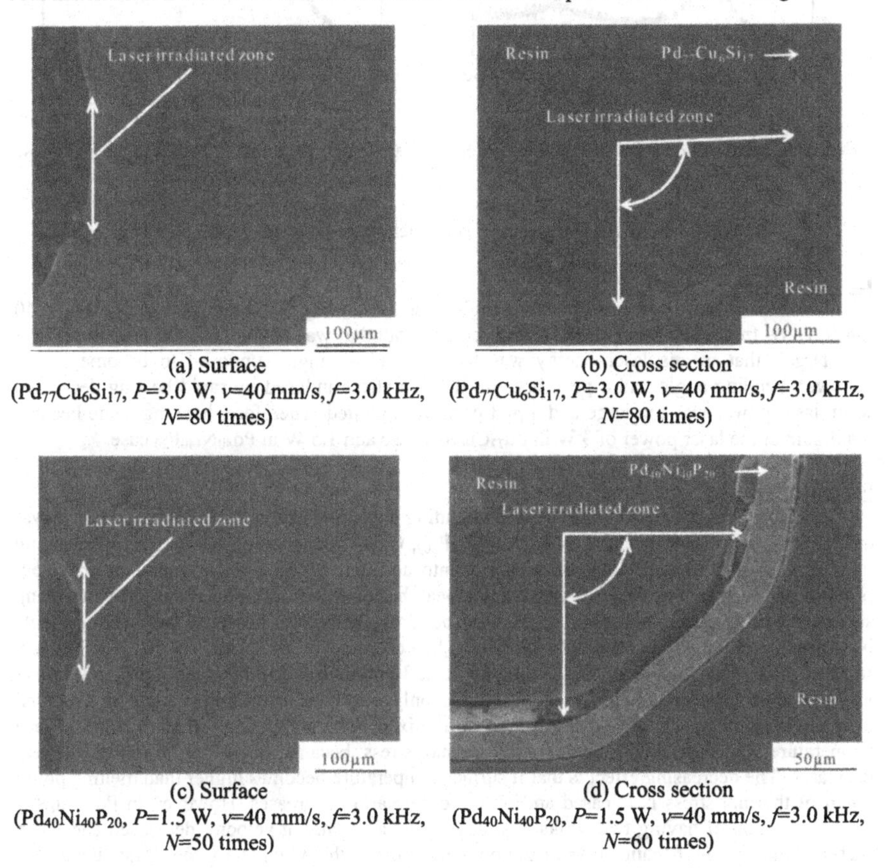

(a) Surface
($Pd_{77}Cu_6Si_{17}$, P=3.0 W, v=40 mm/s, f=3.0 kHz, N=80 times)

(b) Cross section
($Pd_{77}Cu_6Si_{17}$, P=3.0 W, v=40 mm/s, f=3.0 kHz, N=80 times)

(c) Surface
($Pd_{40}Ni_{40}P_{20}$, P=1.5 W, v=40 mm/s, f=3.0 kHz, N=50 times)

(d) Cross section
($Pd_{40}Ni_{40}P_{20}$, P=1.5 W, v=40 mm/s, f=3.0 kHz, N=60 times)

Figure 1. SEM images of surface and cross section of formed specimen.

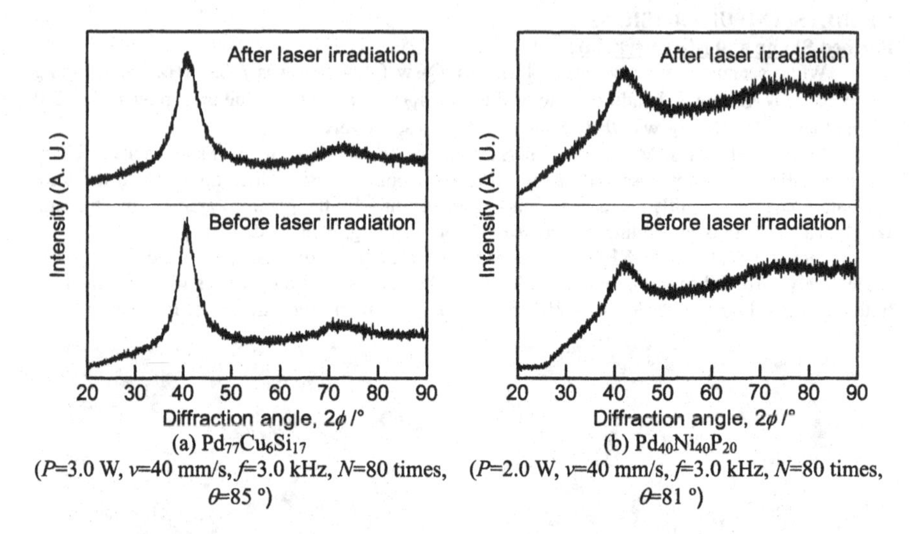

(a) $Pd_{77}Cu_6Si_{17}$
(P=3.0 W, v=40 mm/s, f=3.0 kHz, N=80 times, θ=85 °)

(b) $Pd_{40}Ni_{40}P_{20}$
(P=2.0 W, v=40 mm/s, f=3.0 kHz, N=80 times, θ=81 °)

Figure 2. Micro-XRD patterns of specimen before and after laser forming.

Laser Power

Effect of laser power on bending angle is shown in Fig. 3. Scanning velocity was v=40 mm/s, Q-sw frequency was f=3.0 kHz and scanning number was N=50 times. When laser power was larger, that means laser energy was larger, bending angle supposed to become larger, however, bending angle had a peak because thermal stress induced by rapid heating decreased when laser power was too large and a part of surface melted. Therefore, bending angle became maximum at the laser power of 3 W in $Pd_{77}Cu_6Si_{17}$ case and 1.5 W in $Pd_{40}Ni_{40}P_{20}$ case.

Scanning Velocity

Figure 4 expresses relation between bending angle and scanning velocity at laser power of P=3.0 W ($Pd_{77}Cu_6Si_{17}$) and 1.5 W ($Pd_{40}Ni_{40}P_{20}$), Q-sw frequency of f=3.0 kHz and scanning number of N=50 times. Taking laser energy into account, bending angle would be inversely proportional to scanning velocity. But it had a peak because thermal stress reduced by occurring surface melting when scanning velocity was too slow. Therefore, it seems that bending angle becomes maximum at v=30-45 mm/s in $Pd_{77}Cu_6Si_{17}$ case and v=30-40 mm/s in $Pd_{40}Ni_{40}P_{20}$ case. In $Pd_{77}Cu_6Si_{17}$ case, two peaks of bending angle at 30 mm/s and 45mm/s, and those at 35 mm/s and 40 mm/s became smaller. In metal cases, only one peak of bending angle is observed because increasing and decreasing effects are mixed. The increasing effect is that surface temperature becomes much higher and thermal stress became larger as scanning velocity decreases. The decreasing effect is that if surface temperature becomes higher than melting point, a part of thermal stress is released and plastic deformation decreases. However, in $Pd_{77}Cu_6Si_{17}$ case, the reason of having two peaks is assumed that as scanning velocity decreased and when surface temperature became crystallization temperature, the volume of the crystallized part reduced and a part of thermal stress was reduced, bending angle decreased. Scanning velocity became much smaller surface was melted and much stress release happened and bending angle decreased much more. So one peak at scanning velocity of 45mm/s is the border of super cooled liquid and crystallization, and the other peak at 30mm/s is the border of crystallization and melting. This phenomenon may be the characteristic for materials like metallic glass showing volume reduction at the elevated temperature, and further investigation is needed.

Q-sw Frequency

Figures 5 and 6 show effects of Q-sw frequency on bending angle for $Pd_{77}Cu_6Si_{17}$ and $Pd_{40}Ni_{40}P_{20}$, respectively. Forming was carried out under scanning velocity of v=40 mm/s, scanning number of N=50 times, laser power of P=3.0 W for $Pd_{77}Cu_6Si_{17}$ and P=1.5 W for $Pd_{40}Ni_{40}P_{20}$. In the case of $Pd_{77}Cu_6Si_{17}$, bending angle was almost constant when frequency was from 2.5 to 5.0 kHz and in the other range, bending did not confirmed or became unstable due to deformation by not temperature gradient mechanism but buckling mechanism.

On the other hand, in the case of $Pd_{40}Ni_{40}P_{20}$, bending angle was larger as frequency was smaller when frequency was from 3 to 12 kHz. When frequency was greater than 12 kHz, bending was also unstable due to buckling and bending angle became smaller as well as the $Pd_{77}Cu_6Si_{17}$ case.

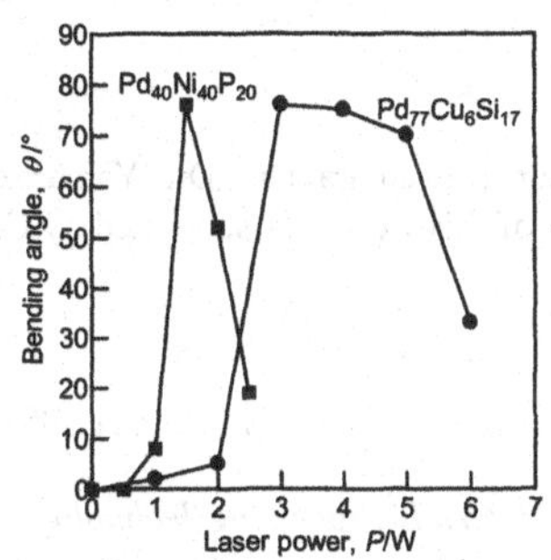

Figure 3. Relation between bending angle and laser power. (v=40 mm/s, f=3.0 kHz, N=50 times)

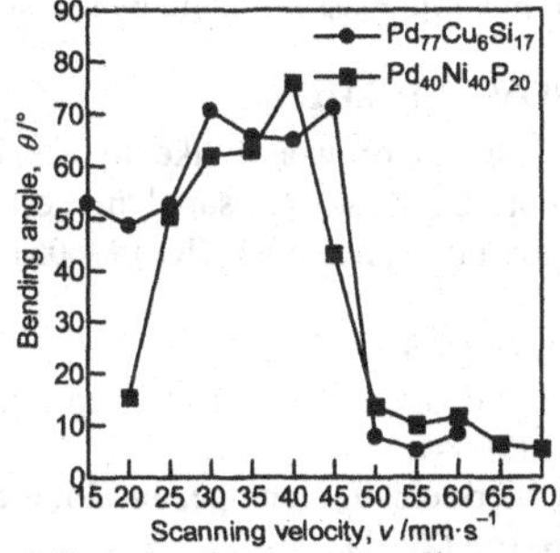

Figure 4. Relation between bending angle and scanning velocity. (P=3.0 W($Pd_{77}Cu_6Si_{17}$), 1.5 W($Pd_{40}Ni_{40}P_{20}$), f=3.0 kHz, N=50 times)

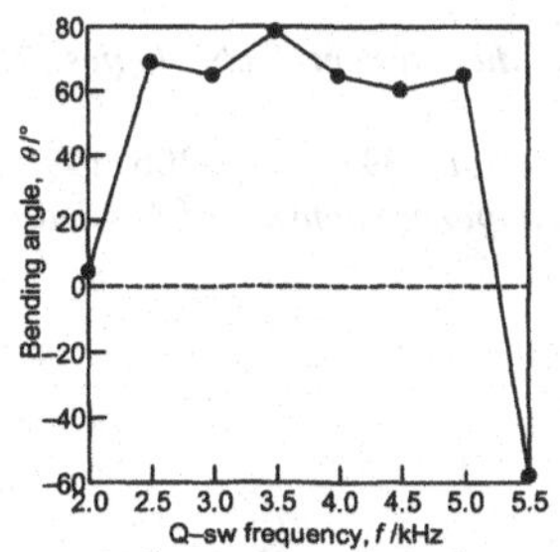

Figure 5. Relation between bending angle and Q-sw frequency of $Pd_{77}Cu_6Si_{17}$. (P=3.0 W, v=40 mm/s, N=50 times)

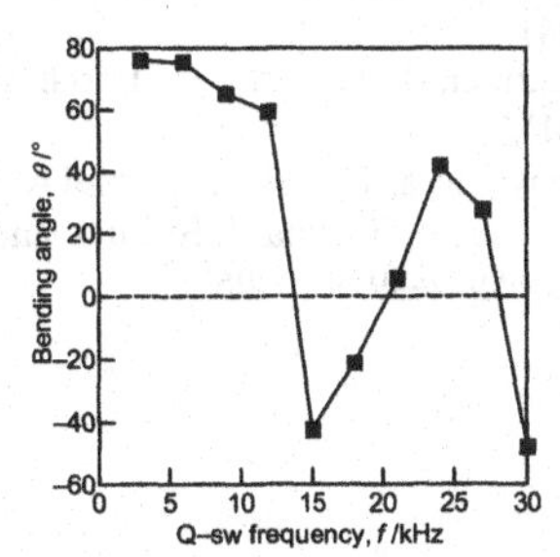

Figure 6. Relation between bending angle and Q-sw frequency of $Pd_{40}Ni_{40}P_{20}$. (P=1.5 W, v=40 mm/s, N=50 times)

CONCLUSIONS

Thin films of palladium based metallic glasses were successfully bent using laser forming. The following experimental results were obtained.

1) Thin film of $Pd_{77}Cu_6Si_{17}$ was bent until θ=89.0 ° and that of $Pd_{40}Ni_{40}P_{20}$ was bent until θ=86.5 °.
2) Both thin films of $Pd_{77}Cu_6Si_{17}$ and $Pd_{40}Ni_{40}P_{20}$ did not crystallize and were amorphous after laser forming.
3) When laser power was changed, bending angle had a peak at P=1.5 W in $Pd_{77}Cu_6Si_{17}$ case and at P=3.0 W in $Pd_{40}Ni_{40}P_{20}$ case.
4) When scanning velocity was varied, bending angle had a broad peak and almost constant at v=30-45 mm/s in $Pd_{77}Cu_6Si_{17}$ case and v=30-40 mm/s in $Pd_{40}Ni_{40}P_{20}$ case.
5) When Q-sw frequency was changed, bending angle was maximum and almost constant at from 2.5 to 5.0 kHz in $Pd_{77}Cu_6Si_{17}$ case, but in $Pd_{40}Ni_{40}P_{20}$ case, bending angle was larger as frequency was smaller.

Further works should be required to clarify the mechanism of occurring plastic deformation of metallic glasses in laser forming process. Possibility of applying to other metallic glasses and forming into three-dimensional shape, and mechanical properties of laser formed thin films of metallic glasses should be investigate, too.

ACKNOWLEDGMENTS

The author would like to acknowledge Professor Kawamura and Dr. Yamasaki of Kumamoto University for supplying test materials. A part of this work was supported by Grant-in-Aid for Young Scientists (R) 19760513.

REFERENCES

1. J. D. Majumdar, A. K. Nath, and I. Manna, *Materials Science and Engineering,* **A 385,** 113-122 (2004).
2. M. Merklein, T. Hennige, and M. Geiger, *Journal of Materials Processing Technology,* **115,** 159-165 (2001).
3. M. Geiger, M. Merklein, and M. Pitz, *Journal of Materials Processing Technology,* **151,** 3-11 (2004).
4. D. J. Chen, and S. C. Wu, *Journal of Materials Processing Technology,* **152** (1), 62-65 (2004).
5. E. Gartner, J. Fruhauf, U. Loschner, and H. Exner, *Microsystem Technologies,* **7,** 23-26 (2001).
6. Y. Kawamura, T. Nakamura, and A. Inoue, *Scripta Materialia,* **39** (3), 301-306 (1998).
7. Y. Yamauchi, S. Hata, J. Sakurai, and A. Shimokohbe, *Japanese Journal of Applied Physics,* **45** (7), 5911-5919 (2006).

Droplet Formation at Microfluidic T-junctions

Yu Xiang[1] and David A. LaVan[2]*
[1] Yale University, New Haven, CT, USA
[2] Ceramics Division, Materials Science and Engineering Laboratory, National Institute of Standards and Technology, Gaithersburg, MD 20899, USA * david.lavan@nist.gov

ABSTRACT

Analysis of droplet formation in microfluidic systems is important to understand the operation of these devices, and to permit optimal design and process control. Droplet formation in microfluidic T-junction devices was studied using experimental and numerical methods. The simulations agree well with experimental data from polydimethylsiloxane (PDMS) devices; they show that droplet pinch-off is controlled not by viscous stress, but rather caused by pressure buildup after channel blocking due to the second phase. The period of droplet formation is dependent on velocity of the flow, but not viscosity or interface tension of the fluids. Analysis using dimensionless period, which is equivalent to dimensionless droplet length, shows that dimensionless period is controlled primarily by water fraction but is also dependent on velocity following a power-law relationship. Higher values of capillary number tend to extend the distance for droplet pinch-off. Droplet length does depend on flow velocity at low velocities, but reaches a relatively constant length at higher flow velocities. The coefficient of variation of droplet volume/length increases with increasing capillary number.

INTRODUCTION

Microfluidic devices have been recognized for a tantalizing potential to provide novel synthesis and research tools[1, 2]. Multiphase flows at the micro scale, particularly droplet-based microfluidics, have been of particular interest[3-6]. There are two major techniques reported to generate micro droplets in microfluidic channels: flow-focusing[7, 8] and T-junctions[9-11].

Droplet generation at T-junctions and the underlying mechanism have been carefully investigated. Although many believe that the balance between capillary force and viscous force controls the formation of droplets[9-13], it has been argued that viscous force in such microfluidic devices is too weak to break up droplets[14]. Instead, analytical models of the droplet formation based on simplified geometric assumptions were proposed[14, 15]. Recently, Stone and coworkers[16] showed experimentally that the buildup of pressure in the oil phase due to blocking of the channel by the intruding phase is indeed the major force to compete with interfacial tension. They predicted that droplet size is a linear function of the flow rate ratio.

It is difficult to simulate multi-phase flows, especially to track moving interfaces and deforming topology[17]. Recently, Osher and coworkers[18, 19] have reported a novel method employing a level set function to track the motion of moving boundaries or interfaces. Zhou et al.[20] reported simulations of the generation of droplets in axisymmetric microfluidic flow-focusing devices using this method with adaptive meshing. Microfluidic devices were made from PDMS for testing. A numerical model of the device was created using a conservative form[21] of the level set method using Comsol Multiphysics software. Further numerical details are in the supplemental materials, along with details of the experimental methods used.

Channel geometry not only affects the production of droplets, but even changes the underlying mechanism. For example, the width ratio of the branch and main channel D_{branch}/D_0 [16], the aspect ratio of channel cross-section[16] and the angle of the T-junction[23], have been reported to alter the properties of multi-phase flows. A 2D numerical model (corresponding to the top-view) was created to match the fabricated devices: width, $D_{branch} = D_0 = 100$ μm and height, $h = 100$ μm. Unless otherwise mentioned, the density of the oil phase was 1.7×10^3 kg m^{-3}, density of the water phase, 1.0×10^3 kg m^{-3}; viscosity of the oil phase 21.1 × 10^{-3} Pa·s, viscosity of the water phase 1.0 × 10^{-3} Pa·s; interfacial tension of oil/water 1.2 × 10^{-3} N/m^1. The contact angle of water on the oil-wetted wall has been shown to have very weak influence on the size of droplet[24]; the contact angle was measured as 144 degrees.

NUMERICAL THEORY AND METHOD

In this work, a reinitialized level set method is employed to solve the motion of interfaces. The 0.5 contour of the level set function was used to represent the water oil interface (0 to be in the oil phase, and 1 in the water phase). Equations (1)-(3) describe the motion of interfaces in two phase flows.

$$\frac{\partial \varphi}{\partial t} + \vec{u} \cdot \nabla \varphi = 0 \quad (1)$$

$$\rho\left(\frac{\partial \vec{u}}{\partial t} + \vec{u} \cdot \nabla \vec{u}\right) - \mu \cdot \nabla^2 \vec{u} + \nabla p = \rho \vec{g} + \vec{F}_\sigma \quad (2)$$

$$\nabla \cdot \vec{u} = 0 \quad (3)$$

Here, ρ is the local density of the flow, μ the local dynamic viscosity, $\vec{u}$ the velocity, p the pressure. $\vec{F}_\sigma$ represents the interface tension effect, which reads

$$\vec{F}_\sigma = \nabla(\sigma \cdot \kappa \cdot \delta(\varphi)) \quad (4)$$

where σ is interface tension coefficient, κ the local curvature of the interface, $\delta(\varphi)$ a Dirac delta function to be non-zero only within the interfacial layer.

In practice, slight modifications have been made to (1) to preserve smoothness of the level set function and the thickness of the interfacial layer, and eventually to improve the accuracy of the solution. The governing equation of the advection of level set function is modified from Eqn. (1) to become

$$\frac{\partial \varphi}{\partial t} + \vec{u} \cdot \nabla \varphi = \gamma\left[\varepsilon \nabla^2 \varphi - \nabla \cdot \vec{f}(\varphi)\right] \quad (5)$$

where $\vec{f}(\varphi) \equiv \varphi(1-\varphi)\vec{n}$ is an artificial compression flux in the normal direction of the interface, and has non-zero values only within the interfacial layer; $\varepsilon \nabla^2 \varphi$ a small amount of added artificial viscosity, introduced to prevent discontinuities at the interface[21]; γ is an artificial parameter, which can be tuned to stabilize the thickness of interface and minimize oscillation in the level set function[22].

Equation (5) can be rewritten as

$$\frac{\partial \varphi}{\partial t} + \vec{u} \cdot \nabla \varphi = \gamma \nabla \cdot (\varepsilon \nabla \varphi - \varphi(1-\varphi)\vec{n}) \quad (6)$$

Equation (6) is solved along with Eqn. (2-3) using Comsol Multiphysics. The 0.5 contour of the level set function φ is considered as the interface between the two phases of fluids. When solving the equations, γ is often on the same order of magnitude of the maximum velocity in the flow field; and ε, the thickness of the interfacial layer, was often equal to the largest value of mesh size within the solved flow field.

To evaluate interface tension force $\vec{F}_{st} = \nabla(\sigma \cdot \kappa \cdot \delta(\varphi))$ in (2), the $\delta(\varphi)$ function is approximated by:

$$\delta = 6|\nabla \varphi| \cdot |\varphi(1-\varphi)| \quad (7)$$

Using level set function, density can be expressed in a global form:

$$\rho = \rho_1 + (\rho_2 - \rho_1)\varphi \quad (8)$$

along with the dynamic viscosity:

$$\mu = \mu_1 + (\mu_2 - \mu_1)\varphi \quad (9)$$

where $\rho_{1,2}$ and $\mu_{1,2}$ are the densities and viscosities of fluid 1 and 2, respectively.

RESULTS and DISCUSSION

Mechanism of droplet formation

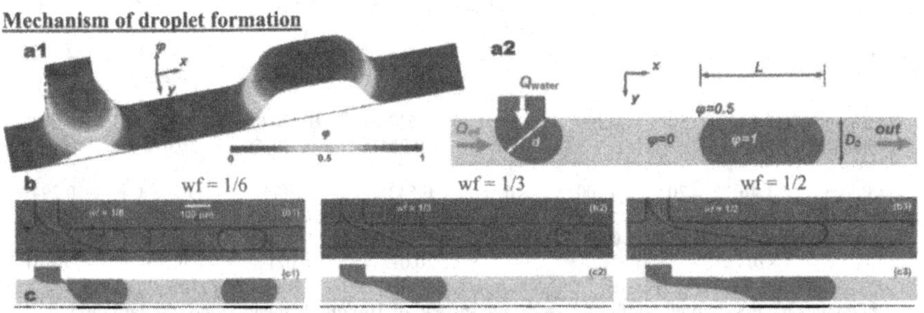

Figure 1. Model and experimental results for droplet formation. Channel width, Do = 100 μm.

The underlying mechanism of droplet formation is controlled by the balance of the hydraulic pressure drop across the interface, Δp, the viscous stress, τ, and the interfacial tension, Ω [25] given non-dimensionally as:

$$\Delta p + \tau \equiv \Omega, \qquad (10)$$

Interfacial tension stabilizes a developing droplet and prevents detachment of the droplet from the water stream. Both viscous stress and pressure drop try to snap droplets off from the water stream. Figure 1(b) and (c) present photographs taken during experiments and corresponding simulation results.

Several dimensionless numbers can be employed to describe the formation of droplets. Capillary number, $Ca \equiv U\mu / \sigma$, the relative value of viscous stress compared to interfacial tension, depends on the velocity of the carrier fluid flow, U, the interfacial tension coefficient at the water/oil interface, σ, and the dynamic viscosity of the carrier fluid, μ. In this study, experiments were conducted at very low Ca values ($Ca \approx 10^{-3}$). ($U \approx 10^{-3}$ m/s, $\mu \approx 10^{-3}$ Pa·s, and $\sigma \approx 10^{-3}$ N/m). At such low Ca values, the viscous stress is extremely low; it is the hydraulic pressure difference that dominates the left side of the balance equation, Eqn. (10), and competes with the interfacial tension to generate discrete droplets in channels.

Water fraction, wf, is defined as the ratio of flow rate of the aqueous phase to the total flow rate of the carrier fluid and the aqueous phase[10], or, $wf \equiv Q_{water} / (Q_{water} + Q_{oil})$, where Q_{water} is the volumetric flow rate of aqueous phase and Q_{oil} the volumetric flow rate of the carrier fluid (oil) stream. The ratio of flow rates is defined by $\Gamma = Q_{water} / Q_{oil}$, or, $\Gamma = wf / (1 - wf)$.

Droplet volume, V, is proportional to its end-to-end linear length L, neglecting a small volume associated with the curvature on the ends of the droplets. To evaluate the variation in size of droplets, simulations of the formation of as many as nine droplets in series were performed at different flow rates, water fractions, interfacial tensions and viscosities (data shown in Table I).

Figure 2 shows the coefficient of variation, $CV = st.\ dev.\ / \ mean$, of droplet volume as a function of Ca. At larger Ca (a regime that is hard to test experimentally), viscous stress becomes more significant compared to interfacial tension, although the growth is still governed by the pressure difference. The increase in the viscous effect alters the uniformity of droplet size. This trend had been reported by Tice et al, whose experiments showed a higher variation of droplet length at higher total flow rate[10].

The growth of a droplet starts when the water stream enters the main channel, and ends at pinch-off. This process is composed of two steps: elongation and detachment, $L = L_{elongation} + L_{detach}$. The growth rate during elongation is determined mainly by U_{water}. $L_{elongation}$, or the intrusion length of water into oil, is often of the same order of magnitude of the width of the main channel, $L_{elongation} \approx O(D_0)$.

Table I. Simulation results for the formation of a series of droplets

Carrier Fluid (oil)	*Ca*	U_{total} (*mm/s*)	*wf*	N, Number of droplets simulated	Pinch-off length (10^{-4} *m*)**	Period of droplet formation (s)**	Length of droplets (10^{-4} *m*)**	Volume of droplets (*nL*)**
Perfluorohe-xylethanol	8.75 $\times 10^{-3}$	0.75	0.333	9	0.383 ± 0.050	0.823 ± 0.096	2.924 ± 0.293	2.770 ± 0.293
Perfluoro-decaline*	3.64 $\times 10^{-3}$	50	0.200	6	0.622 ± 0.064	0.00518 ± 0.000756	1.126 ± 0.072	0.262 ± 0.018
Perfluoro-decaline*	1.46 $\times 10^{-3}$	20	0.200	6	0.542 ± 0.086	0.0171 ± 0.00197	1.268 ± 0.064	0.294 ± 0.0083
Hexadecane	6.00 $\times 10^{-5}$	1.5	0.667	6	0.145 ± 0.012	0.486 ± 0.021	5.525 ± 0.261	5.371 ± 0.261

*Devices have a channel width of $D_0 = 50\ \mu m$ instead of 100 μm. **mean ± one standard deviation

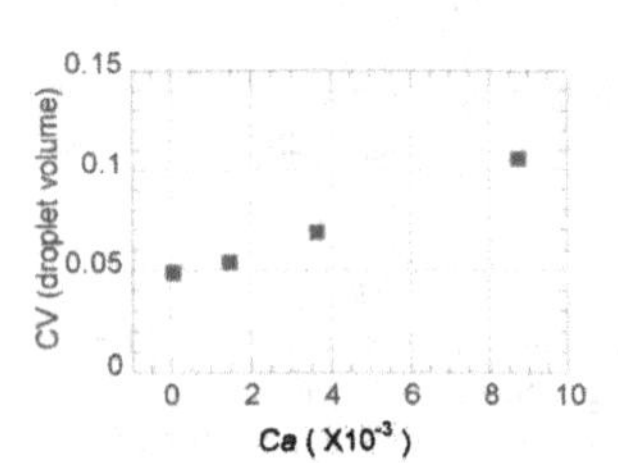

Figure 2. Coefficient of variation of simulation droplet volume as a function of capillary number.

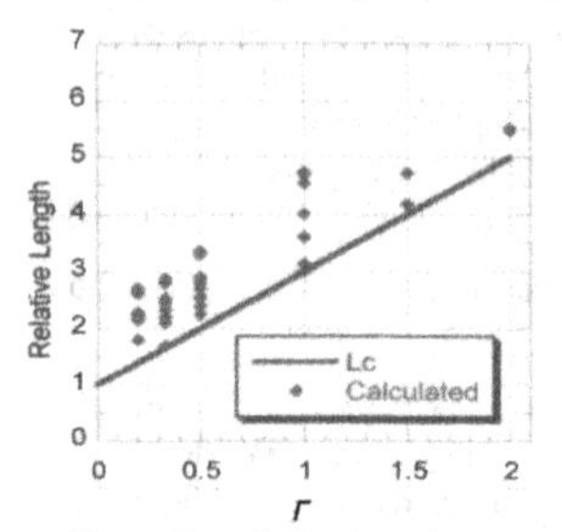

Figure 3. Relative length as a function of flow rate ratio.

A stable aqueous droplet in oil phase at equilibrium sees a higher hydraulic pressure inside of the droplet than outside. The pressure difference across the interface is the difference in pressure between the oil and water phase, $\Delta p = p_{water} - p_{oil}$, and must be balanced by the interfacial tension:

$$\Delta p = p_{water} - p_{oil} = \Omega \qquad (11)$$

At a T-junction, once the water phase blocks the channel, oil can only flow through the thin wetting layer between water and the channel wall; pressure in the oil phase will quickly build up and reduce Δp; which leads to an imbalance with interfacial tension, resulting in the squeezing of the neck of the water thread {with a neck width of *d*, as denoted in Figure 1(a2)}, until the neck breaks off and a droplet detaches.

Negative hydraulic pressure is present near the pinch-off point at the moment of break-off, causing oil from the surrounding area to quickly fill the void. The rate of squeezing is estimated by U_{oil}; the time is $\Delta t_{detach} = d/U_{oil}$ (d is the width of the droplet neck as shown in Figure 1a2); the droplet length grows at a rate determined by flow velocity of water, or $L_{detach} = U_{water} \cdot \Delta t_{detach} = d \cdot \Gamma$. Combined with an estimation of $L_{elongation}$ being D_0, it leads to an estimate of the total length of a droplet [16], of $L = D_0 + d \cdot \Gamma$ which can be rewritten as $\tilde{L} = 1 + \alpha\Gamma$, where $\tilde{L}$ is droplet length relative to the channel width D_0, and α is the ratio of d/D_0. The droplet size is a linear function of the flow rate ratio, with a constant α determined by device geometry. Figure 3 shows the calculated relative droplet length at different flow rate ratios Γ, when *Ca* values are kept low (smaller than 10^{-2}). The general linear trend is clearly seen. The value of $\alpha = 2$ is in agreement with Garstecki et al.[16] Data at the same Γ value were calculated at different U_{water} levels. As shown in Figure 4, as U_{water} increases, the droplet length quickly decays and approaches a characteristic droplet length, L_c, which is only a function of Γ. The linear fit (blue line) in Fig. 3 is, in fact, the $L_c(\Gamma)$ function.

Period of droplet formation

Droplet formation period increases with water fraction, at a given capillary number, due to the conservation of mass. Increasing *Ca* (by increasing flow velocity) will shorten the period; we have examined the parameters of Ca individually: Fig 5(a) shows that changes in the viscosity of the oil, the oil/DI interfacial tension or changes in both (data not shown) do not cause significant changes to the period when flow velocity is fixed. The same data is plotted against Ca in Fig 5(b).

The dependence of period on flow velocity is shown in Figure 4; the period drops quickly as velocity increases (starting below 0.5 *mm/s*). This relationship was found to follow a power law with an exponent close to 7/6 for the studied geometry. For wf = 1/3, $T = k_0 / U_{water}^{1.169}$ with R^2 equal to 0.9968 (N=13) (T in seconds; U_{water} in m/s, $k_0 = 4.98\times10^{-5}$).

Because droplet period is so strongly dependent on flow velocity, changes in velocity will result in different periods even at the same Ca. Changes in viscosity and interfacial tension change Ca but do not change period. Table II provides results corresponding to flow conditions at the same wf value but at different values of flow velocity, viscosity and interfacial tension.

Period, T, is composed of elongation and detachment: $T = \Delta t_{elongation} + \Delta t_{detach}$. During elongation, at a low Ca and moderate wf, the water thread extends almost straight into the oil without significant deformation, until it touches the opposite sidewall. A rough estimate of elongation time can be given by $\Delta t_{elongation} = D_0/U_{water}$. The detachment time is estimated by $\Delta t_{detach} = d/U_{oil}$, relating period to both flow velocity and the ratio of flow rates,

$$T = \frac{D_0}{U_{water}}\left(1 + \frac{d}{D_0}\Gamma\right) \qquad (12)$$

If using D_0/U_{water} as the time scale, one can convert T to a dimension-less period, $\tilde{T} = 1 + \frac{d}{D_0}\Gamma$, which is also the relative length of droplets.

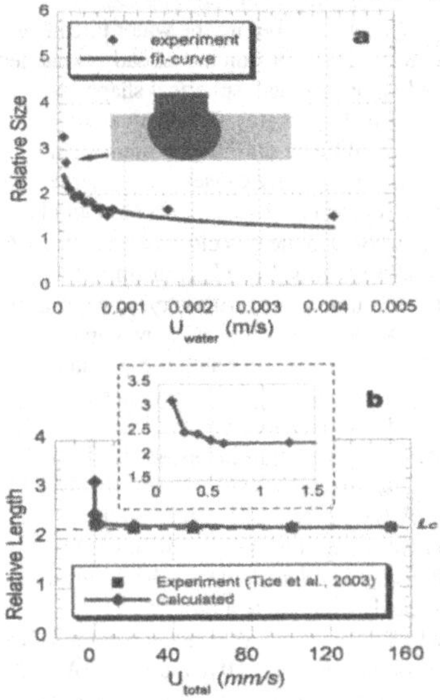

Figure 4. Droplet size and length as a function of velocity

At a given flow rate ratio, Γ, period, T, should be proportional to $1/U_{water}$ according to Eqn. (12); however, at extremely low U_{water} level when Γ is fixed, interfacial tension causes the water thread to expand dramatically during the elongation step, leading D_0/U_{water} to under-estimate $\Delta t_{elongation}$, which is why T scales with $U_{water}^{-\beta}$ (β>1) but not with U_{water}^{-1}.

Table II. Calculated periods at different flow conditions

Carrier Fluid (Oil)	(A) hexadecane	(B) hexadecane	(C) perfluorohexylethanol
μ, measured viscosity (oil), (Pa s)	3 x10^{-3}	3 x10^{-3}	21.1x10^{-3}
σ, measured interface tension (water/oil), (N/m)	25 x10^{-3}	25 x10^{-3}	1.2 x10^{-3}
U_{oil}, oil velocity, (m/s)	73.2 x10^{-3}	0.5x10^{-3}	0.5x10^{-3}
U_{water}, water velocity, (m/s)	36.6 x10^{-3}	0.25x10^{-3}	0.25x10^{-3}
Ca, Capillary Number	8.79 x10^{-3}	6 x10^{-5}	8.79 x10^{-3}
T, calculated period, (s)	0.004	0.84	0.82

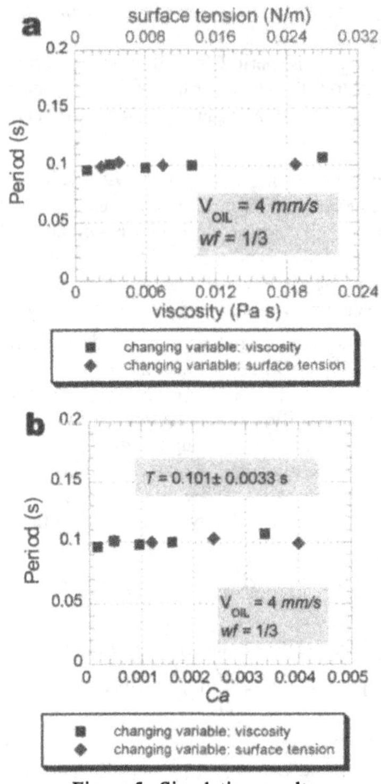

Figure 5. Simulation results for period as a function of (a) viscosity and (b) capillary number.

Length of droplets

From Eqns. (11) and (12), it is easy to see $L = U_{water} \cdot T$. Recall that $T = k_0 / U_{water}^{\beta}$ (where $\beta = 1.169$), so the length of a droplet, L, is found to be weakly dependent on water velocity, $L = k_0 / U_{water}^{\beta-1}$, in accordance with the experimental droplet length data shown in Fig.4(a). The explanation arises from the invalidity of the assumption that $L_{elongation} = D_0$. At extremely low U_{water} levels, before the water thread can block the channel, surface tension has already expanded the water thread, approaching a spherical shape, as shown in the inset of Figure 4(a). The equivalent elongation length will be larger than the main channel width, D_0, which consequently causes L to be larger than the characteristic length $L_c = D_0 + d \cdot T$. It is still not clear as to why the power law exponent is close to -0.17 (≈ -1/6). Similar power laws relating L to U_{water} in similar geometries have been reported, with varying exponents, such as -0.30 by Xu et al.[26] and -0.25 by Van der Graaf et al. [24] The power law relationship also predicts that droplet length is essentially constant once a threshold flow velocity has been reached. Tice et al.[10] reported constant droplet length independent of flow velocity with U_{water} varying between 4 and 30 *mm/s* at *wf* = 0.2; this observation is correct because all these velocities are above the threshold; the difference in droplet length became undetectable for them; however our model can easily interrogate a broad velocity range and identify where droplet length is not constant. Under similar device conditions ($D_0 = 50\ \mu m$, $\sigma = 14\ mN/m$, $\mu_{OIL} = 5.1\ mPa\ s$), but at a broader range of velocities, the droplet length varies as shown in Fig. 4(b).

CONCLUSIONS

The droplet grows at a rate given by the velocity of the impinging flow (U_{water} in our example); in the detachment step, the neck of a developing droplet is squeezed at a rate governed by the carrier flow velocity (U_{oil}). Droplet generation frequency is dependent on the impinging flow fraction and velocity, but not significantly on viscosity or interfacial tension. The dependence of the period on velocity is described by a power law relationship with an exponent of 7/6. The length of droplets varies significantly at low flow velocity, and approaches a constant characteristic length associated with a corresponding water fraction at higher flow velocities. Modeling of droplet fission further shows that pressure buildup due to channel blocking is the major contributing force for droplet break-off.

ACKNOWLEDGMENTS

The full description of the procedures used in this paper requires the identification of certain software and their suppliers. The inclusion of such information should in no way be construed as indicating that such software is endorsed by NIST or is recommended by NIST or that it is necessarily the best software for the purposes described.

REFERENCES

[1] G. M. Whitesides, "The origins and the future of microfluidics," *Nature (London, United Kingdom)*, vol. 442, pp. 368-373, 2006.

[2] T. Thorsen, J. Maerkl Sebastian, and R. Quake Stephen, "Microfluidic large-scale integration," *Science*, vol. 298, pp. 580-4, 2002.

[3] D. R. Link, S. L. Anna, D. A. Weitz, and H. A. Stone, "Geometrically Mediated Breakup of Drops in Microfluidic Devices," *Physical Review Letters*, vol. 92, pp. 054503/1-054503/4, 2004.

[4] A. Guenther, S. A. Khan, M. Thalmann, F. Trachsel, and K. F. Jensen, "Transport and reaction in microscale segmented gas-liquid flow," *Lab on a Chip*, vol. 4, pp. 278-286, 2004.

[5] J. D. Tice, R. F. Ismagilov, and B. Zheng, "Forming droplets in microfluidic channels with alternating composition and application to indexing concentrations in droplet-based assays," *Abstracts of Papers, 228th ACS National Meeting, Philadelphia, PA, United States, August 22-26, 2004*, pp. CHED-106, 2004.

[6] H. Song, D. L. Chen, and R. F. Ismagilov, "Reactions in droplets in microfluidic channels," *Angewandte Chemie, International Edition*, vol. 45, pp. 7336-7356, 2006.

[7] P. Garstecki, I. Gitlin, W. DiLuzio, G. M. Whitesides, E. Kumacheva, and H. A. Stone, "Formation of monodisperse bubbles in a microfluidic flow-focusing device," *Applied Physics Letters*, vol. 85, pp. 2649-2651, 2004.

[8] S. L. Anna, N. Bontoux, and H. A. Stone, "Formation of dispersions using \"flow focusing\" in microchannels," *Applied Physics Letters*, vol. 82, pp. 364-366, 2003.

[9] T. Thorsen, R. W. Roberts, F. H. Arnold, and S. R. Quake, "Dynamic Pattern Formation in a Vesicle-Generating Microfluidic Device," *Physical Review Letters*, vol. 86, pp. 4163-4166, 2001.

[10] J. D. Tice, H. Song, A. D. Lyon, and R. F. Ismagilov, "Formation of Droplets and Mixing in Multiphase Microfluidics at Low Values of the Reynolds and the Capillary Numbers," *Langmuir*, vol. 19, pp. 9127-9133, 2003.

[11] J. D. Tice, A. D. Lyon, and R. F. Ismagilov, "Effects of viscosity on droplet formation and mixing in microfluidic channels," *Analytica Chimica Acta*, vol. 507, pp. 73-77, 2004.

[12] V. Cristini and Y.-C. Tan, "Theory and numerical simulation of droplet dynamics in complex flows - a review," *Lab on a Chip*, vol. 4, pp. 257-264, 2004.

[13] M. D. Menech, "Modeling of droplet breakup in a microfluidic T-shaped junction with a phase-field model," *Phys Rev E*, vol. 73, pp. 031505, 2006.

[14] P. Guillot and A. Colin, "Stability of parallel flows in a microchannel after a T junction," *Physical Review E: Statistical, Nonlinear, and Soft Matter Physics*, vol. 72, pp. 066301/1-066301/4, 2005.

[15] J. H. Xu, G. S. Luo, G. G. Chen, and J. D. Wang, "Experimental and theoretical approaches on droplet formation from a micrometer screen hole," *Journal of Membrane Science*, vol. 266, pp. 121-131, 2005.

[16] P. Garstecki, M. J. Fuerstman, H. A. Stone, and G. M. Whitesides, "Formation of droplets and bubbles in a microfluidic T-junction-scaling and mechanism of break-up," *Lab on a Chip*, vol. 6, pp. 437-446, 2006.

[17] W. B. J. Zimmerman, *Process modelling and simulation with finite element methods*. Singapore ; Hackensack, NJ: World Scientific, 2004.

[18] F. Losasso, R. Fedkiw, and S. Osher, "Spatially adaptive techniques for level set methods and incompressible flow," *Computers & Fluids*, vol. 35, pp. 995-1010, 2006.

[19] S. Osher and J. Sethian, "Fronts propagating with curvaturedependent speed: algorithms based on Hamilton-Jacobi formulations.," *Journal of Computational Physics*, vol. 79, pp. 12-49, 1988.

[20] C. Zhou, P. Yue, and J. J. Feng, "Formation of simple and compound drops in microfluidic devices," *Physics of Fluids*, vol. 18, pp. 092105/1-092105/14, 2006.

[21] E. Olsson and G. Kreiss, "A conservative level set method for two phase flow," *Journal of Computational Physics*, vol. 210, pp. 225, 2005.

[22] Comsol, *Comsol 3.3a Release Notes*, 2007.

[23] L. Menetrier-Deremble and P. Tabeling, "Droplet breakup in microfluidic junctions of arbitrary angles," *Physical Review E: Statistical, Nonlinear, and Soft Matter Physics*, vol. 74, pp. 035303/1-035303/4, 2006.

[24] S. Van der Graaf, T. Nisisako, C. G. P. H. Schroeen, R. G. M. Van der Sman, and R. M. Boom, "Lattice Boltzmann Simulations of Droplet Formation in a T-Shaped Microchannel," *Langmuir*, vol. 22, pp. 4144-4152, 2006.

[25] T. Ward, M. Faivre, M. Abkarian, and H. A. Stone, "Microfluidic flow focusing: Drop size and scaling in pressure versus flow-rate-driven pumping," *Electrophoresis*, vol. 26, pp. 3716-3724, 2005.

[26] J. H. Xu, S. W. Li, J. Tan, Y. J. Wang, and G. S. Luo, "Preparation of highly monodisperse droplet in a T-junction microfluidic device," *AIChE Journal*, vol. 52, pp. 3005-3010, 2006.

Mater. Res. Soc. Symp. Proc. Vol. 1139 © 2009 Materials Research Society 1139-GG03-20

Controlling the Wrinkling of the Bilayer Thin Films Electrothermally

Shravan Chintapatla[1], John F. Muth[1] and Leda M. Lunardi
[1]Department of Electrical and Computer Engineering, North Carolina State University, 2410 Campus Shore Drive, Raleigh, NC 27606, U.S.A.

ABSTRACT

Wrinkling of thin sheets under strain is a universal phenomenon. The amplitude and period of the wrinkles formed in a thin sheet clamped at both ends are dependent on its strain and material parameters. In our study, wrinkling is observed in microscale for double clamped thin films (L>W>>t) consisting of 200nm deposited low stress silicon nitride bridges fabricated by bulk micromachining. A bilayer system is formed with 30nm aluminum evaporated on to these bridges. At room temperature the bridges are essentially flat. When an electrical current passes through the aluminum layer electrothermal heating results in thermal expansion that wrinkles the bilayer. In addition we investigated various dimensions of the bridges and their correlation to the amplitude and the number of wrinkles. The observations are compared to wrinkling theory.

INTRODUCTION

Thin film materials are used in wide range of applications including microelectronics, coatings and medical devices. Wrinkling and buckling is a common feature for films under stress. Wrinkling is a well studied phenomenon occurring in thin isotropic elastic sheets (L>W>>t) when subjected to a strain greater than the critical value of the material.[1,2] Stretching of the substrate makes large amplitude, long wavelength deformations prohibitive and hence the material wrinkles with the wavelength set by a compromise between bending and stretching energies. As a stress driven instability, the study of buckling can be traced back to the classical Euler buckling of an elastic column and its dependence on the constraints that result in a sine like solution. Understanding wrinkling can be useful in a variety of applications such as diffraction gratings,[3] metrological tools,[4] and stretchable electronics.[5] Controlling the wrinkling phenomenon electronically potentially enables other applications such as controlling the finesse in Fabry-Perot microcavity devices.[6] In the present study, we show that one can make a simple MEMS device, in which the periodic wrinkling is controlled electrothermally and can be reversed. Furthermore, we study some of the properties of the materials to identify the origin of its wrinkling.

EXPERIMENT

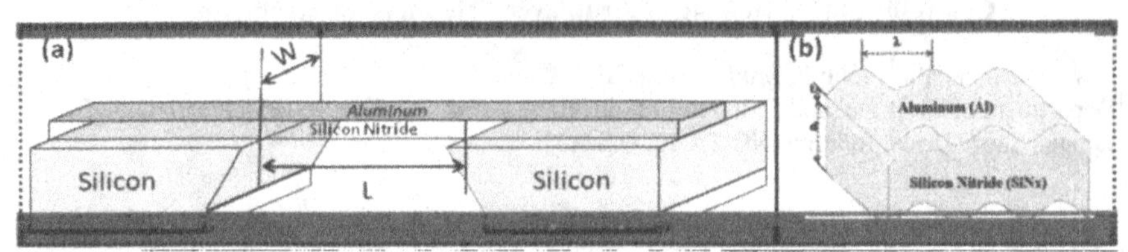

Figure 1. Schematic of the free standing double clamped microbridge structure (a) finished MEMS device and (b) Aluminum/Silicon nitride double layer showing wrinkles of wavelength λ..

A double sided polished 4-inch silicon wafer was cleaved and deposited with low stress silicon nitride by LPCVD. Double clamped microbridge structures are fabricated by patterning and then releasing a bi-layer of aluminized silicon nitride membrane through a series of the following processing steps: The silicon wafer with 200nm of low stress silicon nitride (σ=250MPa, E=250GPa)[7] was patterned on one side and by selectively plasma etching the nitride film, thus acting as a mask bulk micromachining using KOH as the wet etchant and forming silicon nitride membranes. Afterwards, the membrane is patterned and 30-40 nm thermally evaporated aluminum film deposited, which acts as an etch mask for dry etching (RIE), the silicon nitride and releasing the bilayer microbridges. Figure 1 displays the fabricated device. An optical microscope image of an array of microbridges is shown in figure 2.

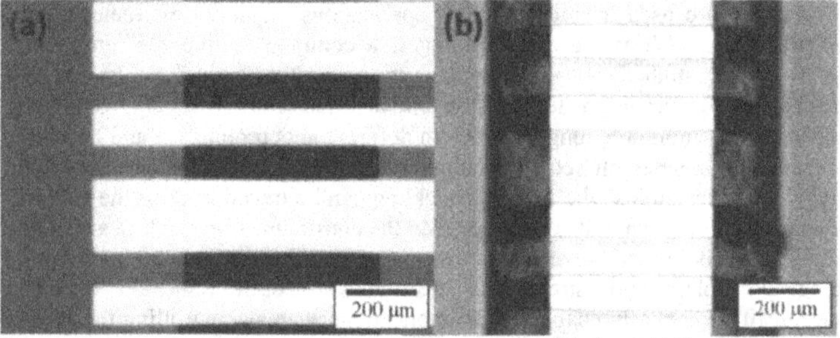

Figure 2: An optical microscope image of an array of fabricated bilayer bridges of several dimensions with 30nm of Aluminum on 200nm of silicon nitride (a) Top view and (b) Bottom view through the silicon trench formed by the KOH etchant.

DISCUSSION

The bilayer microstructure is composed of a silicon nitride as the lower layer and a top layer of thin aluminum film. The silicon nitride has a measured initial residual stress of 250 MPa, from the deposition process, defined as biaxial stress as shown in Figure 3(a). Given clamped boundary conditions with L>>W>>t, wrinkling periods[1] are observed as shown in the figure 3(b). After the initial Aluminum top layer deposition, the wrinkling disappears for any length to width ratio of the devices.

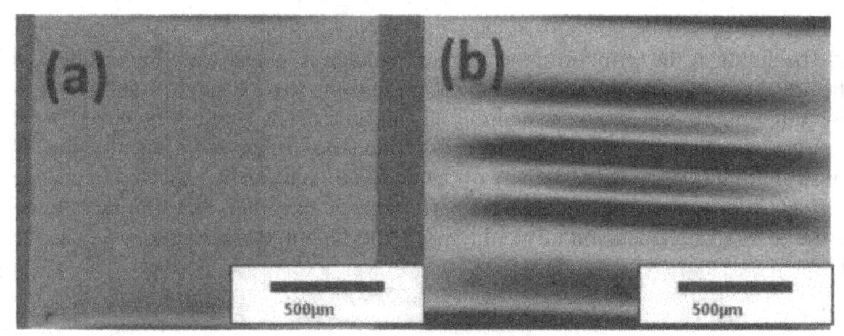

Figure 3. A silicon nitride microbridge of dimensions (a) L=500μm, W=500μm and (b) L=1,500μm >> W=750μm >> t=200nm, with 3 wrinkles.

To study the wrinkling process, a voltage is applied across the bridge with a current flowing through the aluminum. Thermal expansion due to joule heating causes the microbridge to be under compressive stress because of its double clamped condition. Wrinkling occurs when the critical stress is exceeded, as shown in the figure 4 for a typical 460μm x 500μm microbridge. As the voltage increases, the characteristic wavelength of the wrinkles increases as well. Above 3 V, all bridges show fracture behavior and irreversible damage can occur.

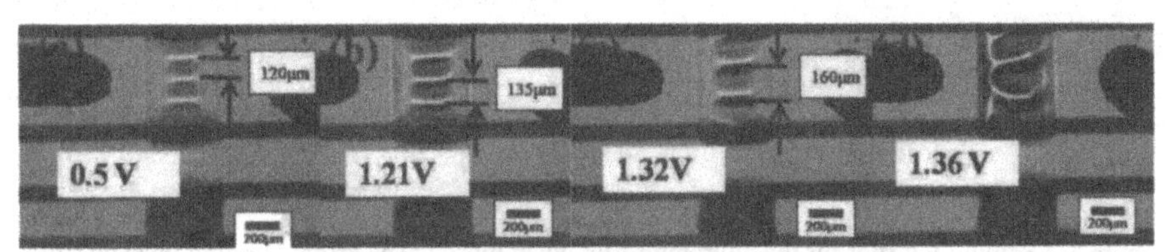

Figure 4. Electrothermal wrinkling phenomenon observed in 460μm x 500μm microbridge (left to right for increasing voltage) (a) 3 wrinkles (b) 3 wrinkles with increased amplitude (c) 2 wrinkles (d) state before fracturing.

Analytically the detailed behavior of wrinkling depends on the directions of the acting stresses, modulus of the materials and geometric constraints. The wavelengths of the wrinkles, measured in figure 4 can be predicted using the empirical Cerda's formula[1]:

$$\lambda \sim \left(\frac{B}{K}\right)^{1/4} \tag{1}$$

where B is the bending modulus of the aluminum film and K is the effective elastic foundation of stiffness (for silicon nitride film). Here silicon nitride is assumed not to be affected by heating it is only the bending modulus of the aluminum that changes with increasing bias on the microbridge. The bending modulus is given by.

$$B = \frac{Et^3}{\left[12\left(1-\nu^2\right)\right]} \tag{2}$$

where ν is the Poison's ratio and E is the Young's modulus of the aluminum film which is temperature sensitive, becoming stiffer with increasing voltage bias.

To understand the behavior of the wrinkles under applied voltage it is necessary to consider the material properties of the bridge structure. In our case, we assume the mechanical and physical properties of the aluminum i.e. the Young's modulus (E) and resistivity more influenced by the joule heating[8] than the silicon nitride material properties . Since the surface and grain structure of the thin films can strongly influence the resistivity of the material[9] and can be expected to change under heating current – voltage measurements were performed on the 30nm thin film aluminum. Figure 5displays the resistance values, initially as high as ~500 Ω, but decreasing to ~ 5 Ω as the film is annealed.

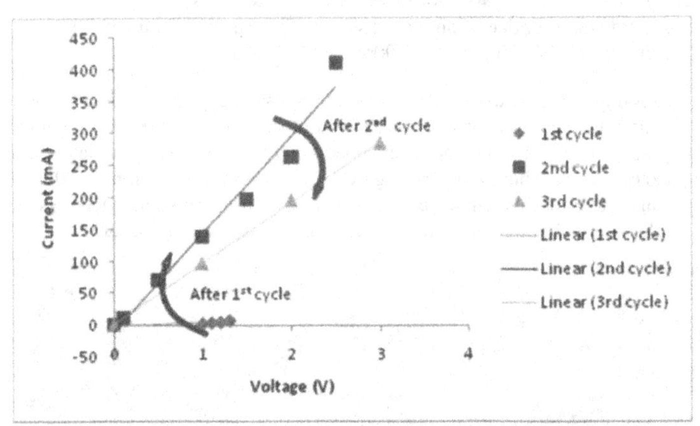

Figure 5. : Sample I-V plot for the aluminum film when biased. The arrow pointing up shows the transition from the 1st cycle to the 2nd cycle and from there to the 3rd cycle is shown by the arrow pointing down

As shown in Figure 5, the resistance behaves linearly with increasing bias (1st cycle) and the resistance is ~ 500 Ω. Joule heating annealing the film results in grain texture changes that lower the resistance to ~ 5 Ω (2nd cycle). Leaving the film heated at high temperatures for a long period of time, the film becomes more resistive (3rd cycle) with the film being destroyed it by fracturing if the applied voltage is over 3V. The change in resistance is attributed to changes in the microstructure of the aluminum[10, 11]. An increase in the Young's modulus is also observed. The resistance measurements are also useful in calculating the approximate temperature of the bridge using the strain to be calculated by the thermal expansion coefficient (TCE) of aluminum multiplied by change in temperature (ΔT) due to joule heating of the microbridge. The stress is then calculated by multiplying strain with Young's modulus. Using this methodology, the wavelengths measured using the reticule of an optical microscope and Cerda's equation one finds good qualitative agreement between theory and measurement as shown in Figure 6.

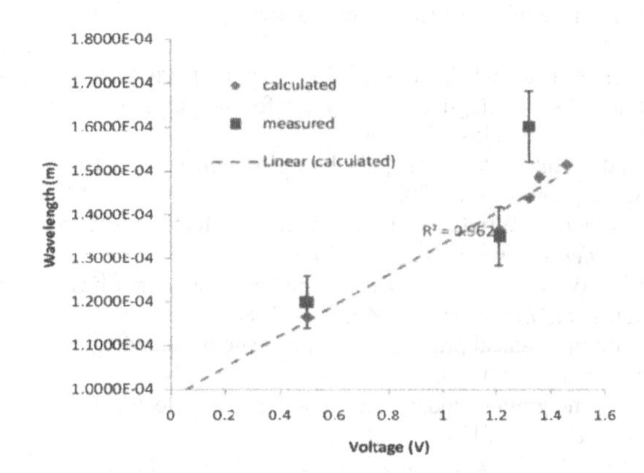

Figure 6. Plot of calculated and measured wavelengths for a 460μm x 500μm MEMS microbridge. The dashed line is the linear fit considering 5% margin error from the optical reticule.

CONCLUSIONS

A simple bulk micromachined MEMS device was constructed for studying the wrinkling of an aluminum/silicon nitride thin film bilayer. Applying voltage resulting in joule heating causes the aluminum/silicon nitride bilayer to form periodic wrinkles on the previously flat double clamped microbridges. The wrinkles can be controlled by the voltage bias. Good qualitative agreement between optical measurements of the wavelength of the wrinkles and Cerda's formula was found when the material properties of the aluminum thin film and how they changed under joule heating were considered. The wrinkling process was observed to be reproducible and reversible for voltages less 0.5 V. For voltages greater than 3 Volts the bridges were found to fracture. The ability to electronically control the formation of wrinkling in thin films may be useful in the construction of novel MEMs devices.

REFERENCES

1. E. Cerda and L. Mahadevan, "Geometry and Physics of Wrinkling," *Physical Review Letters* Vol. 90 (7), 1-4, (2003).
2. E. Cerda, K. Ravi-Chandar and L. Mahadevan, "Wrinkling of an elastic sheet under tension," *Nature,* Vol. 419, 579-580, (2002).
3. R.A. Guerrero, J.T. Barretto, J.L.V. Uy, I.B. Culaba and B.O.Chan, "Effects of spontaneous surface buckling on the diffraction performance of an Au-Coated elastomeric grating," *Optics Communications*, 270, 1-7, (2006).
4. C.M. Stafford, C. Harrison, K.L. Beers, A. Karim, E.J. Amis, M.R. Vanlandingham, H. Kim, W. Volksen, R.D. Miller and E.E. Simony, "A buckling-based metrology for measuring the elastic moduli of polymeric thin films," *Nature*, 3, 545-550 (2004).
5. S. P. Lacour, S. Wagner, Z. Huang and Z. Suo, "Strechable gold conductors on elastomeric substrates," *Applied Physics Letters*, 82 (15), 2404-2406, (2003).
6. S. Chintapatla and J. F. Muth, "Electrothermal Wrinkling of Silicon Nitride Membrane Mirror in a Fabry-Perot Cavity," *IEEE/LEOS Conference Proceedings,* (2007).
7. X. Ren and D.C.C. Lam, "Characterizations of elastic behaviors of silicon nitride thin films with varying thickness," *Materials Science and Engineering A*, 467, 93-96, (2007).
8. M.A. Haque and M.T.A. Saif, "Thermo-mechanical properties of nano-scale freestanding aluminum films," *Thin Solid Films*, 484, 364-368, (2005).
9. J.M. Camacho and A.I. Oliva, "Surface and grain boundary contributions in the electrical resistivity of metallic nanofilms," *Thin Solid Films*, 515, 1881-1885, (2006).
10. D.T. Read, Y. Cheng, R. Keller and J.D. McColskey, "Tensile properties of free standing aluminum thin films," *Scripta Materialia*, 45, 583-589, (2001).
11. D.S. Gardener and P.A. Flinn, "Mechanical Stress as a Function of temperature in Aluminum Films," *IEEE Transactions of Electron Devices*, 35(12), 2160-2169, (1988).

Mater. Res. Soc. Symp. Proc. Vol. 1139 © 2009 Materials Research Society 1139-GG03-22

Thermal Bubble Nucleation in Nanochannels: Simulations and Strategies for Nanobubble Nucleation and Sensing

Manoj Sridhar[1], Dongyan Xu[2], Anthony B. Hmelo[1], Deyu Li[2], Leonard C. Feldman[1,3]

[1]Department of Physics and Astronomy, Vanderbilt University, Nashville, TN, USA.
[2]Department of Mechanical Engineering, Vanderbilt University, Nashville, TN, USA.
[3]Institute of Advanced Materials, Devices and Nanotechnology, Rutgers University, New Brunswick, NJ, USA.

ABSTRACT

Progress in the state of the art of nanofabrication now allows devices that may enable the experimental sensing of bubble nucleation in nanochannels, and the direct measurement of the bubble nucleation rate in nanoconfined water and other fluids. In this paper we report on two aspects in achieving this goal: 1) new molecular dynamics simulations of nanobubble formation in nanoconfined argon and water model systems and 2) an ultrasensitive nanofluidic device architecture potentially able to detect individual nanobubble nucleation events.

INTRODUCTION

The classical description of homogeneous bubble nucleation in water suggests that the critical radius is a sensitive function of temperature and pressure, but of the order of one micron at 373 K. Thus we ask a simple question: Can fluids such as water boil when confined at length scales smaller than this critical dimension? Clearly, our question is concerned with confinement effects on pre-critical density fluctuations in the bulk fluid away from confining walls. We report progress on our efforts to simulate nanobubble nucleation, to understand the nucleation rate, and recent innovations in detector technology that will enable us to directly sense individual nucleation events. Potential extensions of this device include in-situ reaction monitoring, the investigation of two-phase flow, the stability of fluids with respect to the solid-liquid transition, and the granularity of matter at the nanoscale. Such nanoscale fluidic devices are ultimately limited by fundamental noise considerations, possibly different than those of other electronic systems. One important phenomenon is the influence of spontaneous thermal bubble nucleation, which has recently been postulated as a source of detector noise in nanofludic measurements [1]. Thus the investigation of nanobubble nucleation is of intrinsic interest and has direct applicability to the ultimate performance of nanofluidic devices.

MOLECULAR DYNAMICS SIMULATIONS

Thermal bubble nucleation in nanoconfined spaces has attracted considerable attention in recent years [2-7]. Most of this work has focused on using canonical-ensemble (NVT) molecular dynamics (MD) simulations to study nanoscale bubble formation. Such simulations may be regarded as bubble cavitation rather than spontaneous thermal bubble nucleation. Our goal is simulate thermal bubble nucleation inside nanochannels using an isobaric-isothermal ensemble (NPT) MD approach. However we first present the results of NVT simulations of argon and water systems to validate our approach and verify that our models are working properly by comparing our results against those previously published by other researchers.

NVT simulations

Details of our simulation system and intermolecular potentials can be found elsewhere [8]. Using a Leonard-Jones potential for a canonical ensemble of 2606 argon atoms at a temperature of 100 K, we found that bubbles were formed in the argon system when the average density of the fluid was less than $0.8\rho_{0,argon}$, where $\rho_{0,argon}$ is the saturated liquid argon density at atmospheric pressure (1397 kg/m^3), in agreement with the molecular dynamics results published by several others [5,6,9]. Figure 1a shows a typical argon atom configuration when the density of the liquid was $0.84\rho_{0,argon}$. No bubble was observed in this case over the entire simulation period of 3 ns. Figure 1b shows a typical argon atom configuration when the density of argon was $0.70\rho_{0,argon}$ and a stable bubble was observed in this case. The nucleation rate for the bubble in Figure 1b was 4×10^{28} cm^{-3}s^{-1}, which is close to values reported in the literature [2,9]. We conclude that our molecular dynamics code and interatomic potential reproduces previously published and accepted results.

Figure 1. Argon NVT (a) $T = 100$ K, $\rho = 0.84\rho_{0,argon}$, (b) $T = 100$ K and $\rho = 0.70\rho_{0,argon}$.

We performed NVT MD simulations for a canonical ensemble of 2008 water molecules using a simple point charge with polarization (SPC/E) model at a temperature of 450 K. For the water NVT systems, we observed stable bubbles when the average fluid density, $\rho < 0.74\rho_{0,water}$.

Figure 2. Water NVT (a) $T = 450$ K, $\rho = 0.95\rho_{0,water}$. (b) $T = 450$ K, $\rho = 0.66\rho_{0,water}$.

Figure 2a shows the water molecule configuration when the density is not low enough and hence, no bubble is observed. Figure 2b shows the configuration when the density is low enough to form a stable bubble.

NPT simulations

We report the initial results of molecular dynamics simulations of thermal bubble nucleation in argon and water systems nanoconfined between parallel silicon walls using an isothermal-isobaric (NPT) ensemble to determine the conditions under which nanobubble nucleation may be expected. No bubbles were observed for either system in an external pressure range of 0.01 - 0.1 MPa, even for temperatures much higher than the boiling temperature of the respective liquids at 0.1 MPa. A typical snapshot of the argon NPT system is shown in Figure 3, where the argon atoms were sandwiched by two silicon plates in the z-direction. First, the NPT system was modeled under a constant pressure of 0.1 MPa. In all simulations, we equilibrated our system until the top silicon plate oscillated within a distance of less than 1 Å. This ensured that the overall pressure in our system was kept constant.

Figure 3. The simulation box with confining silicon plates used for argon NPT case. The x and y dimensions are 5.43 nm. The z dimension depends on the net pressure of the system.

At a pressure of 0.1 MPa, we report that no bubble was observed in the NPT system up to $T = 120$ K. We also simulated argon NPT systems at a reduced pressure of 0.01 MPa and found that no bubbles were observed up to $T = 120$ K, similar to our observations at $P = 0.1$ MPa. Furthermore, for water NPT systems, no bubbles were observed over a temperature range of T = 450 – 540 K at external pressures of 0.1 or 0.01 MPa. We observe that, for both argon and water NPT systems, the average density of the fluid was roughly independent of pressure, indicating that the fluids were behaving like liquids with low compressibility even at temperatures much higher than the normal boiling point of the respective fluids. Thus, our NPT simulations suggest that bubble formation is very difficult inside nanochannels and may require temperatures much higher than the superheat limit of the fluids. Details of the nanobubble nucleation depend sensitively on how the liquid-solid interactions are modeled. Our simulation results suggest limits on the nanochannel length scale and conditions under which nanobubble nucleation can be expected. Additional MD simulations of nanoconfined Ar and water are in progress to further our understanding of bubble nucleation inside nanochannels, and will be reported shortly [8].

MOSFET-BASED FLUIDIC SENSORS

Recent advances in micro and nanofabrication technologies have enabled us to fabricate and test micro and nanofluidic devices for a variety of applications. Experimental sensing of bubble nucleation in a nanochannel reactor requires an advanced detection scheme. We report on the development of a new ultrasensitive sensor that has been used to detect the translocation of small particles through a sensing fluidic channel, and demonstrate it on the micro- and nanoscale. The device is an enhancement of a traditional coulter counter, and couples a fluidic circuit to the gate of a MOSFET that detects particles by monitoring the MOSFET drain current modulation instead of the modulation in the ionic current through the sensing channel. We demonstrate the application of the device concept as a particle sensor for polystyrene and glass beads on a variety of length scales, and extrapolate its use for the detection of spontaneous thermal bubble nucleation.

Microfluidic sensors

Figure 4 shows a schematic of the MOSFET-based microfluidic sensor that we have developed and reported recently [10, 11]. In this novel sensing scheme, we monitor the MOSFET drain current as a function of time and detect particle translocations based on the temporal modulations of the MOSFET drain current.

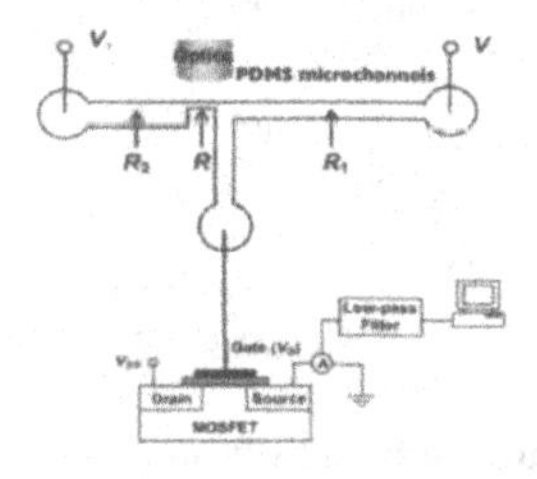

Figure 4. Schematic of experimental setup used (not to scale).

Using this MOSFET-based microfluidic device, we are able to sense particles with a volume ratio of 0.006% defined as the ratio of the particle volume to the sensing channel volume, which represents a ten-fold improvement over the current state-of-the-art [12]. This result, illustrated in Figure 5 below, shows the MOSFET drain current as a function of time when microbeads of two different sizes are present in the buffer mixture.

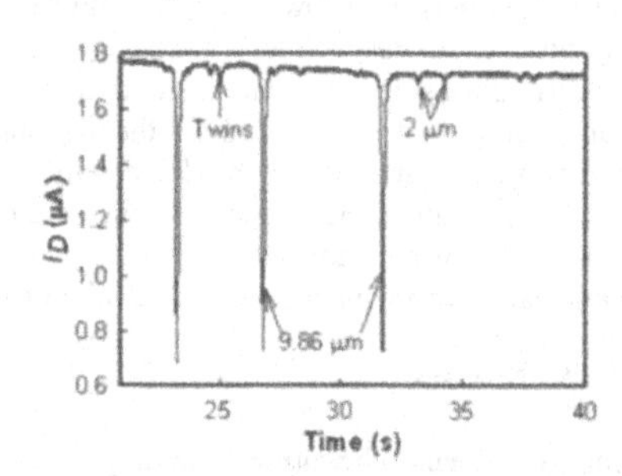

Figure 5. MOSFET drain current as a function of time for a mixture of 9.86 μm and 2 μm diameter PS microbeads.

Nanofluidic sensors

Next we have extended this sensing scheme down to the nanoscale. Figure 6 shows a schematic of the MOSFET-based nanofluidic sensor, which is very similar to its microfluidic counterpart. A scanning electron microscope (SEM) image of the sensing channel is shown in Figure 7 and the sensing channel was measured to be 5 μm long, 350 nm wide and 500 nm deep.

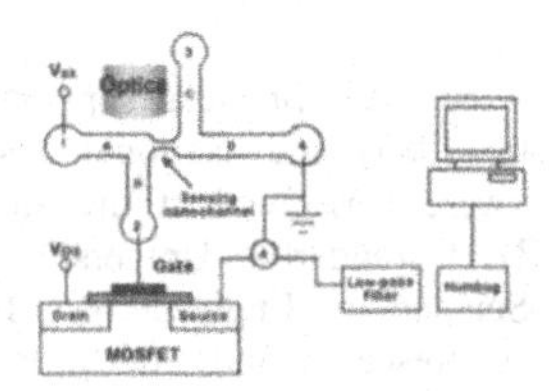

Figure 6. Schematic of the experimental setup for MOSFET-based nanofluidic sensor.

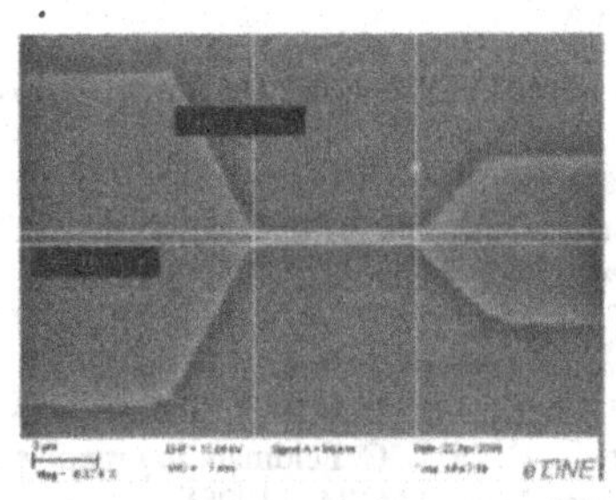

Figure 7. SEM image showing the dimensions of the sensing nanochannel.

Results of our preliminary investigations with this nanofluidic device have shown that we are able to detect 210 nm nanobeads with a volume ratio of 0.55% as shown in Figure 8. With further refinement we expect to achieve a minimum volume ratio closer to that observed for our MOSFET-based microfluidic sensor, i.e. 0.006%.

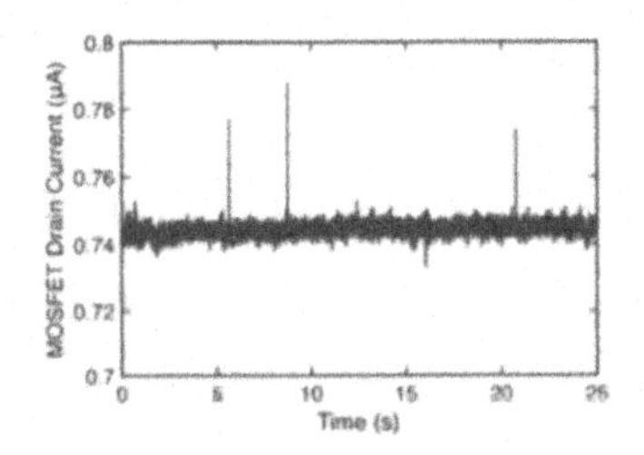

Figure 8. MOSFET drain current vs. time for 210-nm diameter PS nanobeads.

CONCLUSIONS

We believe that this novel MOSFET-based sensing scheme will enable thermal bubble nucleation detection inside nanochannels. Ongoing efforts in our research group are directed towards the fabrication of an experimental device with on-board thermal heaters that will allow further experimental tests of the predictions of our MD simulations.

ACKNOWLEDGMENTS

The authors wish to acknowledge the financial support from the US National Science Foundation (Award No. CBET-0643583) and National Institute of Health (Award No. 5R01HG002647). Computations were performed at the Advanced Computing Center for Research and Education (ACCRE) at Vanderbilt University. Finally, the authors thank the Vanderbilt Institute of Nanoscale Science and Engineering (VINSE) as well as the Vanderbilt Institute for Integrative Biosystems Research and Education (VIIBRE) for access to their nanofabrication facilities.

REFERENCES

1. R. M. M. Smeets, U. F. Keyser, M. Y. Wu, N. H. Dekker, C. Dekker, *Phys. Rev. Lett.* **97**, 088101 (2006).
2. T. Kinjo, K. Ohguchi, K. Yasuoka, M. Matsumoto, *Comp. Mat. Sci.* **14**, 138 (1999).
3. T. Kinjo, G. T. Gao, X.C. Zeng, *Prog. Theor Phys. Suppl.* **138**, 732 (2000).
4. S. Maruyama, T. Kimura, *Int J Heat Technol* **8**, 69 (2000).
5. S. Park, J. G. Weng, C. L. Tien, *Microscale Thermophys Eng* **4**, 161 (2000).
6. Y. W. Wu, C. Pan, *Microscale Thermophys Eng* **7**, 137 (2003).
7. G. Nagayama, T. Tsuruta, P. Cheng, *Int J Heat Mass Tran* **49**, 4437 (2006).
8. M. Sridhar, D. Xu, A.B. Hmelo, D. Li, L. C. Feldman, *in preparation* (2009).
9. T. Kinjo, M. Matsumoto, *Phase Equil.* **144**, 343 (1998).
10. D. Xu, Y. Kang, M. Sridhar, A. B. Hmelo, L. C. Feldman, D. Li, D. Li, *Appl. Phys. Lett.* **91**, 013901 (2007).
11. M. Sridhar, D. Xu, Y. Kang, A. B. Hmelo, L. C. Feldman, D. Li, D. Li, *J. Appl. Phys.* **103**, 104701 (2008).
12. R. W. DeBlois, C.P. Bean, *Rev. Sci. Instrum.* **41**, 909 (1970).

Mater. Res. Soc. Symp. Proc. Vol. 1139 © 2009 Materials Research Society 1139-GG03-25

Rapid Cell Manipulation by Rotating Nanowires

Hansong Zeng, Joshua Ebel, and Yi Zhao
Department of Biomedical Engineering, The Ohio State University, 1080 Carmack Road, Columbus, OH 43210, U.S.A.

ABSTRACT

In this paper, we report the rotational maneuver of ferromagnetic nanowires and their applications in cell manipulation. The experimental results show that when the nanowires contact with the bottom surfaces, the rotation under a modest external magnetic field can generate rapid lateral motion; while the floating nanowires do not exhibit substantial lateral displacements. Cell manipulation using skeletal myoblasts C2C12 showed that the living cells can be moved efficiently on the bottom surface by the rotational maneuver of attached nanowires. We also demonstrate the use of rotational maneuver of nanowires for creating three-dimensional multicellular clusters. This work is expected to add to the knowledge of nanowire-based cell manipulation and complement a full spectrum of controlling strategies for efficient use of nanowires in micro-total-analytical-systems.

INTRODUCTION

The development of miniaturized tools has been a great technical challenge for precise manipulation of living cells, which is significant for a wide range of biomedical applications, such as cell differentiation [1], and cell-based diagnosis [2]. During the past decades, a broad array of non-invasive cell manipulation strategies has been developed, including microfluidic approaches [3], dielectrophoretic approaches [4], optomechanical approaches [5], and etc.

Magnetic nanoparticles have been proven efficient for cell manipulation, separation, and patterning [6-8]. Among these nanoparticles, one-dimensional nanowires with large aspect ratios have recently drawn extensive interests due to their larger magnetic moments. The characteristic cross-sectional dimensions of these nanowires range from tens of nm to several hundreds of nm, while the length is in the order of a few μm or more. Such unique geometries provide a micro/nano interface to wire the nanowires with the "bigger" surroundings. In this paper, we report the rotational maneuver of magnetic nanowires under a rotating magnetic field. Such rotation brings nanowires out of the lateral plane, and leads to a larger displacement. The rotational maneuver thus allows rapid manipulation of nanowires or cell-nanowire couples on the lateral plane. It also allows three dimensional assemblies of cell-nanowire couples, which may open new perspective for various applications in tissue engineering.

EXPERIMENT

Nanowire synthesis

Figure 1a shows the electroplating process for nickel nanowires synthesis. An anodic alumina membrane (Figure 1b) (Fisher Scientific, Pittsburgh, PA) was used as the template for growing nanowires. The GaIn eutectic alloy (Sigma-Aldrich, St. Louis, MO) was applied to one side of the membrane surface as the seed layer. Afterwards, the copper plate was electrically connected with a nickel wire and immerged into the nickel plating solution (Ni Pure, Technic Inc., Cranston, RI). During the electroplating, the copper plate carrying the membrane served as the anode and the nickel wire served as the cathode. Nickel was growing in the nanopores of the anodic alumina membrane under a DC bias of 1.5 V. The diameter of the nanowires was determined by the pore size in the anodic alumina membrane. The length of the nanowires was controlled by the electroplating time (Figure 1c). After completion of the electroplating process, the anodic alumina membrane was detached from the copper plate. The GaIn eutectic alloy was removed using concentrated nitric acid. The alumina membrane was dissolved in NaOH solution (20 wt %). The released nanowires were collected and suspended into phosphate buffer solution (PBS). Figure 1d shows the SEM pictures of released nanowires. The diameter of the released nanowires was measured as about 200 nm (Figure 1e).

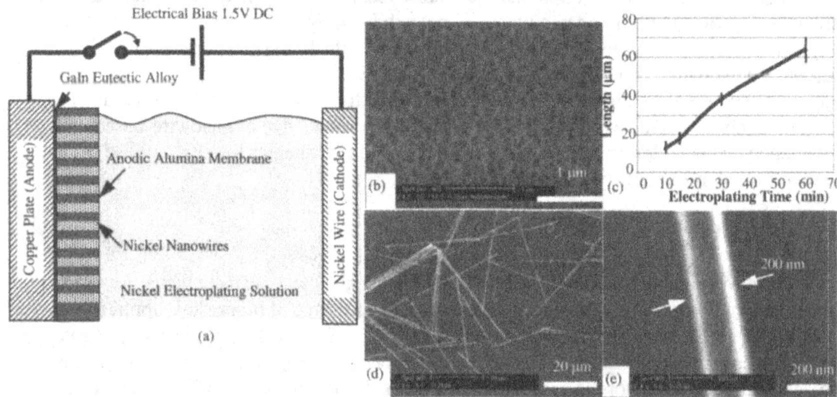

Figure 1. Nickel nanowires were synthesized by electroplating using a porous alumina anodic membrane. (a) The electroplating was performed under an electrical bias of 1.5 V DC; (b) The nanowires were growing into the pores of the alumina anodic membrane; (c) The length of the nanowires was controlled by the electroplating time. (d) and (e) show the SEM micrographs of the as-fabricated nickel nanowires. The diameter of the nanowires is about 200 nm.

Cell culture and incubation with nanowires

The nanowires were sterilized in ethanol (70 wt %), and precipitated by centrifugation at 900rpm for 5 min. These wires were then resuspended into the culture medium [Dulbecco's Modified Eagle Medium (DMEM; Invitrogen, CA) supplemented with 10% fetal bovine serum, 2 mM glutamine, and 1% penicillin/streptomycin], kept in incubator at 37 °C, and sonicated for 5 mins to avoid aggregation before being mixed with cells.

Cell manipulation by the rotational maneuver of nanowires was demonstrated with skeletal myoblasts C2C12 (ATCC, MD). After the cells proliferated to reach a desired density, they were trypsinized and resuspended in the culture medium and mixed with the PBS containing slightly over-numbered nanowires. The mixture was stirred thoroughly. Optical observation showed most cells were well attached to one or more nanowires. This is attributed to the cell affinity for binding with hydrophilic surfaces of native nickel oxide layer on the nanowires [9]. The nanowire disengagement from the cell was not observed throughtout the entire experiment. The cell-nanowire couples were then aspirated and added to the culture medium in a 100 petri-dish at $1\times10^5/cm^2$.

Experiment setup

The rotational maneuver of the nanowires was studied under a rotating magnetic field. The experimental setup was illustrated in Figure 2a, where the petri-dish containing nanowire suspension was placed on the stage of a measuring microscope. The external magnetic field was generated by a rectangular magnet, which is 4.75 cm in length, 2.22 cm in width, and 0.95 cm in thickness. It was positioned on the stage close to the petri-dish. The distance from the point of interest to the center of the magnet was about 3 cm. The magnet was able to rotate around the lateral axis (y axis) at a controllable rate. Similarly, the rotation of nanowires along y-axis was realized by placing a second permanent magnet perpendicular to the first one. The nanowires or the cell-nanowire couples can thus be rotated along any arbitrary axis in the lateral surface. The rotational maneuver and the lateral displacement of the nanowires were continuously monitored using the microscope. Given the fact that the characteristic dimensions of the permanent magnet were orders larger than the dimensions of an individual nanowire, the field gradient at a subject nanowire is negligible. The lateral traveling velocity due to the field non-uniformity is orders lower than that due to the rotating field, and is neglected.

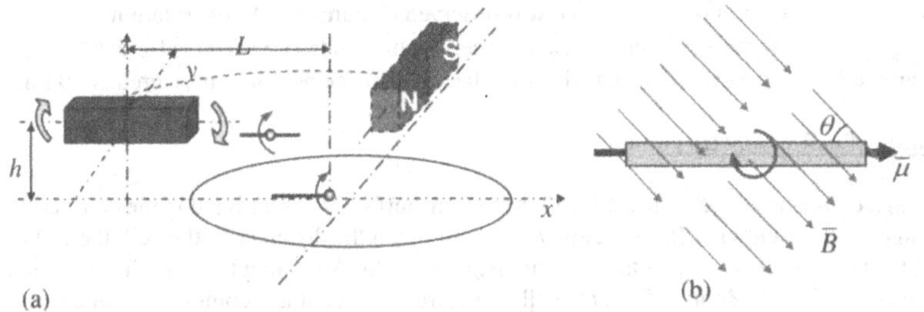

Figure 2. The roatational maneuver of the nanowires was studied using a rotating external magnetic field. (a) The field was generated using permanent magnets positioned close to the point of interest; and (b) the Interaction between the external field and the induced magnetic dipole moment rotates the nanowire out of the lateral plane.

Manipulation of nanowires

Figure 3 is the experiment results showing the rotation of nanowires under two suspension conditions (suspending in the medium and contacting with the bottom surface). The medium containing nanowires was added to a petri-dish and shaked gently, followed by immediate observation using the optical microscope. The focal plane was carefully tuned to the level where the subject nanowire suspended. The nanowire rotation was seen from the change of the projected length in response to the rotating magnetic field (Figure 3a). The suspending nanowire did not exhibit substantial lateral displacement during the observation, as evidenced by the small lateral displacement with respect to the stationary reference line.

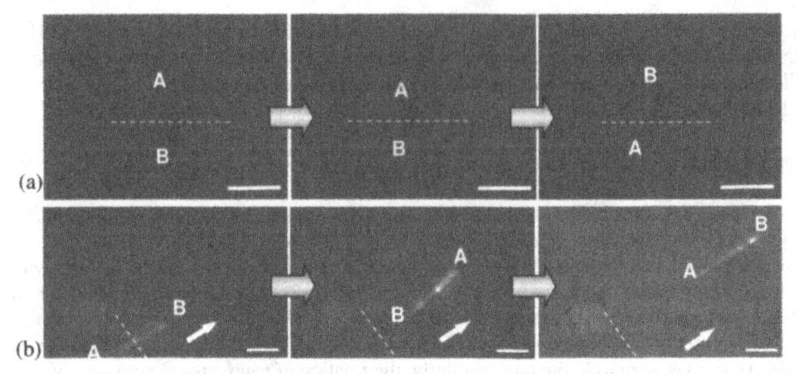

Figure 3. Experimental validation of the nanowire rotation under two different suspending conditions. (a) A floating nanowire rotating around its midpoint did not exhibit substantial lateral displacement; and (b) A nanowire contacting with the bottom surface had lateral displacement under a rotating field. The spacebars indicate 10 μm.

Under the second suspending condition, the nanowire suspension was kept stationary in the petri-dish for 10 mins to allow the sedimentation. The focal plane was adjusted to the bottom surface. After the nanowires lay on the surface, the rotating external field was applied. The rotation of the nanowire around one endpoint was observed (Figure 3b). As the rotation proceeds, the nanowire moved laterally on the surface. It took less than 5 seconds for a 20 μm long nanowire to finish one cycle, during which the lateral displacement was measured at 40 μm.

Manipulation of cell-nanowire couples

When a cell-nanowire couple settles to the bottom surface, the rotational maneuver varies with the length of the nanowire. If the nanowire is shorter than the diameter of the cell, the cell is rotating on the bottom surface under the rotating magnetic field. Assuming the non-slip condition between the cell and the bottom surface, the cell-nanowire movement is a combination of the translational motion along the lateral direction and the rotation around the midpoint of the nanowire (the center of the cell) (Figure 4a).

If the nanowire is longer than the diameter of the cell, the rotational maneuver becomes a little more complicated since the endpoints of the nanowire contact with the bottom surface two times during each cycle, as shown in Figure 4b. The process can be explained by two phases. In the first phase, the cell-nanowire couple rotates around the midpoint of the nanowire until one endpoint of the wire touches the bottom surface. In the second phase, the nanowire-cell couple rotates around the contacting endpoint, lifting the cell slightly off the surface until the cell contacts the bottom surface with the other side. The rotation proceeds as these two phases alternate, moving the cell-nanowire couple on the lateral direction. Figure 4c shows the motion of a cell-nanowire couple under a rotating magnetic field. For such a cell-nanowire couple, the lateral displacement of over 1mm was achieved within 5 mins using a modest magnetic field ranged from 500 G to 2000 G, with the rotating rate of about 20 rpm.

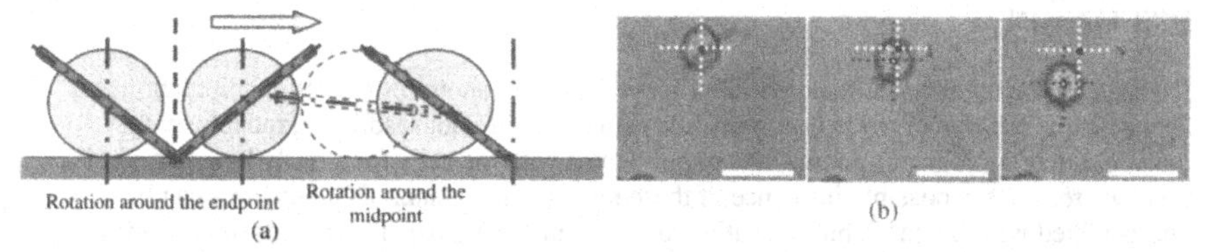

Figure 4. The rotational maneuver of a cell-nanowire couple contacting with the bottom surface. (a) The rotation consists of two phases: (1) rotation around the midpoint plus the lateral movement; and (2) rotation around the endpoint lifting the cell out of the lateral surface; (b) The experimental observation of cell manipulation on the lateral surface under a rotating magnetic field. The spacebars indicate 20μm.

Nanowire clustering and 3D cell-nanowire assembly

It is known that ferromagnetic nanowires are able to attract and connect with each other when they are brought approached. In this work, we explored the use of rotational maneuver of ferromagnetic nanowires for creating multicelluar assemblies. Figure 5a shows the manipulation of a cell-nanowire couple to approach another couple under the rotating magnetic field. By continuing the process, the cluster served as a seed for picking up other cell-nanowire couples and forming a large rod-shaped cell-nanowire assembly (Figure 5b-d). The assembly can be wrapped by rotating the field towards the longitudinal axis of the rod-shape cluster to form more spherical geometry. Optical examination showed the cluster holds the cells in a three dimensional fashion. The thickness of a representative rod shaped cluster is over 200 μm. The lateral dimension is over 1 mm. The size and the geometry of the cluster can be adjusted by controlling the traveling route of the seed cluster for selective picking up and wrapping.

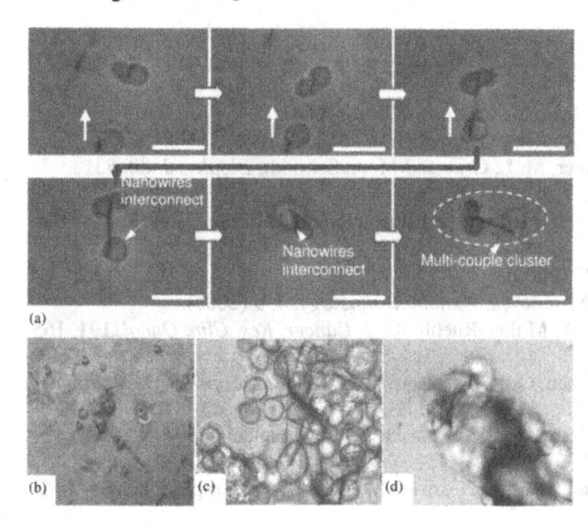

Figure 5. Multicellular clustering using the rotational maneuver. (a) A cell-nanowire couple was brought approached to other couples to develop a large cluster upon the nanowire connection; and (b-d) Three-dimensional cell-nanowire clusters were formed by picking up all the cell-nanowire couples on the lateral surface. The spacebars indicate 20μm.

DISCUSSION

Despite the promising potential of ferromagnetic nanowires for cell manipulation and analysis, one critique of these nanomaterials is that they are not readily degradable by the metabolism of biological systems, and may co-exist with the cells for a fairly long period of time. To reduce the possible influence of these metallic nanomaterials, nanowires can be encapsulated within a more biocompatible polymer material [10]. This avoids direct contact of metallic nanomaterials with the cells while retaining the manipulability. One can also fabricate one-dimensional nanomaterials using biodegradable nanomaterials with ferromagnetic or paramagnetic properties for cell manipulation [11].

CONCLUSIONS

This work explores the use of rotational maneuver of nanowires for cell manipulation and clustering under an external magnetic field. The rotational maneuver and manipulation efficiency are interpreted by considering two different suspending conditions of the nanowires. The movement of cell-nanowire clusters were experimentally validated. It thus facilitates the integration of magnetic nanomaterials with cellular objects. This work is expected to open new prospective for micropattern-free cell manipulation, and help the scaffold-free fabrication of three-dimensional cell assemblies in various studies of tissue engineering.

ACKNOWLEDGMENTS

The authors would thank Dr. Sudha Agarwal for her generous supply of C2C12 skeletal myoblasts.

REFERENCES

1. R. Mcbeath, D. M. Pirone, C. M. Nelson, K. Bhadriraju and C. S. Chen, *Dev. Cell.* **6**, 483 (2004).
2. R. Giordano, L. Lazzari and P. Rebulla, *Vox Sang.* **87**, 65 (2004).
3. C. Yi, C. Li, S. Ji and M. Yang, *Analytica Chimica Acta.* **560**, 1 (2006).
4. J. Voldman, R. A. Braff, M. Toner, M. L. Gray and M. A. Schmidt, *Biophys J.* **80**, 531 (2001).
5. M. Ozkan, M. Wang, C. Ozkan, R. Flynn and S. Esener, *Biomed. Microdevices.* **9**, 435 (2007).
6. M. Zahn, *J. Nanopart. Res.* **3**, 73 (2001).
7. A. Prina-Mello, Z. Diao and J. M. Coey, *J. Nanobiotechnology.* **4**, 9 (2006).
8. J. Kandzia, M. J. Anderson and W. Muller-Ruchholtz, *J. Cancer. Res. Clin. Oncol.* **101**, 165 (1981).
9. A. Hultgren, M. Tanase, C. S. Chen, G. J. Meyer and D. H. Reich, *J. appl. Phys.* **93**, 7554 (2003).
10. R. Gunawidjaja, C. Jiang, S. Peleshanko, M. Ornatska, S. Singamaneni and V. V. Tsukruk, *Adv. Funct. Mater.* **16**, 2024 (2006).
11. S. T. Tan, J. H. Wendorff, C. Pietzonka, Z. H. Jia and G. Q. Wang, Chemphyschem, 6, 1461 (2005)

Mater. Res. Soc. Symp. Proc. Vol. 1139 © 2009 Materials Research Society 1139-GG03-31

Novel Technique to Determine Elastic Constants of Thin Films

K.J. Martinschitz[1], R. Daniel[2], Ch. Mitterer[2] and J. Keckes[1,*]

[1]Erich Schmid Institute for Materials Science, Austrian Academy of Science and Department of Materials Physics, University of Leoben, Austria
[2]Department of Physical Metallurgy and Materials Testing and Christian-Doppler Laboratory for Advanced Hard Coatings, University of Leoben, Austria
*jozef.keckes@mu-leoben.at

ABSTRACT

A new X-ray diffraction technique to determine elastic moduli of polycrystalline thin films deposited on monocrystalline substrates is demonstrated. The technique is based on the combination of $\sin^2\psi$ and X-ray diffraction wafer curvature techniques which are used to characterize X-ray elastic strains and macroscopic stress in thin film. The strain measurements must be performed for various *hkl* reflections. The stresses are determined from the substrate curvature applying the Stoney's equation. The stress and strain values are used to calculate *hkl* reflection dependent X-ray elastic moduli. The mechanical elastic moduli can be then extrapolated from X-ray elastic moduli considering macroscopic elastic anisotropy of the film. The derived approach shows for which reflection and corresponding value of the X-ray anisotropic factor Γ the X-ray elastic moduli are equal to their mechanical counterparts in the case of fibre-textured cubic polycrystalline aggregates. The approach is independent of the crystal elastic anisotropy and depends on the fibre texture type, the texture sharpness, the amount of randomly oriented crystallites and on the supposed grain interaction model. The new method is demonstrated on a fiber textured Cu thin film deposited on monocrystalline Si(100) substrate. The advantage of the new technique remains in the fact that moduli are determined non-destructively, using a static diffraction experiment and represent volume averaged quantities.

INTRODUCTION

The reliability and the performance of thin films used *e.g.* in microelectronics or as protective coatings on working tools is closely related to their mechanical properties [1,2]. Advanced characterization techniques operating on very small scale have been used to assess phenomena like residual stress, yield stress, hardness, indentation modulus *etc.* [1].

Recently, it has been demonstrated that X-ray elastic strain and macroscopic stress in polycrystalline thin films can be rapidly determined by a simultaneous application of $\sin^2\psi$ and X-ray diffraction (XRD) substrate curvature techniques [3-5]. The experimental strain and stress values were used to quantify X-ray elastic constants and stress factors [3-5].

Mechanical elastic constants can be extrapolated from X-ray elastic constants considering macroscopic elastic anisotropy of the sample. In the case of cubic polycrystalline aggregates with macroscopic elastic isotropy (quasi-isotropic materials) which obey the Hill grain interaction model [6,7], it was demonstrated that X-ray elastic constants correspond to their mechanical counterparts for $\Gamma_{hkl} = 0.2$ whereby Γ_{hkl} is the X-ray anisotropic factor according to the Reuss grain-interaction model with

$$\Gamma_{hkl} = \frac{h^2k^2 + k^2l^2 + l^2h^2}{(h^2 + k^2 + l^2)^2} \tag{1}$$

where *hkl* are indices of the reflection measured [8]. According the Reuss model, X ray elastic anisotropy is often expressed as a function of $3\,\Gamma_{hkl}$ and this formalism will be applied hereafter [9].

Provided grain interaction mechanism and elastic anisotropy of the sample are known, mechanical elastic constants can be extrapolated from X-ray elastic constants. The aim of this work is to demonstrate a practical application of the new self-consistent XRD technique to determine elastic moduli of polycrystalline thin films deposited on monocrystalline substrates. At first a certain type of $3\Gamma_{hkl}$ selection rule is presented whereby *hkl* in $3\Gamma_{hkl}$ denotes a reflection for which mechanical and X-ray elastic constants are equal. As a next step, out-of-plane Young's moduli of 111 fiber-textured Cu thin films are determined using XRD under static conditions.

METHODOLOGY

According to the X-ray diffraction Hooke's law, the elastic strain $\{\varepsilon_{33}\}_{\psi}^{hkl}$ of a fiber textured film can be described as follows [7]

$$\{\varepsilon_{33}\}_{\psi}^{hkl} = \langle\sigma\rangle \left[\left(\{S_{3311}\}_{\psi}^{hkl} + \{S_{3322}\}_{\psi}^{hkl}\right) + \left(\{S_{3333}\}_{\psi}^{hkl} - \{S_{3311}\}_{\psi}^{hkl}\right)\sin^2\psi\right] \tag{2}$$

where $\{S_{33ij}\}_{\psi}^{hkl}$ represent X-ray elastic constants of a *hkl* reflection measured at a tilt angle ψ and $\langle\sigma\rangle$ represents an in-plane stress imposed on the film. The tilt angle ψ is defined as an angle between the sample normal and the diffraction vector [10]. In the case of experimental $\sin^2\psi$ dependences, the term $\{S_{3311}\}_{\psi}^{hkl} + \{S_{3322}\}_{\psi}^{hkl}$ corresponds to the intercept on the $\{\varepsilon_{33}\}_{\psi}^{hkl}$ axis and the term $\{S_{3333}\}_{\psi}^{hkl} - \{S_{3311}\}_{\psi}^{hkl}$ is proportional to the plot at the tilt angle ψ. Since for $\psi=0$ $\{S_{3311}\}_{\psi}^{hkl} = \{S_{3322}\}_{\psi}^{hkl}$, the knowledge of $\{S_{3311}\}_{\psi}^{hkl} + \{S_{3322}\}_{\psi}^{hkl}$ and $\{S_{3333}\}_{\psi}^{hkl} - \{S_{3311}\}_{\psi}^{hkl}$ parameters opens a possibility to quantify X-ray compliance $\{S_{3333}\}_{\psi}^{hkl}$ which is inversely proportional to the diffraction Young's modulus $\{E\}_{\psi}^{hkl}$.

Supposing the Hill grain interaction model [6], out of plane XRD $\{E\}_{\psi=0}^{hkl}$ and mechanical $\{E\}_{\psi=0}^{\mathrm{M}}$ moduli of numerous materials were numerically calculated and compared. As a parameter for the numerical calculations, various *uvw* fibre textures (characterized by the parameter $3\,\Gamma_{uvw} = \left(u^2v^2 + u^2w^2 + v^2w^2\right)/(u^2 + v^2 + w^2)^2$) with the fibre axis oriented perpendicular to the substrate surface were considered. Moreover, it was supposed that the fibre textures can possess various sharpness expressed by the parameter ψ_{FWHM} which represents the full width at half maximum of the peak in the centre of *uvw* pole figures. It was moreover supposed that the films possess a different amount of isotropic oriented crystallites (ISO) from the range 0-100%. The aim of the numerical calculation was to find $3\,\Gamma_{hkl}^{*}$ parameter for which out of plane XRD

$\{E\}^{hkl}_{\psi=0}$ and mechanical $\{E\}^{M}_{\psi=0}$ moduli are equal. In order to express the dependence of $3\,\Gamma^{*}_{hkl}$ on the $3\,\Gamma_{uvw}$, on ψ_{FWHM} (in degrees) and on ISO generally, the following empirical equation was derived:

$$3\,\Gamma^{*}_{hkl} = A + 3\,\Gamma_{uvw}\left(1 - \frac{A}{0.6}\right) \qquad (3)$$

where $A = (\psi_{FWHM} * 8.8 + ISO*5.8 - \psi_{FWHM}\ ISO*0.083)/1000$.
Next the procedure to determine out of plane mechanical moduli will be demonstrated.

EXPERIMENT

A 600 nm thick Cu thin film was deposited on Si(100) using magnetron sputtering in laboratory scale deposition system. In order to induce a measurable substrate curvature and to avoid a substrate plastic deformation, a monocrystalline Si(100) wafer with the thickness of 140 μm and lateral dimensions of 30 x 8 mm^2 was chosen for the deposition. The film was deposited in argon atmosphere at room temperature and then annealed at 400 °C for 10 minutes in order to increase the residual stress (and substrate curvature) magnitude.
The substrate curvature, elastic strain and texture in the film were characterized in laboratory conditions using a Seifert 3000 PTS four-circle diffractometer. The setup comprised Cu K_α radiation, polycapillary optics on the primary side, Soller slits, a graphite monochromator and a scintillation detector on the secondary side. For the elastic strain and curvature characterization, the beam size of 3.0 and 0.5 mm in diameter were chosen. The rectangular sample was glued just with one of their narrower side on a sample holder to allow for the bending when the strain and the curvature were characterized in the diffractometer.

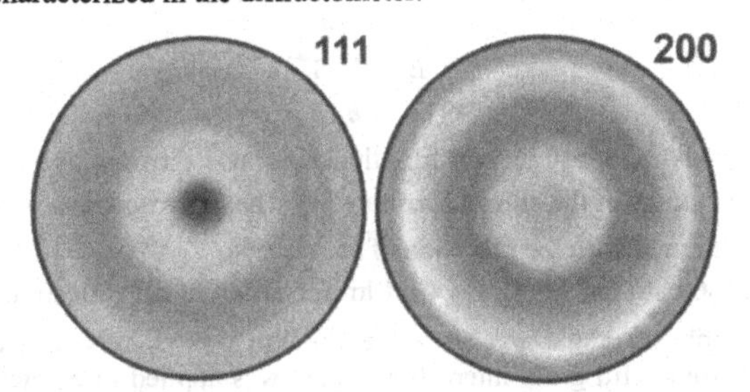

Figure 1. Experimental Cu 111 and 200 pole figures document a 111 fibre texture in Cu thin film.

RESULTS

The texture in Cu thin film was characterized using pole figure measurements whereby a relatively sharp 111 fiber texture was identified (Fig. 1). The analysis of the data showed that the width at half maximum ψ_{FWHM} of the peak in the centre of 111 pole figure is 14 degrees (Fig. 1 left) and the amount of randomly oriented crystallites (ISO) corresponds to 10 %. The experimental parameters $3\Gamma_{uvw}$, ψ_{FWHM} and ISO were used to determine $3\Gamma^{*}_{hkl}$=0.89 using Eq. 3.

The volume-averaged macroscopic stress in Cu thin film was determined using XRD substrate curvature method [11,12]. The quantification of the curvature was performed by the measurement of rocking curves of Si 400 reflections at different sample positions x. In Fig. 2, the plot of rocking curve relative angular position expressed through $\Delta\omega$ is presented as a function of the relative sample position Δx.

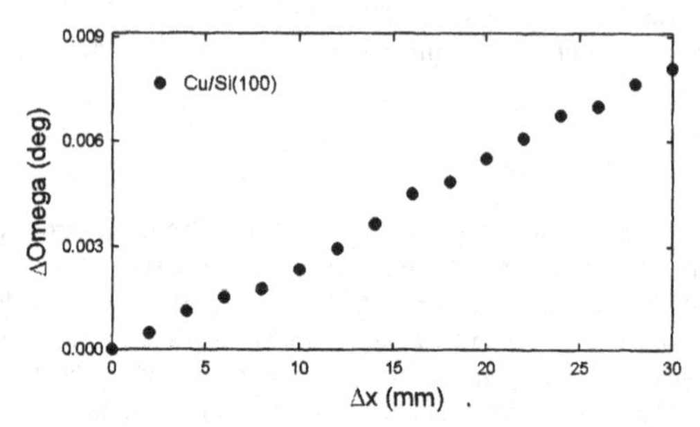

Figure 2. The plot of $\Delta\omega$ dependence on Δx. The results document a radius of the curvature R of 2.193 m which corresponds to the tensile stress of 275.9 ± about 20 MPa .

The data in Fig. 2 were used to calculate the radius of the curvature according $R \cong (\partial\Delta\omega / \partial\Delta x)^{-1}$ whereby $\partial\Delta\omega / \partial\Delta x$ represent the slope of the linear dependency. Applying R, it was possible to determine the macroscopic in-plane isotropic residual stress $\langle\sigma\rangle$ in the films using the Stoney formula [11]

$$\langle\sigma\rangle = \frac{E}{6(1-\nu)} \frac{h_s^2}{h_f} \frac{1}{R}, \qquad (4)$$

where h_s and h_f stands for substrate and film thickness, respectively, and the term $E/(1-\nu)$=181 GPa is the biaxial modulus of the silicon substrate [1]. The macroscopic stress $\langle\sigma\rangle$ of 275.9 MPa in Cu film was determined with the precision of about ±5%.

In Fig. 3, X-ray elastic strains $\{\varepsilon_{33}\}_{\psi}^{hkl}$ in Cu film for different *hkl* reflections are presented as a function of the sample tilt angle ψ. Since the Cu film was polycrystalline, the methodology based on the anisotropic Hill grain interaction model was applied to extract mechanical elastic constants applying Eq. 2. This type of analysis was performed in order to evaluate (*i*) $\{S_{3311}\}_{\psi=0}^{hkl} + \{S_{3322}\}_{\psi=0}^{hkl} = 2\{S_{3311}\}_{\psi=0}^{hkl}$ from the intercepts on the $\{\varepsilon_{33}\}_{\psi}^{hkl}$ axis and (*ii*) $\{S_{3333}\}_{\psi\to 0}^{hkl} - \{S_{3311}\}_{\psi\to 0}^{hkl}$ from the slopes in Fig. 3. In Fig. 4, the fitted parameters are shown as a function of $3\Gamma_{hkl}$ whereby the numerical dependence is presented in Table 1. Those parameters differ for various *hkl* reflections what is the consequence of the crystal elastic anisotropy. By easy calculus it was possible to derive also a dependence of $\{S_{3333}\}_{\psi\to 0}^{hkl}$ on $3\Gamma_{hkl}$ (Table 1). Considering the macroscopic elastic anisotropy and by applying the $3\Gamma^{*}_{hkl}$=0.89 (Eq. 3) one could determine an inverse out-of-plane XRD elastic modulus $\{S_{3333}\}_{\psi=0}^{3\Gamma^*}$ which is equal to the mechanical compliance $\{S_{3333}\}_{\psi=0}^{\mathrm{M}}$ and inversely proportional to the out-of plane Young's modulus of the film $\{E\}_{\psi=0}^{\mathrm{M}}$. Finally the out-of-plane modulus of the film was found to be 169.40 MPa. Considering the experimental uncertainties, it is supposed that the modulus was determined with an error of ±15%.

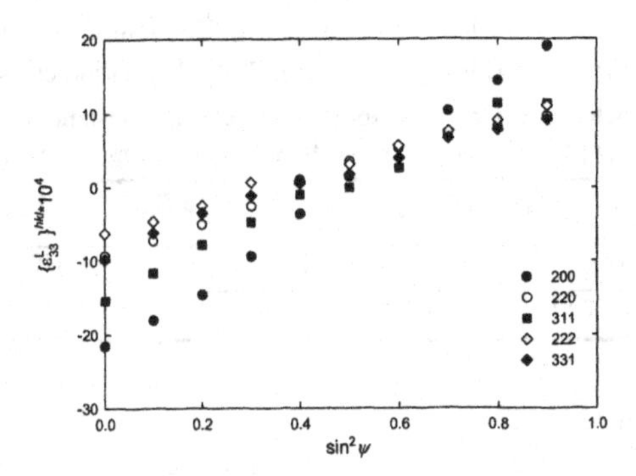

Figure 3. X-ray elastic strains $\{\varepsilon_{33}\}_{\psi}^{hkl}$ in Cu thin film as a function of the sample tilt angle ψ. The strains were determined with a precision better than ±10%.

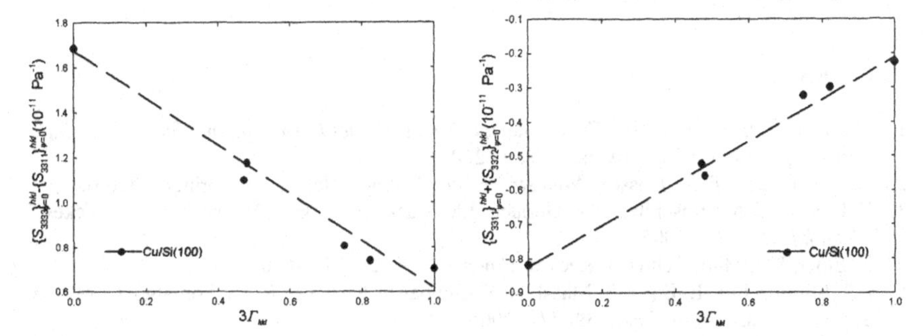

Figure 4. X-ray elastic constants $\{S_{3333}\}_{\psi\to0}^{hkl} - \{S_{3311}\}_{\psi\to0}^{hkl}$ and $\{S_{3311}\}_{\psi=0}^{hkl} + \{S_{3322}\}_{\psi=0}^{hkl}$ obtained by fitting the Eq. 2 to the data from Fig. 3 and by evaluating the slopes (left) and the intercepts (right) under the consideration of the macroscopic stress $\langle\sigma\rangle$=275.9 MPa (Fig. 2). The numerical data are presented in Table 1.

Conclusions

The aim of this contribution was to show for which *hkl* reflection the X-ray elastic strain is equal to the mechanical elastic strain in the case of equi-biaxially strained fiber-textured thin films obeying the Hill grain interaction model. Moreover, a new experimental method to determine elastic moduli of thin films using a static diffraction experiment was introduced. The method is based on the combination of XRD substrate curvature and $\sin^2\psi$ techniques which enable to determine stress and X-ray elastic strains in thin films using XRD.

Table 1. An experimental algorithm to determinate out-of-plane mechanical moduli of fiber textured thin films is presented. The macroscopic stress $\langle\sigma\rangle$ was determined using the curvature measurement (Fig. 2, Eq. 4). The elastic strains $\{\varepsilon_{33}\}_{\psi}^{hkl}$ dependences on $\sin^2\psi$ (Fig. 3) were analyzed in order to evaluate the intercepts $\{S_{3311}\}_{\psi=0}^{hkl}+\{S_{3322}\}_{\psi=0}^{hkl}=2\{S_{3311}\}_{\psi=0}^{hkl}$ and slopes $\{S_{3333}\}_{\psi\to0}^{hkl}-\{S_{3311}\}_{\psi\to0}^{hkl}$ for $\psi\to0$ (Fig. 4). The factor $3\Gamma_{hkl}^{*}$ (Eq. 2) indicates for which value of the X-ray anisotropic factor the X-ray and mechanical elastic constants are equal. The out-of-plane Young's modulus $\{E\}_{\psi=0}^{\mathrm{M}}$ is inversely proportional to the compliance $\{S_{3333}\}_{\psi=0}^{\mathrm{M}}$.

$\langle\sigma\rangle$	$\{S_{3311}\}_{\psi=0}^{hkl}+\{S_{3322}\}_{\psi=0}^{hkl}$	$\{S_{3333}\}_{\psi\to0}^{hkl}-\{S_{3311}\}_{\psi\to0}^{hkl}$	$3\Gamma_{hkl}^{*}$	$\{S_{3333}\}_{\psi=0}^{\mathrm{M}}$	$\{E\}_{\psi=0}^{\mathrm{M}}$
(MPa)	(10^{-11} Pa^{-1})	(10^{-11} Pa^{-1})		(10^{-11} Pa^{-1})	GPa
275.9	$0.62*3\Gamma_{hkl}-0.82$	$-1.04*3\Gamma_{hkl}+1.65$	0.89	0.5903	169.40

ACKNOWLEDGMENTS

This work was supported by the Austrian NANO Initiative via a grant from the Austrian Science Fund FWF within the project "StressDesign - Development of Fundamentals for Residual Stress Design in Coated Surfaces".

REFERENCES

1. S. Suresh, L.B. Freund, *Thin Film Materials Stress, Defect Formation and Surface Evolution.* Cambridge: Cambridge University Press (2003).
2. A. Cavaleiro, J. T. De Hosson, *Nanostructured Coatings.* New York: Springer (2006).
3. E. Eiper, K.J. Martinschitz, J.W. Gerlach, J.M. Lackner, I. Zizak, N. Darowski, J. Keckes, *Z. Metallkd.* **96**, 1069 (2005).
4. E. Eiper, K.J. Martinschitz, J. Keckes, *Powder Diffr.* **21**, 25 (2006).
5. K.J. Martinschitz, E. Eiper, S. Massl, H. Kostenbauer, R Daniel, R., G. Fontalvo, C. Mitterer, and J. Keckes, *J. Appl. Cryst.* **39**, 777 (2006).
6. R. Hill, *Proc. Phys. Soc. London, Sect A* **65**, 349 (1952).
7. P. van Houtte and L. De Buyser, *Acta metall. mater.* **41**, 323 (1993).
8. F. Bollenrath, V. Hauk, E.H. Müller, *Z. Metallkd.* **58**, 76 (1967).
9. A. Reuss, *Z. Angew. Math. Mech.* **9**, 49 (1929).
10. I.C. Noyan, J.B. Cohen, *Residual Stress Measurement by Diffraction and Interpretation.* Springer. (1987).
11. G.G. Stoney, *P. Roy. Soc. Lond. A* **82**, 172 (1909).
12. J. Keckes, E. Eiper, K.J. Martinschitz, H. Koestenbauer, R. Daniel, C. Mitterer, *Rev. Sci. Instrum.* **78**, 036103 (2007).

Mater. Res. Soc. Symp. Proc. Vol. 1139 © 2009 Materials Research Society 1139-GG03-32

Surface Treated PDMS by UV-Vis Light Applied to Micro-fluidic Device

Seisuke Kano[1], Sohei Matsumoto[1] and Naoki Ichikawa[1]
[1]National Institute of Science and Technology (AIST), JAPAN

ABSTRACT

Hydrophobic property of PDMS surface was improved by the 400 W UV-Vis lamp light irradiation in the atmospheric condition for several ten minutes. As a result of this surface treatment, the surface became to hydrophilic character for one month long. This surface treatment technique applied to PDMS micro-fluidic device and verified valve-less switching. The UV-Vis light irradiated to PDMS micro fluidic pattern with partly covered by aluminum foil. Finally inlet and outlets were connected 0.5 mm diameter tubes. The syringe pumps injected the distilled water into the inlet of the PDMS micro channel at the flow rates of 0.5, 5.0, and 50 μl / min for the both width channel. As results of water injection water flowed only the UV-Vis treated channel at the flow rates of 0.5 and 5.0 μl / min. On the other hand, the water flowed for all channels at the flow rate of 50 μl / min. This result was observed from 5.0 μl/min flow again for both width devices which dried by air. These results were occurred by the difference of the flow conductance and wettability. The mechanism of this hyrophilicity of PDMS was reported to form Si-O in the surface by means of glassy surface. From the IR spectra, the Si-O-Si peak shifted to higher wave number for UV-Vis irradiated PDMS than the untreated PDMS comparing with the other IR peaks. This result showed that the Si-O-Si network bonding of PDMS changed to the O-Si-O bonding around the surface.

INTRODUCTION

Polydimethyl siloxane, PDMS, has a highly potential as using to micro fluidic device, because of its economical, ecological, and flexible properties. However its surface showed hydrophobic property, therefore its application was limited. Lots of researchers try to surface treatments to control the wettability [1] by plasma irradiation [2-4], VUV light irradiation [5, 6], and so on. And also discussing the mechanism of hydrophilic surface of PDMS was reported as forming O-Si-O near the top surface by means of glassy surface. The problems of theses surface treatment were the hydrophilic property disappeared in a few hours. The reason was unknown yet but one of the possibilities the O-Si-O binding recover to Si-O network connection of PDMS origin. In many cases the treatments might be limited on the top surface of the PDMS by lower energy than that of concreting glassy O-Si-O bindings. The PDMS can store gas inside the body, that is, the body has channels connect to the surface. This structural property might be induced above recovery phenomena of the Si-O bindings. Therefore in our study the surface of PDMS was treated by irradiation of strong power light and changing the bindings inside the body.

In our research, several types of UV lights including pulsed laser beam irradiated to the PDMS surface. It was succeeded to obtain long term hydrophilic surface of PDMS by the irradiation of high power UV-Vis lamp light for one hour in the atmospheric condition [7]. The authors considered that this technique could be applied to produce valve-less devices to select a flow channel by the flow rate control and several types of bio-chip devices. In this research, this technique applied to simple flow device and verified to select the flow channel by the difference of the wettability.

EXPERIMENT

Surface Treatment by UV-Vis Light

The PDMS was spin-coated on the 35 × 50 mm^2 slide glass and heated at 60 °C for one hour in ambient atmospheric conditions. Part of the PDMS film was irradiated using UV lamp light for several minutes (2, 5, 10, 20, and 60 minutes) at 20 mm distance from the lamp which power of 400 W and wavelengths of 200–600 nm. After UV light irradiation, the wettability of the treated PDMS film surface was observed using a contact-angle tester with distilled water at just after irradiation, 2, 4, 8, 16, and 30 days later after irradiation comparing with glass surface and untreated PDMS surface.

Valve-less Micro-fluidic Device Test

The PDMS was poured on the Si wafer which the flow channel formed by SU-8, and heated at 60 °C for one hour. The PDMS had 2 mm thickness and the channel had 200 or 100 μm width and 100 μm depth. There was an inlet and separated to four channels with outlets finally, shown in figure 1 (a). The parts of the channels were masked by an aluminum foil (figure 1 (b)) and irradiated UV-Vis lamp light for 20 minutes at the 20 mm far from the lamp which electrical power of 400 W (in figure 2). The effects of the UV-Vis irradiation were evaluated by the contact angle measurements and IR spectroscopy. The contact angles were changed from 30 degrees at just after UV-Vis irradiation. From the IR spectra, the Si-O-Si peak shifted to higher wave number for UV-Vis irradiated PDMS than the untreated PDMS comparing with the other IR peaks. This result showed that the Si-O-Si network bonding of PDMS changed to the Si-O bonding around the surface.

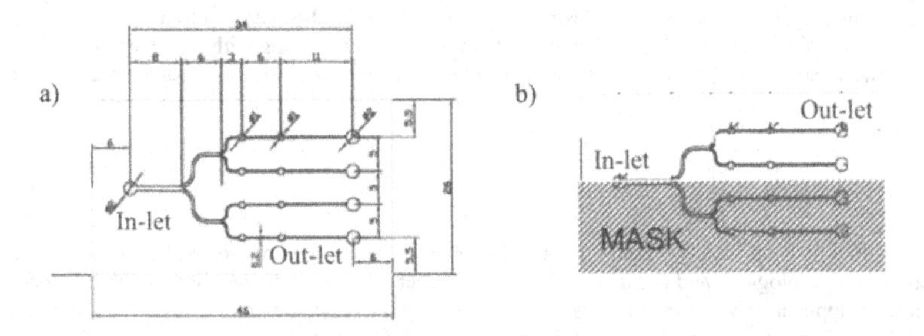

Figure 1. Schematic drawing of (a) channel pattern and (b) masked area for UV-Vis irradiation.

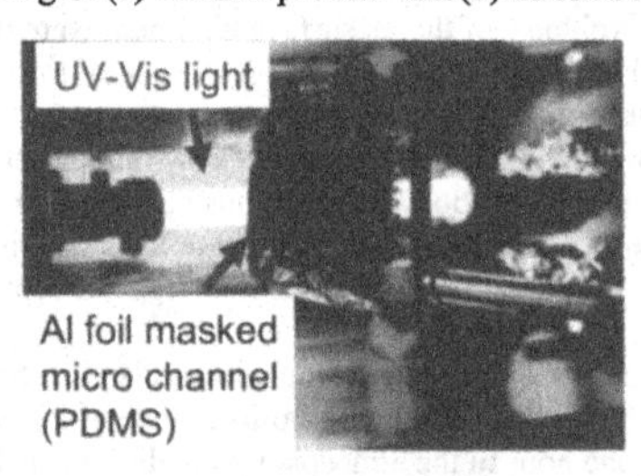

Figure 2. View of UV-Vis irradiation process and sample setup.

RESULTS

Surface Treatment by UV-Vis Light

The UV-Vis light irradiated PDMS surface was observed using a contact-angle tester with distilled water. Figure 3 shows long term measurements of the PDMS and glass surface contact angles. The PDMS surface shows hydrophobic property. This result was observed in 2 minutes UV-Vis irradiation case. On the other hand, hydrophilic surface of glass shows the contact angle kept lower than 90 degrees for over one month. The one hour UV-Vis irradiation case, the contact angles of the PDMS were kept under 90 degrees for one month like glass surface. From this result the UV-Vis treatment could be change the PDMS surface from hydrophobic to hydrophilic property until one month later after irradiation.

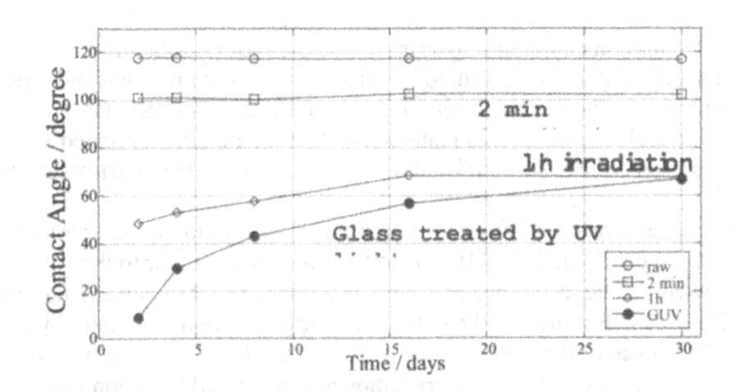

Figure 3. Contact angles of PDMS, UV irradiated PDMS, and glass.

Valve-less Micro-fluidic Device Test

The PDMS was treated by the UV-Vis lamp light and partly untreated inside the micro-channels. As results of water injection, the water flowed only UV-Vis treated channel at the flow rates of 0.5 and 5.0 µl / min in figure 4 (a). On the other hand, the water flowed for all channels at the flow rate of 50 µl / min in figure 4 (b). After the flow test at the rate of 50 µl / min, the channels dried with dry air and the water flowed again at the flow rates of 5.0 µl / min. In this case, the water flowed only UV-Vis light treated channels.

These results indicated that this UV-Vis lamp light treatment could be applied to valve-less micro channel device and selection of the fluid pass way by the changing of the flow rate or flow pressure for long term until one month.

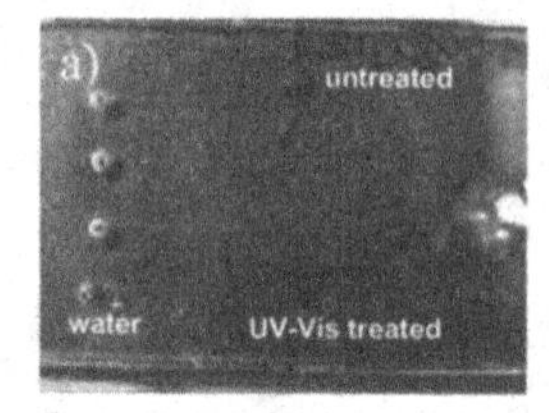

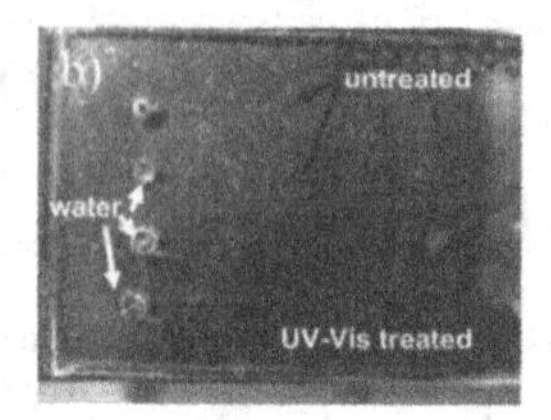

Figure 4. Micro channel devices of 200 μm width and 100 μm depth; (a) water poured at 5.0 μl /min, and (b) water poured at 50 μl /min. In the case of 5.0 μl /min the water run from UV-Vis treated channel only and the case of 50 μl /min the water run from both channels.

DISCUSSION

The UV-Vis irradiation could hold hydrophilic property of PDMS surface until one month after irradiation. The surface character was measured by micro-Raman spectroscopy and micro-FTIR spectroscopy. Figure 5 shows the Raman spectra from the untreated PDMS surface, UV-Vis lamp light irradiated PDMS for several time, and glass plate. From the spectra, only untreated PDMS and two minutes UV-Vis irradiation surface show several peaks from Si or C. The other cases, the spectra show broaden patterns.

Hydrophilic property observed from the 5 minutes UV-Vis irradiation or longer time irradiation. The Raman spectra of the over 5 minutes UV-Vis irradiations were similar to the spectra of the glass. The Raman spectrum showed drastic difference between the untreated PDMS surface and the UV-Vis irradiated PDMS. These results indicated the PDMS structure decomposing easily by the UV-Vis irradiation. Typical difference of these spectra the peaks of the PDMS induced from the original binding of the PDMS disappeared after UV-Vis light irradiation and generated Si-O bonding. These differences might be related to the hydrophilic property.

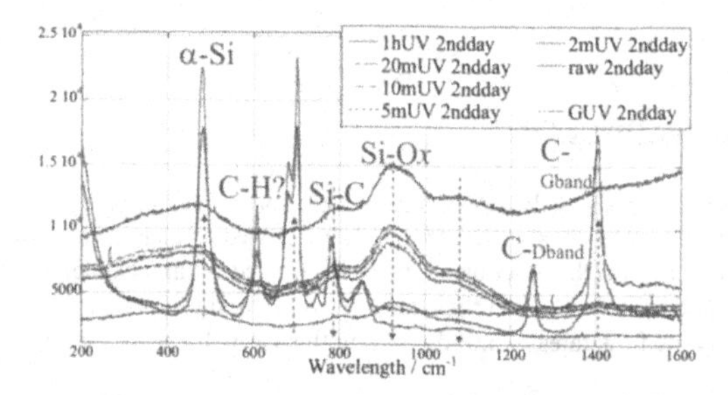

Figure 5. Raman spectra of UV-Vis treated and untreated PDMS surfaces. The α-Si, C-H, and C-bands peaks are disappeared from the UV-Vis treated PDMS. On the other hand, the Si-Ox broaden peak appeared in the UV-Vis treated PDMS.

In figure 6, the FTIR spectra are shown from the untreated PDMS surface and the UV-Vis lamp light irradiated PDMS. From the spectra, the Vas (Si-O-Si) absorption peak of the UV-Vis irradiated PDMS at 1110 cm^{-1} shifted to higher wave-number than that of untreated PDMS.

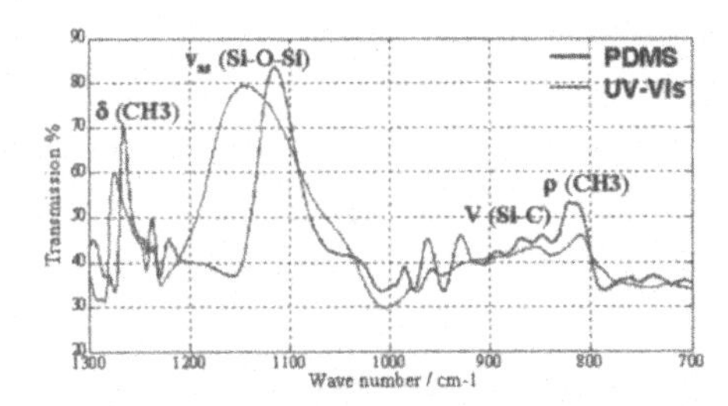

Figure 6. FTIR spectra of the UV-Vis treated and untreated PDMS surface. The Si-O-Si peak of UV-Vis treated PDMS shifted to higher wave number and broader than that of untreated PDMS.

The Raman and the FTIR spectra of UV-Vis irradiated PDMS surface showed glassy peaks. These Si-Ox bonds would be induced long term hydrophilic property for UV-Vis irradiated PDMS.

CONCLUSIONS

The PDMS surface was treated by the UV-Vis lamp light which power of 400 W at the distance of 20 mm for several ten minutes in the ambient atmospheric condition. As a result of this surface treatment, the surface became to hydrophilic character for one month long. This surface treatment technique applied to PDMS micro-fluidic device and verified valve-less switching. The syringe pumps injected the distilled water into the inlet of the PDMS micro channel at the flow rates of 0.5, 5.0, and 50 μl / min for the UV-Vis treated and untreated channels. As results of water injection water flowed only the UV-Vis treated channel at the flow rates of 0.5 and 5.0 μl / min. On the other hand, the water flowed for all channels at the flow rate of 50 μl / min. These results were occurred by the difference of the flow conductance and wettability.

ACKNOWLEDGMENTS

Part of this study supported by Dr T. Sato to treat the PDMS with UV pulsed laser.

REFERENCES

1. H. Makamba, J. H. Kim, K. Lim, N. Park and J. H. Hahn, *Electrophoresis*, 24, 3607 (2003).
2. S. Bhattacharya, A. Datta, J. M. Berg, et al., *J Microelectromechanical Systems*, 14 (3), 590 (2005).
3. M. Manca, B. Cortese, I. Viola, A. S. Arico, R. Cingolani and G. Gigli, *Langmuir*, **24 (5)**, 1833 (2008)
4. T. G. Vladkova, I. L. Keranov, P. D. Dineff, S. Y. Youroukov, I. A. Avramova, N. Krasteva and G. P. Altankov, *Nucl. Inst. Meth*, **B 236**, 552 (2005).
5. "UV/ozone modification of poly (dimethyl siloxane) microfluidic channels," Y. Berdichevsky, J. Khandurina, A. Guttman, Y.-H. Lo, Sensors and Actuators B, **97**, 102 (2004).
6. T. Yamamoto and T Fujii, *Proceedings of uTAS 2004*, **2**, 133 (2004).
7. S. Kano, S. Matsumoto, T, N. Ichikawa, *Proceedings of uTAS 2007*, **2**, 1037 (2007).

Mater. Res. Soc. Symp. Proc. Vol. 1139 © 2009 Materials Research Society 1139-GG03-33

Metal Wafer Bonding for MEMS Applications

V. Dragoi[1], G. Mittendorfer[1], F. Murauer[1], E. Cakmak[2] and E. Pabo[2]
[1]EV Group, DI E. Thallner Str. 1, 4782 – St. Florian/Inn, Austria
[2]EV Group Inc., 7700 S. River Parkway, Tempe, AZ 85284, U.S.A.

ABSTRACT

Metal layers can be used as bonding layers at wafer-level in MEMS manufacturing processes for device assembly as well as just for electrical integration of different levels. One has to distinguish between two main types of processes: metal diffusion bonding and bonding with formation of an interface eutectic alloy layer or an intermetallic compound. The different process principles determine also the applications area for each. From electrical interconnections to wafer-level packaging (with emphasis on vacuum packaging) metal wafer bonding is a very important technology in MEMS manufacturing processes.

INTRODUCTION

Wafer bonding plays an important role in Micro-Electro-Mechanical Systems (MEMS) applications as a technique used for joining two substrates. The large device variety required the development of various wafer bonding processes.

Among the wafer bonding processes currently used for industrial applications can be mentioned direct bonding (also named fusion or molecular bonding - adhesion appears between the molecules on the two surfaces) [1], anodic bonding (used to bond a Si wafer to a glass wafer - adhesion appears through an oxide layer grown at the interface) [2], adhesive bonding (uses intermediate layers, typically polymers) [3, 4], eutectic bonding (adhesion appears through an eutectic alloy layer grown at the interface) [5] or intermetallic bonding [6] and thermocompression bonding (metal bond - adhesion occurs between two metal surfaces pressed together under heating) [7].

Wafer bonding process selection is based on various criteria related to the materials (substrates types, bonding temperature) as well as to the desired application (type of bond – mechanical connection, electrical or thermal conductivity of interface, optical properties and device working temperature).

This paper presents a review of wafer boding processes based on the use of metal layers and introduces the process selection criteria for metal wafer bonding.

BASICS

Apart from direct (fusion) bonding for which adhesion between two substrates is generated through molecular bonds, processes using "bonding layers" are extensively used in MEMS manufacturing . The choice of intermediate layers for wafer bonding is made based on required processing temperatures as well as on other materials characteristics (e.g. specific outgasing, transparency, thermal/electrical/optical conductivity).

Wafer bonding using metal bonding layers is an important technique used for applications requiring good thermal conductance and applications in which electrical conductivity is required (e.g. for 3D TSV – Through-Si Vias - applications).
Two main types of principles are governing metal bonding: alloying (eutectic alloy or intermetallic compound formation), and metal diffusion by thermo-compression bonding.
In numerous situations there is a state of confusion surrounding the two types of metal bonds and this impacts mainly on further experiments design.
From wafer bonding process and equipment perspective there are major differences between the two processes (table I).

Table I. Main process characteristics for eutectic and metal diffusion wafer bonding.

Parameter	Eutectic wafer bonding	Metal thermo-compression wafer bonding
Process temperature	Au:Sn ~300°C (280°)* Au:Si ~380°C (363°C)* Au :Ge ~380°C (361°C)* Al:Ge ~440°C (424°C)* Au:In ~510°C (495°C)*	300°C – 500°C (Au-Au, Cu-Cu, Al-Al)
Temperature variations	+(10°C - 15°C) from eutectic point	100°C – 200°C (correlating with contact force)
Surface quality	Low	High
Contact force	Low	High
Atmosphere	Inert or reducing	Inert or reducing
Liquid phase	Yes	No

*Temperature values in parenthesis show the eutectic temperatures.

Eutectic wafer bonding

Eutectic alloy is formed at the bonding interface in a process which goes through a liquid phase: for this reason, eutectic bonding is less sensitive to surface flatness irregularities, scratches, as well as to particles compared to the direct wafer bonding methods (metal layer can incorporate particles with diameter lower than the eutectic layer thickness).

Some of the main eutectic alloys used for wafer bonding applications are listed in table 1. For a successful eutectic bonding process it is very important that bonder assures a good temperature uniformity across the entire wafer surface and also to control very well the temperature value (avoid overshooting the set point) in order to have a reliable process.

Experimental results showed that good quality interfaces are obtained when temperature is raised to a value lower than the eutectic temperature (heating simultaneously from top/bottom), maintained constant for short time to reach uniform heating of both wafers, than increased again by heating both heaters to a temperature exceeding the eutectic point with 10-20°C (depending on specific process conditions and on substrates restrictions) followed by cooling down to a temperature below the eutectic temperature.

Eutectic wafer bonding does not require application of high contact force. Due to the liquid phase formed during the process, high contact force results always in metal squeezing out of the interface, resulting in poor interface layer uniformity as well as contamination of the bond tools and bond chamber. The role of the light contact force required is just to ensure good contact of the two wafers and good contact of the two heaters of the bonder with wafers' back sides.

Eutectic wafer bonding is a good candidate to high-vacuum applications as this process has a very low specific outgasing due to the use of only high purity components. The liquid melt formed during process can only enhance the high vacuum compatibility by allowing high quality sealing even on non-perfect surfaces.

Wafer bonding using diffusion soldering

For some applications the process temperatures must be lower than the bonding temperatures of the most usual eutectic alloys (300°C - 400°C). In such situations an alternative process can be used, which results in an inter-metallic compound bonding layer. In literature this process is known under different names among which can be mentioned "diffusion soldering" [8] or Transient Liquid Phase (TLP) bonding [6, 8].
This bonding process is an advanced type of solder bond that can form high-quality hermetic seals at lower temperatures than other bonding technologies. This technique uses one thin layer of metal (typically 1-10μm thick) which during a thermal process diffuses into its bonding partner forming an inter-metallic compound layer with remelting temperature higher than the bonding temperature (table II) [8].

If Cu:Sn system was reported in literature mainly for die bonding ("Chip-to-Wafer" approach, C2W [9, 10]), Au:In was mainly reported for wafer-level vacuum packaging [8, 11]. Same as eutectic wafer bonding, diffusion soldering bonding is attractive for MEMS vacuum packaging as the process is completed at low temperatures (150°C - 300 °C) and can withstand much higher temperatures after bonding (see Table II), bonding layers are made out of metals (low permeability), and they can planarize over surface defects or particles resulting from prior processes.

Table II. Examples of metals which can be used for diffusion soldering (TLP)

Component 1 (thick)	Component 2 (thickness)	Bonding temperature	Diffusion time	Remelting temperature
Cu	Sn (1μm)	280°C	4 min.	>415°C
	Sn (5μm)	300°C	20 min.	>676°C
Au	In (2μm)	260°C	15 min.	>278°C
	In (5μm)	200°C	30 sec.	>495°C
	In (2μm)	160°C - 240°C	10 min.	>495°C
Ag	Sn (2μm)	250°C - 350°C	10 min.	>600°C
	Sn (5μm)	250°C	60 min.	>600°C

* Table adapted from [8]. Times correspond to full diffusion of the specified thickness of Component 2.

Metal thermo-compression wafer bonding

Quite often people are wrongly considering that thermo-compression and eutectic bonding are one single process. In thermo-compression bonding process the two surfaces adhere to each other due to a metal bond established between two metal surfaces pressed together under heating. The bonding mechanism is enhanced by the deformation of the two surfaces in contact in order to disrupt any intervening surface films and enable metal-to-metal contact. By heating the two metal surfaces the contact force applied for the bond process can be minimized. High force uniformity across the bonding area results in high yield.

There are several metals used for metal thermo-compression bonding, as Au [7], Cu [12] or Al [13]. These are considered interesting for wafer bonded MEMS applications mainly due to their availability in main microelectronics applications. Their use for one or another type of applications is conditioned by the type of substrates used (e.g. no Au-containing substrates can be further processed in CMOS lines).

This type of wafer bonding process is used mainly when electric interconnections between the two wafers are required. However, metal thermo-compression bonding is used also in vacuum packaging applications [13].

DISCUSSION

As briefly discussed above, there are numerous metal wafer bonding processes available for MEMS manufacturing. The choice of the appropriate process has to be made very carefully, considering aspects related to substrates, device requirements and process conditions. This chapter aims emphasizing the importance of some criteria which have to be considered in process design.

Material-related criteria

The type of substrates mainly influences the choice of the process type from the two base categories (eutectic/soldering or thermo-compression). The type of substrate may also determine the choice of the metal based on factors as compatibility of bonding material with other materials or processes used for manufacturing process. Thus for example, Cu is always preferred to Au if there is a major concern of Au diffusion in a thermo-compression process.

One important challenge in metal wafer bonding is surface oxidation. As most of the metals used as bonding layers oxidize relatively easy when exposed to ambient conditions it is important to prevent this from happening. For eutectic/diffusion soldering it is possible to protect the fast-oxidizing component by deposition of very thin layer of the other component of the alloy (e.g. a thin layer of Au is evaporated on top of the Sn or In layers for Au:Sn or Au:In systems: the Au layer doesn't change the process kinetics but prevents Sn or In oxidation as it is evaporated in high vacuum).

In case of metal thermo-compression bonding process the situation is different: if for Au surface oxidation is not of concern (in this case organic contamination may be a considerable issue and would require oxygen plasma cleaning prior to bonding), for Cu and Al oxide layer on the bonding surfaces is a major topic.

As Cu layers are typically fabricated in a damascene process, during post CMP cleaning the Cu surface can be coated with an organic layer (e.g. tolyltriazole): this organic layer will protect Cu surface from contact with air and during the bond process will evaporate leaving no residuals on the surface (evaporation temperature ~250°C - 300°C). For Al thermo-compression bonding the oxide layer is one of the main problems. Some groups reported results improvement by increasing the contact force during the process [13].

Besides oxidation it has to be considered that metal diffusion can not be controlled in one direction and will occur also inside the substrate. In order to avoid any undesired diffusion-related failure the bonding metallization has to be deposited on the proper adhesion layer/diffusion barrier layer (typically combinations of Pt, W, Ti or Cr with thicknesses in the range of few tens of nanometers).

Process-related criteria

An important process parameter is the atmosphere composition inside the bond chamber during process: in order to prevent oxidation an inert gas may be used (nitrogen or argon). In order to be even more efficient, a forming gas or formic acid vapor gas atmosphere can be used: besides the oxidation prevention some oxide may be reduced in this case. Considerable improvement of bond results was observed if Cu wafers were first heated at 300°C in forming gas (4% nitrogen content) for few minutes prior to bonding.
Equipment needs to be well set in order to provide good temperature uniformity across the bonding area and very uniform distribution of contact force. Process parameters have to be optimized for producing the best bonding quality. The wafer bonding results can be quantified by inspecting the bonding lines integrity (e.g. by Scanning Acoustic Microscopy - fig. 1), or inspecting integrity and material composition (e.g. by Scanning Electron Microscopy cross section analysis - using EDAX the composition variation inside the intermetallic layer can be determined).

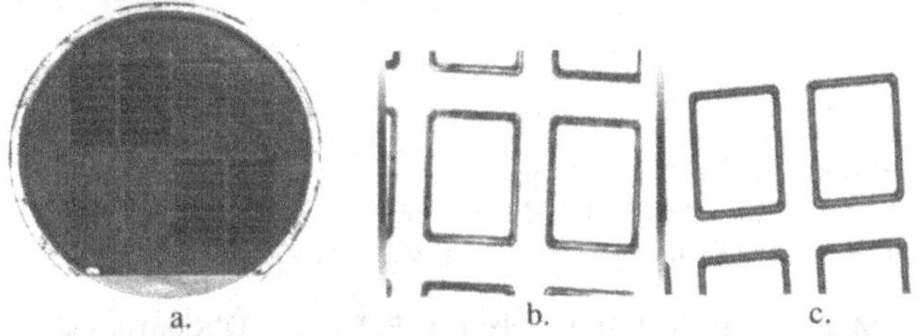

Figure 1. SAM images: a.- 2" diameter patterned InP/GaAs wafer pair bonded with Au:Sn eutectic; b.- Au:In sealing frames bonded with non-optimized process, and c.- Au:In sealing frames bonded with optimized process.

Wafer-to-wafer alignment accuracy

Wafer bonding process is strongly influencing the alignment accuracy. Table III summarizes the main wafer bonding processes used for aligned wafer bonding and shows main process feature with impact on wafer-to-wafer alignment accuracy.

Table III. Main wafer bonding features with impact on wafer-to-wafer alignment.

	Bonding temperature	Interface (bonding)	Interconnection type	Alignment accuracy
Metal thermo-compression	300°C–450°C	Solid	Mechanical+electrical	0.6μm
Eutectic/Soldering	200°C–400°C	Liquid	Mechanical+electrical	1μm

Among the most significant factors influencing alignment accuracy can be mentioned compression of intermediate bond layers which induces shifts (for bonding with intermediate layers as for eutectic bonding, or diffusion soldering bonding), different thermal expansion of the two substrates which induces run-out-type errors and the z-travel range of the wafers when brought in contact (e.g. given by thickness of the spacers used in the bond setup). For such bond processes typically two alignment specifications are defined: post-alignment accuracy (accuracy provided exclusively by the optical alignment equipment) and post-bonding accuracy (the final accuracy measured after bonding, when the bonded interface is already rigid).

CONCLUSIONS

Metal wafer bonding is a very important technology for MEMS and 3D interconnects applications. Apparently simple process, this category of wafer bonding has very strict requirements in terms of substrate preparation (including here the metal bonding layers), handling through the process flow as well as the right choice of the process conditions. Bonding surfaces require special attention as the most used metals are sensitive to oxidation. Diffusion inside the substrate may be another undesired effect which can be prevented by using diffusion barrier layers.

Metal wafer bonding is compatible for high vacuum packaging applications allowing encapsulation of vacuum levels higher than 10^{-3}mbar, required by new applications as high frequency resonators, optical switches or IR sensors.

REFERENCES

1. W. P. Maszara, *J. Electrochem. Soc.* 138 (1), 1991, pp. 341.
2. G. Wallis and D. Pommerantz, *J. Appl. Phys.* **40**, p. 3946, 1969.
3. F. Niklaus, G. Stemme, J.-Q. Lu and R. J. Gutmann, *J. of Appl. Phys.* **99** (2006), pp. 031101-1.
4. J. O'Brien, P. J. Hughes, M. Brunet, B. O'Neill, J. Alderman, B. Lane, A. O'Riordan and C. O'Driscoll, *J. of Micromech. Microeng.* **11** (2001), pp. 353.
5. R. F. Wolffenbuttel and K. D. Wise, *Sensors & Actuators A*, 43 (1994), pp. 223.
6. W. C. Welch III and K. Najafi, in Proc. of IEEE MEMS 2008, pp. 806.
7. C. H. Tsau, S. M. Spearing, and M. A. Schmidt, *J. of Microelectromech. Systems* 13 (6), (2004) pp. 963.
8. G. Humpston, and D. Jacobson, in *Principles of Soldering*, ASM International 2004, pp. 231.
9. C. Schiering, H. Kostner, P. Lindner, and S. Pargfieder, *Advanced Packaging* 5 (2005), pp. 26.
10. S. Pozder, A. Jain, R. Chatterjee, Z. Huang, R.E. Jones, E. Acosta, B. Marlin, G. Hillmann, M. Sobczak, G. Kreindl, S. Kanagavel, H. Kostner, and Stefan Pargfrieder, in Proc. of IEEE Interconnect Technology Conference - IITC 2008, Burlingame, USA, p. 46.
11. Y.-C. Sohn, Q. Wang, S.-J. Ham, and B.-G. Jeong, in Proc. of IEEE Int. Conf. on Electronic Comp. and Techn., Reno - Nevada, USA, 2007, pp. 633.
12. K.N. Chen, C.S. Tan, A. Fan, and R. Reif, *Electrochem. and Solid-State Lett.* 7 (1), (2004), pp. G14.
13. C.H. Yun, J.R. Martin, L. Chen, and T.J. Frey, *ECS Transactions* 16 (8), (2008), pp. 117.

Mater. Res. Soc. Symp. Proc. Vol. 1139 © 2009 Materials Research Society 1139-GG03-35

Effect of Process Variables on Glass Frit Wafer Bonding in MEMS Wafer Level Packaging

S. Sridharan[1], J. Henry[2], J. Maloney[1], B. Gardner[2], and K. Mason[2]
V. Dragoi[3], J. Burggraf[3], E. Pabo[4], and E. Cakmak[4]

[1,2]Ferro Corporation, Electronic Materials System,
[1]7500 E. Pleasant Valley Road, Independence, OH 44136, USA.
[2]1395 Aspen Way, Vista, CA 92083, USA.
[3]EV Group, DI E. Thallner 1, 4782-St. Florian, Austria.
[4]EV Group Inc, 7700 S. River Pkwy, Tempe, AZ 85284, USA.

ABSTRACT

Among different MEMS wafer level bonding processes, glass frit bonding provides reliable vacuum tight seals in volume production. The quality of the seal is a function of both seal glass material and the processing parameters used in glass frit bonding. Therefore, in this study Taguchi L18 screening Design of Experiment (DOE) was used to study the effect of materials and process variables on the quality of the glass seal in 6" silicon wafers bonded in an EVG520IS bonder. Six bonding process variables at three levels and two types of sealing glass pastes were considered. The seals were characterized by Scanning Acoustic Microscopy (SAM), cross sectional Scanning Electron Microscopy (SEM) and Energy Dispersive X-ray Analysis (EDAX). The results were quantified into four responses for DOE analysis. Key results are a) peak temperature has the strongest influence on seal properties, b) hot melt paste has significantly lower defects compared to liquid paste, and c) peak firing temperatures can be as low as 400°C under certain conditions.

INTRODUCTION

Due to rapid advances in consumer and automotive electronics, and mobile communications the Micro Electro-Mechanical Systems (MEMS) market is growing rapidly [1]. Applications for MEMS cover diverse fields, such as, automotive, aeronautics, consumer, defense, industrial, medical and life sciences, and telecommunications. Various MEMS devices require a vacuum level or controlled atmosphere operation in order to ensure either good performance or acceptable operating life. Different wafer bonding technologies are commercially used for MEMS packaging [2-7]. Among these, glass frit bonding [7] offers many advantages such as superior hermeticity, less stringent requirement on roughness of the surfaces being joined and tailored expansion matching to the materials being joined.

In glass frit wafer bonding, sealing glass paste is screen printed into the desired geometry. This layer is dried in an Infrared (IR) oven, and then fired in a furnace to burn the binder and pre-fuse the sealing glass layer. This process is called "glazing". Glazing is usually done in either a belt furnace or box furnace under atmospheric conditions. Then the pre-glazed wafer is placed in a wafer bonder and bonded to a cap wafer in a temperature-time-pressure cycle under desired vacuum in a process called "frit bonding" to produce a hermetically sealed assembly. The quality of the resultant seal (or bond) is a function of both seal glass material used and the process variables employed in both glazing and frit bonding. Understanding the interactions between these variables is important to improve the quality of the seal. Therefore in this study we evaluated the effect of six significant wafer bonding variables, and the effect of two types of pastes (Table I) on the quality of the bond as measured by bond strength, bonding uniformity, size and type of voids in the bond lines, and lead precipitation at the interface, using Taguchi L18 screening design [8].

EXPERIMENTAL DETAILS

Taguchi L18 screening design with seven factors - one material variable (A) which is qualitative and six bonding variables (B to G) that are quantitative– and four responses was used for this study. These are detailed in Table I. The sealing glass pastes used in this study are FX11-036 paste (standard thick film liquid paste) and A401-17B paste (hot melt) manufactured by Ferro Corporation, Cleveland, OH, USA. While the inorganic content of these two pastes is the same, the organic content is different, designed to be liquid in the former case and solid in the latter case at room temperature. Therefore while FX11-036 can be screen printed at room temperature, the hot melt paste A401-17B requires a heated (to 65°C) screen.

Table I: Taguchi L18 design matrix – variables and responses - for this study

	A	B	C	D	E	F	G
Factor	**Paste**	**Back Fill**	**Peak Temp.**	**Peak Time**	**Ramp Rate**	**Bonding Force**	**Cool Down Rate**
Row #	**Type**	**Type**	**°C**	**Min**	**°C/min**	**N**	**°C/min**
1	Regular	N2	400	15	5	3200	5
2	Regular	N2	430	30	18	4000	13
3	Regular	N2	460	45	30	4800	20
4	Regular	None	400	15	18	4000	20
5	Regular	None	430	30	30	4800	5
6	Regular	None	460	45	5	3200	13
7	Regular	Air	400	30	5	4800	13
8	Regular	Air	430	45	18	3200	20
9	Regular	Air	460	15	30	4000	5
10	Hot melt	N2	400	45	30	4000	13
11	Hot melt	N2	430	15	5	4800	20
12	Hot melt	N2	460	30	18	3200	5
13	Hot melt	None	400	30	30	3200	20
14	Hot melt	None	430	45	5	4000	5
15	Hot melt	None	460	15	18	4800	13
16	Hot melt	Air	400	45	18	4800	5
17	Hot melt	Air	430	15	30	3200	13
18	Hot melt	Air	460	30	5	4000	20

Responses
•Voids (score)
-diameter of voids in μm if spherical
or size of largest dimension in μm if non spherical
-Lower score is desired
•Lead precipitates (score)
-None – 0
-Small ppts - 0.5
-Normal ppts – 1.0
-Big ppts – 2.0
-Lower score is desired
•Measure of defects in SEM microstructure (score)
-Measure of severity of voids, de bonding & other unwanted features
-Debond at Si/glass interface - 40
-Large voids - 20
-Small voids - 5
-Mud cracks - 10
-Leafy precipitates - 5
-Lower score is desired
•Measure of defects in SAM scan (score)
-% area looks hazy (not a good bond)
'- Lower score is desired

The responses were quantified as follows:. A defect is an unwanted feature observed in SAM (typically low magnification) or SEM (here at 5600X magnification). Except for de bonding, unusually large voids (in SEM) and completely unresolved bond line (in SAM), not all defects quantified here are detrimental to the performance of the bond, but are useful in predicting the trends towards good bond. To remove the subjectivity as much as possible numerical values were assigned for different defects as listed in Table 1. For example, based on average size of the lead precipitates in the SEM microstructure, the lead precipitates were classified as none (not present), small (≤0.2 μm), normal (0.2 – 0.5 μm) and big (>0.5 μm), and were given scores of 0, 0.5, 1.0, or 2.0, according to the list in Table I. Lead precipitates scores from both bottom and cap wafer/Si surfaces were added to get the total score for the experiment. Since there was a repeat for each experiment, the reported final score was the average of scores for the two experiments, for each response. Quantification of the voids was difficult due to their sporadic occurrence and their shapes. Therefore, often multiple random cross sections were looked at both low and high magnifications before locating a void and assigning void size. For quantification of SEM defects the voids were classified as small (≤5 µm) and large (>5 µm) based on their size, and were given weighted scores of 5 or 20. Since debonding at the Si/glass interface and the presence of large voids are bad for the functionality of the bond, these were given higher weighted scores of 40 and 20 to reflect their severity. Since mud crack appearance and leafy precipitates in the cross section, although undesirable, did not result in necessarily bad bonds, these were given scores of 10 and 5 intermediate to large and small voids. Repetition of a defect was not considered for scoring. Higher scores might either be due to larger size or higher weighting of a given defect or the presence of multiple defects or presence of defects at both interfaces. Therefore the scores should be viewed as means to predict the trends. Nevertheless lower the score better the bond quality.

The pastes were screen printed on 6-inch diameter (thickness: 675 ± 25 µm; weight: 28.14 ± 0.50 gm) <100> boron doped silicon (Si) wafers without any special cleaning. The sealing glass pattern in Figure 1(a) was printed using 325 mesh screen (45° mesh angle/27.9 µm stainless steel wire diameter/30.5 µm emulsion thickness) using 7.62 mm/min squeegee speed and 1.52 mm snap off distance. For both cases, the average dry print thickness was 27 µm, and the average fired thickness was 16.7 µm after glazing and 9 µm after final frit bonding. Three different widths (~260, 330, 370 µm) are inherent in the pattern in Figure 1(a). These printed widths increased by 4 to 6 % in the regular liquid paste case, and exhibited no increase in the hot melt case, after the glazing step. After screen printing, the wafers were dried in an IR belt dryer at 120°C peak temperature. Then the wafers were randomized for further processing steps.

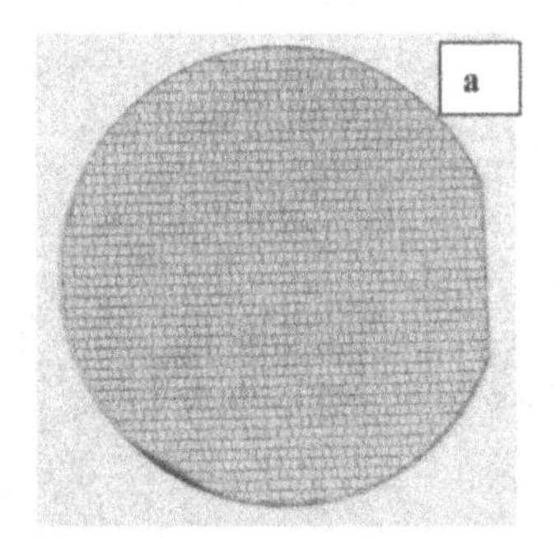

Figure 1: (a) Sealing glass pattern used (courtesy of ST Microelectronics, Italy); (b) Wafer bonder EVG520IS used. (Courtesy of EVGroup, Austria)

The sealing glass patterns on the bottom wafers were fired in the glazing profile shown in Figure 2 [295°C (40 min) binder burn out, and 410°C (15 min) glazing], in 4 hr cycle in a box furnace with adequate air flow. After the glazing step the cap wafers were sealed to the bottom wafers in EVG520IS wafer bonder (cf Figure 1 b). Figure 3 shows the typical glass frit wafer bonding temperature-time-piston force – backfill cycle employed. All the experiments in Table I were done in random order with one repeat for each experiment. The resulting bond lines were examined by both SAM and cross sectional SEM and EDAX analysis for adhesion to the wafers, size and location of voids, cracks, lead and other precipitates. These observations were quantified into four responses according to the criteria outlined in Table I for analysis in DOE.

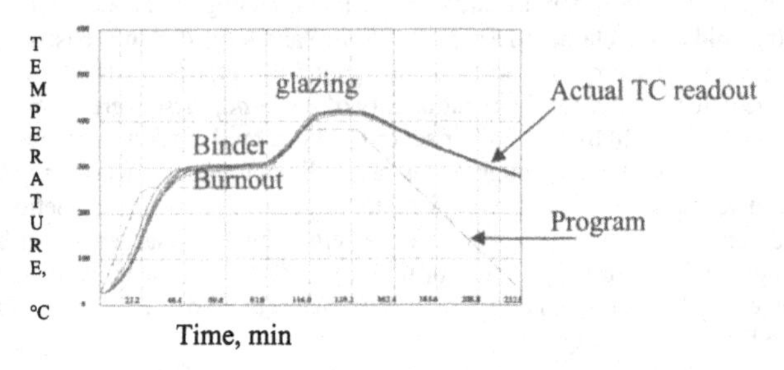

Figure 2: Time-Temperature cycle used in glazing step

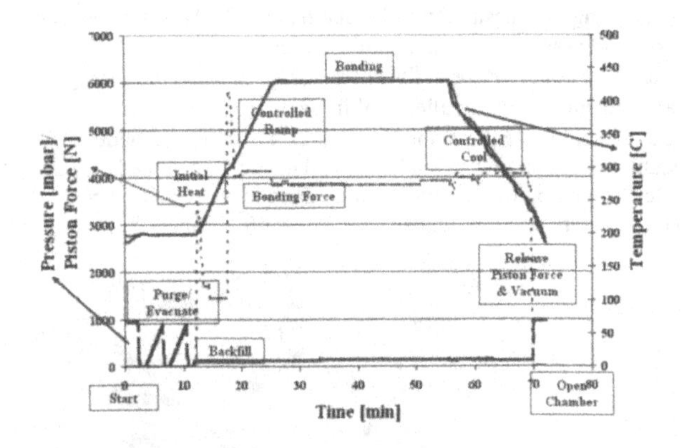

Figure 3: Time-Temperature-Piston Force-Backfill Cycle in Glass Frit Bonding. [For this experiment: Bonding Force= 4000N; Peak Temp = 430°C; Peak Time = 30min; Ramp Rate = 18°C/min; Cool Down Rate = 13°C/min; Backfill = N_2]

RESULTS AND DISCUSSION

Figure 4 shows the SAM surface image and SEM cross sectional images at optimum processing conditions for the two different pastes considered for this study. The SAM images depict the quality of the bond lines across the entire wafer. Sharper and more uniformly dark (distinguishable) bond lines in SAM indicate a better quality bond. The cross sectional SEM pictures in Figure 4 at two different magnifications provide information on voids, debonding tendency, cracks, lead and other precipitates in the seal. The criteria outlined in Table I were utilized to quantify the responses from each experiment, and tabulated in Table II.

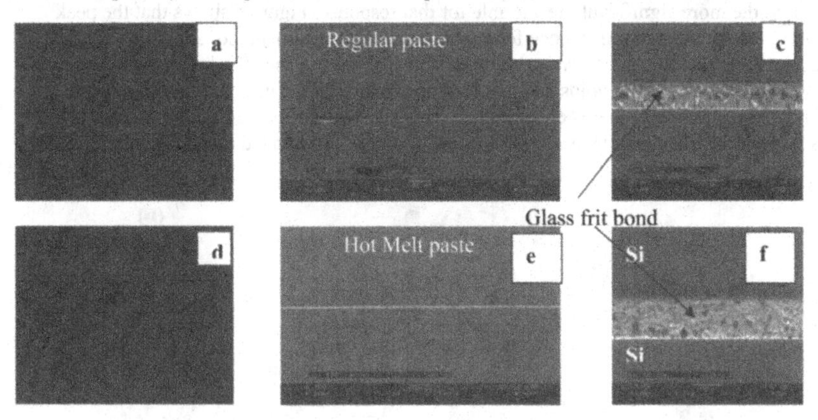

Figure 4: SAM & SEM cross section images; (a) – SAM, (b) – SEM at low mag, (c) – SEM at high magnification for bond line with regular paste (experiment 17 in row 5 of Table II); (d)-SAM, (e)-SEM at low magnification, (f)-SEM at high magnification for bond line with hot melt paste (experiment 30 in row 13 of Table II).

Table II: Responses versus effects for experiments in Table I

	A	B	C	D	E	F	G		Responses			
Factor	Paste	Back Fill	Peak Temp.	Peak Time	Ramp Rate	Bonding Force	Cool Down Rate	Randomized Expt. No.	Void	Lead Precipitates	SEM Defects	SAM Defects
Row #	Type	Type	°C	Min	°C/min	N	°C/min		score	score	score	score
1	Regular	N2	400	15	5	3200	5	13, 18	5.25	0.25	52.5	8.39
2	Regular	N2	430	30	18	4000	13	37, 42	13.75	1.75	22.5	3.75
3	Regular	N2	460	45	30	4800	20	16, 39	12.50	2.00	17.5	9.92
4	Regular	None	400	15	18	4000	20	38,	3.00	1.00	55.0	6.64
5	**Regular**	None	**430**	**30**	**30**	**4800**	**5**	**17, 43**	**5.25**	**1.25**	**5.0**	**2.38**
6	Regular	None	460	45	5	3200	13	22, 20	17.50	1.50	30.0	6.75
7	Regular	Air	400	30	5	4800	13	31, 25	7.50	1.00	27.5	2.84
8	**Regular**	Air	**430**	**45**	**18**	**3200**	**20**	**34, 27**	**6.25**	**1.75**	**15.0**	**0.34**
9	Regular	Air	460	15	30	4000	5	24, 15	15.00	1.75	27.5	10.28
10	Hot melt	N2	400	45	30	4000	13	35, 26	1.25	0.75	32.5	6.12
11	**Hot melt**	N2	**430**	**15**	**5**	**4800**	**20**	**29, 36**	**3.00**	**1.00**	**12.5**	**1.21**
12	Hot melt	N2	460	30	18	3200	5	12, 14	11.25	2.00	15.0	5.13
13	**Hot melt**	None	**400**	**30**	**30**	**3200**	**20**	**30, 44**	**1.00**	**1.00**	**10.0**	**13.12**
14	Hot melt	None	430	45	5	4000	5	28, 40	2.50	1.75	45.0	5.10
15	Hot melt	None	460	15	18	4800	13	23, 41	27.50	1.75	27.5	11.71
16	Hot melt	Air	400	45	18	4800	5	21, 19	2.50	0.50	22.5	8.45
17	**Hot melt**	Air	**430**	**15**	**30**	**3200**	**13**	**45, 33**	**5.00**	**0.25**	**15.0**	**6.27**
18	**Hot melt**	Air	**460**	**30**	**5**	**4000**	**20**	**32,**	**5.00**	**2.00**	**5.0**	**8.60**

Lower the score for the responses better the quality of bonding.

Rows 5 & 8 have overall low scores in liquid paste case.

Rows 11, 13, 17, & 18 have overall low scores in hot melt paste case.

Table II summarizes the responses for each experiment for different factors considered in this study. The lower the numerical score for each response, the better the quality of the frit bonded seal with respect to that defect. A combined minimum for all four responses signifies the best condition for glass frit bonding. Table II clearly shows two such conditions (rows: 5, 8) for liquid paste (FX11-036) and four such conditions (11, 13, 17, 18) for hot melt paste (A401-17B).

The marginal means plot and Pareto chart for the lead precipitates response alone are shown in figure 5 for clarity. The steeper the line in the marginal means plot and the higher the bar in the Pareto chart, the more significant the variable for that response. Figure 5 shows that the peak temperature has the strongest influence for lead precipitation. Although not shown here, similar plots for the other three responses were obtained and analyzed. Basically, lower peak temperature produces smaller voids and less lead precipitates at the interface. Although not plotted here, the interaction (between variables) plots showed that as the peak temperature decreases, peak time has to increase to produce a good glass frit bonded microstructure.

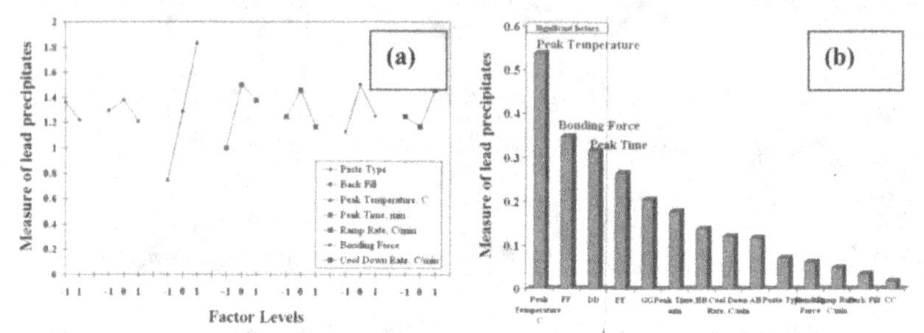

Figure 5: (a) Marginal means plot and (b) Pareto chart for Lead Precipitates response

Table III summarizes the inferences that could be deduced from this study. For example lead precipitates [9, 10] at the glass frit bonded interfaces, which are a consequence of redox reaction at the Si-glass interface [10], can be minimized by lowering the peak temperature [≤430°C], minimizing the time at peak [15 min] and increasing cooling rate. Similarly voids sizes can be reduced by lowering the peak temperature and increasing the cool down rate. Bonding force in the range [3200-4800N] considered for this study has no effect. Similarly the back fill gas (air or nitrogen) has no effect on final seal properties.

Table III: Inferences

	Factor	**Paste**	**Back Fill**	**Peak Temp.**	**Peak Time**	**Ramp Rate**	**Bonding Force**	**Cool Down Rate**
	Row #	**Type**	**to 100 mbar**	°C	Min	°C/min	N	°C/min
Voids	**To reduce void size**	HM	yes	≤ 430 °C	≥ 30 min	------	-----	Faster
Lead ppts	**To reduce lead ppts**	HM	yes	≤ 430 °C	15 min	-----	----	Faster
Defects in SEM	**To reduce debonds, voids & cracks**	HM	yes	≥ 430 °C	30 min OK	Faster	------	Faster
Defects in SAM	**To reduce poor bond lines**	REG	yes	430 °C OK	≤ 30 min	Slower	-----	---

CONCLUSIONS

This study showed that in glass frit bonding of MEMS wafers, among the different frit bonding variables considered, a) peak temperature has the strongest influence on seal properties, b) 400°C peak temperature is possible at certain combination of factors for these glass pastes, and c) for hot melt paste A401-17B defects are significantly lower over wide process conditions.

ACKNOWLEDGEMENTS

The authors gratefully acknowledge the excellent SEM work by Mr. Dave Gnizak of Ferro Corporation, Independence, OH, and the screen printing and firing work by Mr. Jeff Holthus of Ferro Corporation, Vista, CA.

REFERENCES

1. Market Forecast, Yole Development, France, http://www.yole.fr.
2. R. Stengl, T. Tan, and U. Gösele, *J. J. Appl. Phys.*, **28**, 1735 (1989).
3. C. H. Tsau, S. M. Spearing, and M. A. Schmidt, *J. of Microelectromech. Systems*, **11**, 641 (2002).
4. G. Wallis and D. Pommerantz, *J. Appl. Phys.*, **40**, 3946 (1969).
5. V. Dragoi, T. Glinsner, G. Mittendorfer, B. Wieder, P. Lindner, vol. 5116, p. 160, SPIE Proc. Series, Bellingham, WA (2003).
6. J.-W. Yang, S. Hayes, J.-K. Lin, and D. Frear, *J. of Appl. Phys.*, **95**, 6077 (2004)
7. M. Petzold, C. Dresbach, M. Ebert, J. Bagdahn, M. Wiemer, K. Glien, J. Graf, R. Muller-Fiedler, and H. Hofer, Proc. Of "10th Intersociety Conf. on Thermal and Thermomech. Phenom. in Electronics Syst., ITHERM '06", IEEE Proc. Series, 1343 (2006).
8. S.R. Schmidt, and R.G. Launsby, "*Understanding Industrial Designed Experiments*", page 3-47 Air Academy Press & Associates, Colorado Springs, CO 80920, (2005); The DOE software SPC XL2000, licensed from Air Academy Associates LLC was used for this study.
9. R. Knechtel, *Microsyst Technol.*, **12**, 63-68 (2005).
10. B. Boettge, C. Dresbach, A. Graft, M. Petzold and J. Bagdahn, "*Mechanical Characterization and Microstructure Diagnostics of Glass Frit Bonded Interfaces*", *ECS Transactions,* **16 (8)** 441-448 (2008), also known as "*Semiconductor Wafer Bonding 10: Science, Technology, and Applications*".

Mater. Res. Soc. Symp. Proc. Vol. 1139 © 2009 Materials Research Society 1139-GG03-37

Charging processes in silicon nitride films for RF-MEMS capacitive switches: The effect of deposition method and film thickness

U. Zaghloul[1], A. Belarni[1], F. Coccetti[1], G.J. Papaioannou[1,2], R. Plana[1] and P. Pons[1]
[1] University of Toulouse, LAAS-CNRS, 7 avenue du Colonel Roche, 31077 Toulouse, France
[2]University of Athens, Solid State Physics, Panepistimiopolis Zografos, Athens 15784, Greece

ABSTRACT

The paper investigates the dielectric charging in silicon nitride for RF-MEMS capacitive switches. In order to determine the surface potential induced at the SiNx surface by injected charges through asperities, the investigation is performed with EFM Kelvin Probe Microscopy method (KPM). The effect of charging is investigated on dielectric films that have been deposited on silicon wafers and on gold, which have been evaporated on silicon wafers. The SiNx films were deposited with high frequency and low frequency PECVD method. Finally, the effect of dielectric film thickness has been studied by depositing films with different thicknesses.

INTRODUCTION

Capacitive RF MEMS switches are one of the most promising applications in microelectromechanical systems (MEMS). In spite of this, their commercialization is still hindered by reliability problems nonetheless the studies on the subject carried out by several laboratories and companies worldwide. The most important reliability problem is the charging of the dielectric [1]. On the way to solve the problem of dielectric charging, several materials have been investigated among which the most intensively investigated ones are SiO_2 and Si_3N_4. In contrast to the extensive knowledge gathered on their electrical properties, and stemming from semiconductor technology, a comprehensive understanding of the charging problem in dielectric material is still missing. This can be attributed to the fact that in MEMS the dielectric films are deposited on rough substrates, typically with low temperature processes, which lead to significant deviation from material steochiometry and the introduction of hydrogen during growth.

In dielectrics the charging effect arises from dipoles orientation and charge trapping that occurs under the strong electric field during the down state of MEMS switches [2]. The charge injection was reported to follow the Poole-Frenkel effect, a fact that has been supported by the shift of pull-up voltage [3]. In the case of SiNx it has been shown that holes injection introduces metastable traps, which give rise to asymmetrical current-voltage characteristic in symmetrical Metal-Insulator-Metal capacitors [4]. This asymmetry in both symmetrical and asymmetric Metal-Insulator-Metal capacitors was found to be greatly reduced in low temperature (150°C) PECVD SiNx [5]. Moreover the stored charge was found to increase with increasing the dielectric film thickness and the nitride deposition temperature [5].

Up to now, the assessment of dielectric charging has been performed through current measurements in Metal-Insulator-Metal (MIM) capacitors and capacitance and/or voltage measurements in MEMS capacitive switches [2,5,6]. These methods, although extremely useful, lead to results that depend strongly on the nature the device under test. Thus, in MIM capacitors the presence of top electrode introduces interfacial charging and contributes to the discharge of injected charges while in MEMS the injected charges are collected through the bottom electrode in the up-state. Finally, the assessment of dielectric charging with EFM Kelvin Probe Microscopy method has been recently introduced [7,8] and showed to be a promising method.

In view of all these, the aim of the present work is to investigate systematically the dielectric charging in silicon nitride films for RF-MEMS capacitive switches. The investigation took the advantage of KPM method to simulate the charge injection through asperities and then determine the surface potential induced at the SiNx surface by injected charges. Both, the potential distribution and decay time constants have been investigated for charging induced by applying positive or negative bias to the AFM tip. The effect of charging has been investigated on dielectric films that have been deposited on silicon wafers and on gold films evaporated on silicon wafers. Moreover, the SiNx films were deposited with high frequency (HF) and low frequency (LF) PECVD method in order to study the effect of the deposition conditions on the dielectric charging mechanism. The effect of dielectric film thickness has been studied by depositing films with different thicknesses ranging from 100nm to 500nm. Finally, the KPM assessment was performed on the dielectric free surface and the corresponding layer structures.

EXPERIMENTAL DETAILS

All the measurements presented in this article have been performed in ambient air at room temperature using the Digital Instrument 3100 Nanoscope IV. SCM-PIT conductive tips which are Antimony n-doped silicon and coated with Cobalt-chrome have been used. The metallic coating as well as the low spring constant (K=1-5 N/m) and resonance frequency (f_o=60-100 KHz) of these tips make them well suited for sensitive electrical measurements. It should be mentioned that the measured surface potential varies from one tip to another according to the difference in the intrinsic characteristics of each tip. Therefore, the same tip should be used in all measurements in order to be able to compare the results precisely. One important scanning parameter that can be controlled during the surface potential scanning is the lift scan height, which controls the separation between the tip and sample surface. In our measurements, the optimised value of the lift scan height is 20 nm. The dielectric charging has been performed by applying either positive or negative bias of 10 V for 60 seconds in order to study the effect of the bias polarity on the charging process.

Table I. **Layer structure**

Layer	S1	S2
Si_3N_4	100nm - 500nm	100nm - 500nm
Au	500nm	-
Ti	100nm	-
Si	500μm	500μm

The samples were fabricated by depositing SiNx films with PECVD method on silicon wafers or on gold film evaporated on silicon wafer. Moreover, the SiNx films were deposited at 200°C with high frequency (HF) and low frequency (LF) PECVD method in order to study the effect of the deposition conditions on the dielectric charging mechanism. The effect of dielectric film thickness has been studied by depositing films with different thicknesses ranging from 100nm to 500nm. Finally, the nitride films were deposited on metalized (S1) and non metalized (S2) silicon substrates (Table 1). This was done in order to investigate the effect of the formation of space charge region in the Si substrate on the charging process. Practically this attempts to resemble the extension of the nitride film beyond the metallic transmission line of a MEMS capacitive switch.

RESULTS AND DISCUSSION

In MEMS capacitive switches during pull-down state the contacting area is limited due to roughness at the dielectric and suspended electrode surfaces. Thus the charge injection through these areas can be well simulated by the AFM tip. This effect has been clearly depicted through potential distribution images obtained from charged dielectric film of MEMS switches [10]. The images presented a non uniform charge distribution and calculations suggested lateral charge diffusion with a diffusion coefficient of the order of $3\text{-}10\times10^{-11}cm^2/sec$ [10].

The charge injection takes place through different mechanisms among which the ones that contribute to charging phenomena are the Trap Assisted Tunneling and the time dependent Pool-Frenkel effect [11]. The distribution of injected and trapped charges will depend strongly on the intensity and distribution of electric field, the later being strongly dependent on the presence or not of underlying metallic electrodes or semiconducting films as well as the thickness of the dielectric film.

S1 structures

Figure 1 shows the distribution of the surface potential arising from charges injected through the AFM tip and measured with the KPM method. A comparison of the two distributions (fig.1a and b) shows that in the case of the thicker dielectric film the potential distribution is wider. This effect is expected if we take into account that both films are deposited on metal contact. Thus in the case of the thin dielectric film the electric field is more confined while in the case of the thicker one the electric field spreads over a wider area. The relatively small differences between the two distributions have to be attributed to the lower electric field intensity in the thicker dielectric film and the dependence of the density of injected charges on the electric field intensity [11]. Regarding the charge diffusion, the effect is more prominent in the thinner film than in the thicker one. Since the charge diffuses towards laterally and the bottom electrode, which acts as a sink and collects all injected charges, the experimentally observed lateral diffusion will be a function of charge collection time constant and film thickness. In order to determine the charge collection time constant we plotted the time dependence of the peak of potential distribution (fig.1) for different film thicknesses under both positive and negative charge injection. The time dependence of peak potential was found to decrease exponentially following the stretched exponential law, $\exp\left[-\left(\frac{t}{\tau}\right)^{\beta}\right]$, where τ is the process time constant and β ($0\leq\beta\leq1$) the stretch factor. As already discussed in previous publications ([2] and references therein) the stretch factor constitutes an index of charge collection complexity and the lower the value of β the more complex is the charge collection.

A comparison of the plots in Fig.1c,d reveals that the time constant is larger in the thinner dielectric film than in the thicker one. Moreover, the stretch factor approaches is larger in the the thinner material and lower for thicker material. The larger value of stretch factor for the thin (100nm) dielectric film and the lower one for the thicker film (500nm) indicate that the relaxation law is related to the length of the path the charges have to travel before they are collected by the bottom electrode. This leads us to the conclusion that the leading mechanism arises rather from a survival probability of a random walking particle in the presence of a static distribution of random traps [12]. Regarding the fitted relaxation time constants, the differences have to be attributed the fact that the thinner material will be affected more by the defective first layers deposited on the metal electrode surface. Such a material will contain more complex defects with large capture cross section located deep in the band gap. The high value of stretch factor in the thin films also supports the assumption of charge emission from a limited number of deep traps probably located deep in the band gap. In the case of the thick dielectric film the top layers are expected to not be affected by the disordered bottom layers. These defects are expected to exhibit simpler structure and have smaller capture cross section. Finally, the decay time constants and stretch factors are asymmetric when positive or negative bias is applied to the tip during charging. This asymmetry, although not well understood, has been already reported in experiments performed on SiNx MIM capacitors [5] assessed with TSDC method. Here it must be emphasized that the results of HF-SiN were similar to the ones of LF-SiN.

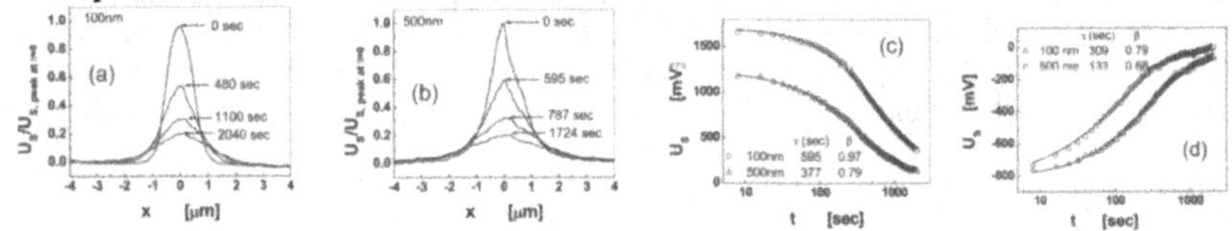

Figure 1. Evolution of surface potential, normalized to peak at t=0 sec, generated from charge injection in (a) a thin and (b) a thick LF-SiN film of structures S1. Time dependence of surface potential for structures S1 (c) with positive charge injection (10 V for 60sec) and (d) with negative charge injection (-10 V for 60 sec)

S2 structures

In samples S2, where the SiNx film has been deposited on bare silicon hence the charge collection by the semiconductor substrate is directly affected by the space charge region at the insulator-semiconductor interface. During the charging process the distribution of electric field will be different from the one of structure S1 and resemble a point contact MIS capacitor. During discharge, the collection of the charge stored in the dielectric film will be affected by the presence and width of the space charge region. Moreover, during discharge the semiconductor surface potential will decrease too leading to shift of Fermi level across band gap at the semiconductor-insulator interface. Taking these into account we are led to the conclusion that the decay of surface potential measured with the KPM method will be also affected by the presence of defects at the semiconductor interface.

Figure 2 shows the distribution of the surface potential in 100nm and 500nm thick SiN S2 structures. Again the comparison of the two distributions (fig.2a and b) shows that in the case of the thicker dielectric film the potential distribution is wider. Here it must be pointed out that in the 100nm thick dielectric the effect of the electric field hence the modulation of the space charge region is expected to be more pronounced. The experimental results show that in the case of the thin dielectric film, in spite of the presence of the space charge region, the electric field is more confined while in the case of the thicker one the electric field spreads over a wider area. Here it must be pointed out that the differences between the two distributions in the case of structures S2, although smaller than in structures S1, are noticeable. The charge diffusion, although present in these structures, is hardly observed within the measurements noise level. This effect is not obvious since the semiconductor space charge region cannot act as an efficient sink as the metal contact. Since the charge diffusion and collection in MEMS switches dielectric film is of primary importance for the device lifetime it becomes obvious that this effects requires further investigation.

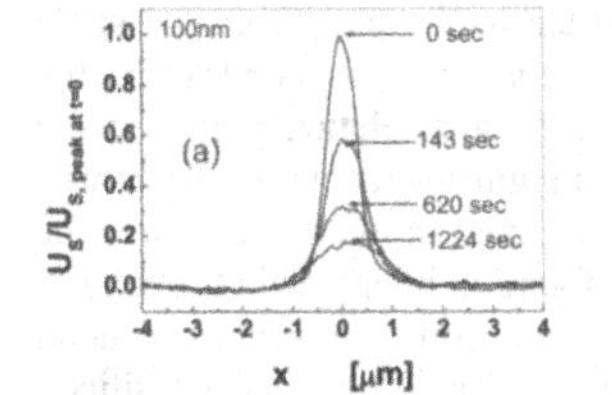

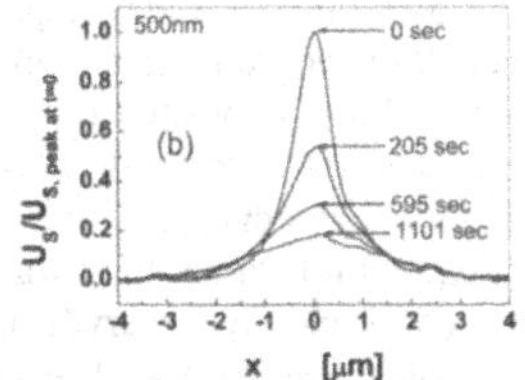

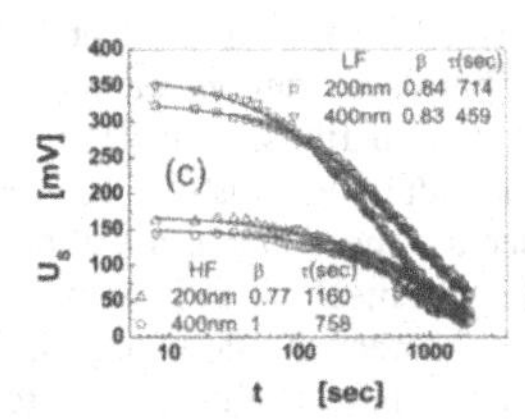

Figure 2. Evolution of surface potential, normalized to peak at t=0 sec, generated from charge injection in (a) a thin and (b) a thick LF-SiN film of structures S2; (c) shows the decay of surface potential maximum for S2 SiNx structures.

The decay of the peak value of the potential distribution for two dielectric film thicknesses obtained with LF and HF SiN is plotted in Figure 2c. The time dependence of peak potential is found to follow the stretched exponential law. Here it is important to notice that the decay times follow the same pattern, which is increase of time constant for thinner dielectric films. The dependence of stretch factor on the film thickness seems to be different from the one of films deposited on metal contact and constitutes a subject for further investigation.

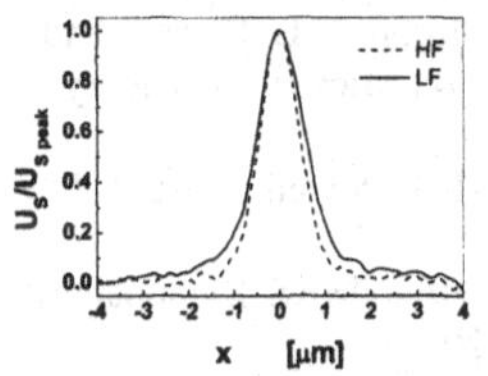

Figure 3. Evolution of normalized surface potential.

In order to compare the effect of material deposition method on the surface potential distribution, hence the charge distribution, we plotted in Fig. 3 the normalized distribution of surface potential for LF- and HF-SiN films with the same thickness (200nm). The differences in potential distribution are clearly small and can be attributed to the higher resistivity of HF SiN [12]. Here it must be emphasized that the LF material is more Si rich than the HF one, N/Si=0.79 and 1.04 respectively. The presence of Si in diluted or nanocluster form is expected to affect the distribution of the electric field. Finally, the observed difference can be also attributed to the modulation of the underlying p-Si substrate depletion region and its relation to the nitride induced band bending.

CONCLUSIONS

The charging of silicon nitride dielectric films has been investigated in view of application in MEMS capacitive switches. Films of different thicknesses deposited under low or high frequency PECVD method were investigated with the Kelvin Probe Microscopy method. In order to simulate the dielectric charging of the insulating film on the coplanar waveguide and on the silicon substrate the silicon nitride films were deposited on metal and silicon substrates.

The investigation revealed that the distribution of surface potential, hence the injected charge, was found to be more confined in thinner than in thicker films, independently of the substrate nature. The surface potential decay resulting from charge collection by the bottom electrode or silicon substrate is strongly affected by both the film thickness and the nature of film substrate. In all cases the potential relaxation follows the stretched exponential law.

In the case of films deposited on metal electrode the charge collection seems to be better described by the survival probability of a random walking particle in the presence of a static distribution of random traps for the thick dielectric films and the emission form deep traps in the case of thin films. When the dielectric film is deposited on silicon the depletion region in the semiconducting substrate and the interface defects seem to play a significant role on the charge collection mechanism. So, although the injected charge is more confined in the thinner films the collection mechanism is more complex. The relaxation time was found to be larger in the thinner films. Moreover, the charge diffusion was clearly observed in thin films deposited on metal contact. The fact that the charge diffusion was hardly observed in films deposited on silicon substrates is an issue that requires further investigation. Finally, the material steochiometry seems to affect the distribution of the injected charge induced surface potential.

REFERENCES

1. J Wibbeler, G Pfeifer and M Hietschold, *Sensors and Actuators A* **71**, 74 (1998)
2. G. J. Papaioannou, M. Exarchos, V. Theonas, G. Wang and J. Papapolymerou, *IEEE Transactions on Microwave Theory and Techniques* **53**, 3467 (2005)
3. S. Melle, D. De Conto, D. Dubuc, K. Grenier, O. Vendier, J.L. Muraro, J.L. Cazaux and R. Plana, Reliability Modeling of Capacitive RF MEMS, IEEE Tans. on Microwave Theory and Techniques **53**, 3482, (2005)
4. K. J. B. M. Nieuwesteeg, J. Boogaard, and G. Oversluizen and M. J. Powell, J. Appl. Phys. **71**, 1290, (1992)
5. R. Daigler, G. Papaioannou, E. Papandreou and J. Papapolymerou, MRS Symposium J: Passive and Electromechanical Materials and Integration, Spring 2008
6. E. Papandreou, M. Lamhamdi, C.M. Skoulikidou, P. Pons, G. Papaioannou and R. Plana, *Microelectronics Reliability* **47**, 1812 (2007)
7. A.Belarni, M.Lamhamdi, P.Pons, L.Boudou, J.Guastavino, Y.Segui, G. Papaioannou and R. Plana, Microelectronics Reliability **48**, 1232, (2008)
8. R.W. Herfst, P.G. Steeneken, J. Schmitz, A.J.G. Mank and M. van Gils, 46th Annual International Reliability Physics Symposium, Phoenix, p.492, (2008)
9. Veeco Instruments Inc, Support Note 231, Revision E ,Electric Techniques on MultiMode TM Systems
10. R. Dianoux, F. Martins, F. Marchi, C. Alandi, F. Comin, and J. Chevrier, Physical Review B **68**, 045403, (2003)
11. R. Ramprasad, Physica Status Solidi (b) **239**, 59, (2003)
12. M.Lamhamdi, J.Guastavino, L.Boudou, Y.Segui, P.Pons, L. Bouscayrol and R. Plana, Microelectronics Reliability **46**, 1700, (2006)

A Perturbation-Based Method for Extracting Elastic Properties during Spherical Indentation of an Elastic Film/Substrate Bilayer

Jae Hun Kim[1], Andrew Gou ldstone[2] and Chad S. Korach[3]
[1]Department of Materials Science and Engineering, Stony Brook University, Stony Brook, NY 11794, U.S.A., [2]Department of Mechanical and Industrial Engineering, Northeastern University, Boston, MA 02199, U.S.A., [3]Department of Mechanical Engineering, Stony Brook University, Stony Brook, NY 11794, U.S.A.

ABSTRACT

Accurate mechanical property measurement of films on substrates by instrumented indentation requires a solution describing the effective modulus of the film/substrate system. Here, a first-order elastic perturbation solution for spherical punch indentation on a film/substrate system is presented. Finite element method (FEM) simulations were conducted for comparison with the analytic solution. FEM results indicate that the new solution is valid for a practical range of modulus mismatch, especially for a stiff film on a compliant substrate.

INTRODUCTION

Measurement of mechanical properties of films deposited on substrates has long been an issue in thin- and thick-film technology. Although indentation is extensively used due to its relative experimental simplicity, analysis is complicated by the inevitable substrate effect. Rules of practice exist that state film properties may be isolated if contact dimensions are small compared to film thickness, but such simplifications are not useful for layers including microstructural size effects, or ultra-thin films. Thus, analyses that consider the relation between film and substrate properties are necessary.

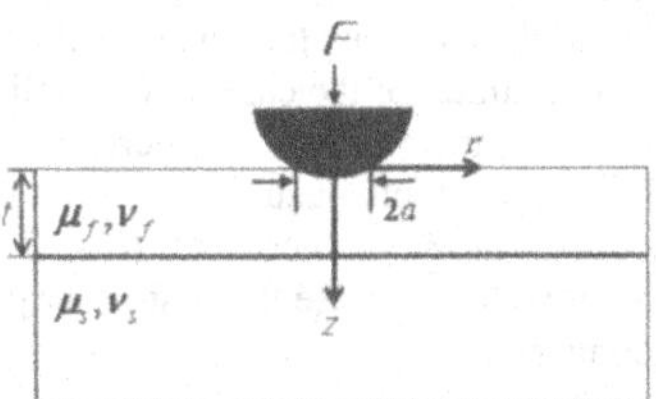

Figure 1. Spherical indentation of an elastic film-substrate bilayer. The force-depth response will be governed by the elastic properties of both constituents.

Consider the indentation of an ideally elastic film/substrate system that is mechanically bonded (Fig. 1). A proper description of the effective modulus μ_{eff} of the film/substrate system, in the context of film and substrate moduli (μ_f and μ_s) is essential to the accurate extraction of the film properties. Since there is no exact solution yet for this problem, a number of approximate models have been proposed to be fitted with data from experiments, and FEM simulations, or analytic solutions for several tip geometries. Those models have shown good yet limited applicability. Most of these approaches were based on the following simple formula:

$$\mu_{eff} = \mu_s + (\mu_f - \mu_s)\phi(a/t, \mu_f/\mu_s) \quad (1)$$

where a is contact radius, t is the thickness of the film and ϕ is a weight function of contact size, tip geometry and modulus mismatch; ϕ approaches 1 when $a/t \to 0$ and 0 when $a/t \to \infty$. Numerous forms of Eq (1) have been proposed for sharp and flat-ended cylindrical geometry. For brevity, they are grouped and listed here.

- Doerner and Nix[1] first introduced an empirical equation of an effective modulus expressed as an exponential form and fit it with experimental data obtained with a sharp tip. King[2] performed a numerical analysis of flat-ended tip indentation and introduced a similar equation to that of Doerner and Nix. Jung et al.[3] developed a power law function based on experimental results from nanoindentation with ceramic films on silicon substrates. These equations have adjustable parameters to be determined by experiment or numerically.
- Gao et al.[4] performed a first order perturbation analysis on the flat-ended cylindrical punch problem, and presented a formula for effective modulus μ_{eff}. Mencik et al.[5] compared this and several empirical equations with experimental data from nanoindentation, and found that Gao et al.'s equation gave the best fit to their experimental data. Xu and Pharr[6] later modified the perturbation solution and comparison with FEM results showed better accuracy for a wide range of modulus mismatch. The perturbation solution is attractive because it is given as a closed form and doesn't include adjustable parameters. But its assumption on flat-ended cylindrical geometry limits the applicability for sharp or spherical tip indentation, particularly when a and t are comparable.
- Different tip geometries have been considered in other investigations. Perriot and Barthel[7] used numerical integration to get effective modulus curves for sharp and spherical tips, and showed they overlap closely, but differ from those for a flat-ended cylindrical tip. Clifford and Seah[8] conducted FEM simulations with spherical tip geometry to get effective modulus curves for compliant polymer films on stiff substrates and proposed a curve-fit equation. Finally, Hsueh and Miranda[9] presented an approximate, analytically derived equation of effective modulus under spherical tip indentation, based on an extension of the Boussinesq Green's function. Using FEM, they showed that their results were valid for a certain range of modulus mismatch, but most accurate for the case of a compliant film on a stiff substrate.

There is an opportunity to revisit the problem of spherical indentation of a bilayer, with particular attention to the stiff film on compliant substrate. From the above investigations it appears that inaccuracies may arise from (a) tip geometry (e.g., flat punch vs spherical) or (b) analysis method. To investigate this, here we use the first-order perturbation solution, and modify it for a spherical tip geometry.

THEORY

A full derivation of the approach is presented in [10], and salient points are shown here. In perturbation analysis, the film/substrate system is treated as a homogeneous material with (initially) properties of the substrate that undergoes a phase transformation to assume film properties in the region $0 < z < t$. During the transformation, the load F is fixed and the displacement h is allowed to change to $h + \delta h$. The force-displacement relation for spherical indentation of an elastic material, is described by Hertzian relation[11] as

$$F = \frac{8\sqrt{R}h^{3/2}\mu}{3(1-\nu)} \quad (2)$$

Eq. 1 contains a single value for shear modulus and Poisson's ratio, but a composite value is used to represent the film/substrate system [8, 9, 12]. Upon the phase transformation, the extra work done by the force F due to the displacement change δh is thus calculated as [13]

$$W = \int F\partial h = \int \frac{8\sqrt{R}h^{3/2}\mu}{3(1-\nu)}\partial h = \frac{2}{5}Fh,$$

$$\partial W = \frac{2}{5}F\partial h \tag{3}$$

The additional work is equal to the strain energy change due to moduli variation in the body, and the energy conservation equation can be written as

$$\frac{2}{5}F\delta h = -\frac{1}{2}\int_{V_f} \delta c_{ijkl} u^o_{i,j} u^o_{k,l} dV \tag{4}$$

where δc_{ijkl} is the change of moduli from substrate material to film material and u^o_i is the known displacement field for the homogeneous substrate material. The right term of Eq. 3 is the energy variation due to a moduli transformation, and can be rearranged (following Gao, et al.) as

$$\frac{2}{5}F\delta h = -\frac{1}{2}e\lambda_s \int_{V_f} \varepsilon^o_{jj}\varepsilon^o_{kk} dV - \frac{1}{2}\frac{\mu_f - \mu_s}{\mu_s}\int_{A_f} \sigma^o_{ij} n^o_i u^o_j dA, \tag{5}$$

where λ_s is the Lame constant of the substrate,

$$e = \frac{\mu_f}{\mu_s}\cdot\frac{(\nu_f - \nu_s)}{\nu_s(1-2\nu_f)}, \tag{6}$$

and σ^o and ε^o are the known stress and strain field for the homogeneous substrate material[4]. The terms V_f and A_f in Eq. 4 indicate the domain volume and the surrounding surface domain of the film region, respectively. The surface domain consists of two planes: $z = 0$ and $z = t$. Using the known values for surface and sub-surface stress distributions under spherical indentation [e.g., 10] perturbation functions I_o and I_1 are found

$$\begin{aligned} I_o = 1 + &\frac{5}{4\pi a^5(1-\nu)} \\ \times\Bigg\{ &\int_0^a \left[\frac{3}{2}\left(-N' + \frac{aM't}{S'}\right)\left(\pi(2a^2 - r^2)(1-\nu) + 4\int_0^t \left(\frac{z(aM - 2S\nu\phi) - NS(1-2\nu)}{S}\right)dz\right)\right] rdr \\ + &\int_a^\infty \left[3\left(-N' + \frac{aM't}{S'}\right)\left(\left(ar\sqrt{1-\frac{a^2}{r^2}} + (2a^2 - r^2)Sin^{-1}\left(\frac{a}{r}\right)\right)(1-\nu)\right.\right. \\ &\left.\left. + 2\int_0^t \left(\frac{z(aM - 2S\nu\phi) - NS(1-2\nu)}{S}\right)dz\right)\right] rdr \\ + &\int_0^\infty \left[2t\left(\frac{N'r}{S'} - \frac{H'rt}{G'^2 + H'^2}\right)\left((3N't^2 + 3r^2t\phi)(1-\nu) - (N'S' + a^3 + 2A'N')(1-2\nu) - 3aM't\nu\right)\right] rdr \Bigg\} \end{aligned} \tag{7}$$

$$I_1 = \frac{30}{\pi}\int_0^t \int_0^\infty (N - z\phi)^2\, rdrdz \tag{8}$$

that can be inserted into the following equation

$$\left[\frac{1-\nu}{\mu}\right]_{eff} = \left[1-\nu_s+(\nu_s-\nu_f)I_1\right]\left[\frac{(1-I_o)}{\mu_s}+\frac{I_o}{\mu_f}\right] \tag{9}$$

which can be further re-arranged to provide *normalized displacement*

$$h_t / h_f = \left[\left(\frac{(1-\nu)}{\mu}\right)_f \Big/ \left(\frac{(1-\nu)}{\mu}\right)_{eff}\right]^{2/3} \tag{10}$$

in which displacement into a composite material may be compared with the corresponding displacement into a homogeneous material with film properties.

RESULTS

Figure 2 illustrates the following normalized displacement curves comparing the current work with FEM results and Hsueh's model[9]. Note that Hsueh's model gives a very similar result to the current solution when the modulus mismatch is small for the case of stiff films, but when the mismatch increases, it shows larger deviation. For $\mu_f / \mu_s = 10$, both Hsueh's and the current solution deviate from the FEM results when $a/t \to \infty$. Hsueh et. al attributed this discrepancy to flexural stresses in the film. The current solution underestimates the modulus in the case of compliant films. We find this acceptable, as Hsueh's solution is sufficiently accurate in those cases. Finally, it is interesting to note that in the case of stiff films, the current solution appears most accurate when a and t are comparable.

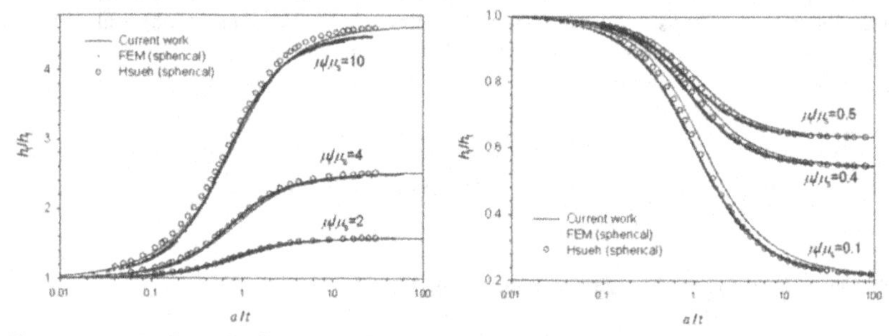

Figure 2. Normalized displacement, h_t / h_f curves for the current work with comparison to FEM results and Hsueh's model: (a) $\mu_f / \mu_s = 2, 4$ and 10, and (b) $\mu_f / \mu_s = 0.1$, 0.2 and 0.5.

ACKNOWLEDGMENTS

JHK and AG gratefully acknowledge the NSF Faculty Early Career Award CMS 0449268 for supporting this work; and CSK would like to gratefully acknowledge the NSF Award CMMI 0626025 for support.

REFERENCES

1. M. F. Doerner and W. D. Nix, J. Mater. Res. **1,** 601 (1986).
2. R. B. King, Int. J. Solids. Struct. **23,** 1657 (1987).
3. Y. G. Jung, B. R. Lawn, M. Martyniuk, H. Huang and X. Z. Hu, J. Mater. Res. **19,** 3076 (2004).
4. H. J. Gao, C. H. Chiu and J. Lee, Int. J. Solids Struct. **29,** 2471 (1992).
5. J. Mencik, D. Munz, E. Quandt, E. R. Weppelmann and M. V. Swain, J. Mater. Res. **12,** 2475 (1997).
6. H. T. Xu and G. M. Pharr, Scripta Mater. **55,** 315 (2006).
7. A. Perriot and E. Barthel, J. Mater. Res. **19,** 600 (2004).
8. C. A. Clifford and M. P. Seah, Nanotechnology **17,** 5283 (2006).
9. C. H. Hsueh and P. Miranda, J. Mater Res. **19,** 94 (2004).
10. J. Kim, C.S. Korach and A. Gouldstone, Accepted in J. Mater. Res (2008).
11. K. Johnson: *Contact Mechanics* (Cambridge University Press, New York, 1985).
12. I. Pane and E. Blank, Surf. Coat. Tech. 200, 1761 (2005).
13. D. Maugis: *Contact, adhesion, and rupture of elastic solids* (Springer, Berlin, 2000).

Mater. Res. Soc. Symp. Proc. Vol. 1139 © 2009 Materials Research Society 1139-GG03-43

Microfabrication of Si/SiO_2–Spherical Shells as a Path to Sub-mm^3 Autonomous Robotic Systems

Vladimir S. Vasilyev[1], J. Robert Reid[2], and Richard T. Webster[2]

[1]S4 Inc., Burlington, MA 01803 U.S.A. (On site contractor for AFRL/RYHA)
[2]Air Force Research Laboratory, Sensors Directorate, Antenna Technology Branch (AFRL/RYHA), Hanscom AFB, MA 01731 U.S.A.

ABSTRACT

A process for forming thin (1-3 μm) stacks of Si/SiO_2 or $SiO_2/Si/SiO_2$ layers into spherical shells 0.5-3.0 mm in diameter is demonstrated as the baseline for realizing sub-mm^3 micro-robots. The fabrication process combines bulk and thin-film micromachining, design of novel masks, and multistage wet and dry etching to release the layers from the substrate. The released layers curl up, self assembling into a spherical shell. The radius of curvature of the released stack is a function of the type, thickness, and residual stresses in the layers. The diameter of the resulting shells is calculated using a mechanical model of the multi-layer stacks. This calculation is compared with measurements of fabricated spheres. The fabrication process is compatible with CMOS circuitry, and future work will focus on realizing spheres with embedded solar cell as a power source and a capacitor for energy storage, which will result in a functional micro-robot.

INTRODUCTION

The development of semi- to fully-autonomous robots that fit in a volume less than one cubic millimeter is a daunting challenge. It requires the integration of energy collection, energy storage, computation, sensing, and actuation in a very compact physical system. Traditional MicroElectroMechanical Systems (MEMS) approaches have tried to do this by first achieving the required functionality and then combining the pieces together to form a robot [1]. Our work takes the opposite approach and focuses on realizing the structure first and then adding functionality into this structure. We have initiated this work by developing a CMOS-compatible process for fabricating sub-cubic-millimeter spherical shells such as those shown in Fig. 1.

These shells are formed from a thin stack of silicon and silicon dioxide layers. These layers

Figure 1. Scanning electron micrograph images showing partially (a) and fully (b,c) released spherical shells composed of thin silicon and silicon dioxide layers.

are initially fabricated on top of a standard thickness (> 0.5 mm) silicon wafer using planar processes. The layers are then released by removing the handle wafer silicon. When released, the residual stresses in the individual layers cause the structure to curl up into the final spherical shape. This fabrication process will eventually evolve to include the integration of active CMOS circuitry and an embedded solar cell to provide the functionality required by the micro-robot. In this paper, we detail the process flow, calculation of the shell diameter, and provide results from our experiments to date. Additional details on the design of the robot have been previously published [2].

METHOD AND EXPERIMENT

Bending model

The radius of curvature, r, for a beam subjected to a bending moment, M, is calculated as

$$r = \frac{EI}{M}, \tag{1}$$

where E is the Young's (or elastic) modulus for the material and I is the moment of inertia of the beam. For multiple layer beams where each layer has different residual stress and material properties, it is necessary to calculate the terms EI and M as the sum of the values for each layer. The value of EI can be calculated as

$$EI = EI_{eff} = \frac{w}{3}\sum_{l=1}^{N} E_l\left(t_l^3 + 3x_l^2 t_l + 3x_l t_l^2\right), \tag{2}$$

where the width, w, of all of the layers is the same; E_l is the Young's modulus of layer l; t_l is the thickness of layer l; N is the number of layers; and x_l is height of the bottom of layer l relative to the bending axis. The position of the bending axis in the layer stack is determined iteratively to minimize the value of EI_{eff}. The bending moment is then calculated by assuming that each layer has a uniform residual stress, σ_l, and summing up the bending moment from each layer as

$$M = M_{eff} = \sum_{l=1}^{N} \frac{\sigma_l w}{2}\left(t_l^2 + 2x_l t_l\right). \tag{3}$$

The results from Eqs. 2 and 3 are used in Eq. 1 to calculate the radius of curvature.

For a two layer device with a stress free top layer and compressively stressed lower layer, as would be expected for bulk Si on a thermally grown SiO_2, the bottom layer releases the stress by expanding, causing the layer stack to bend upwards. As the device curves upwards, the stress in this bottom layer is reduced according to

$$\sigma_{actual} = \sigma_{initial} - \sigma_{reduction} = \sigma_{initial} - \frac{E_l t_l}{R} \tag{4}$$

where R is the radius of curvature of the released shell. The resulting radius for such a two layer shell is found by iterating Eqs. 1-4 to find a consistent solution.

Device designs

The spherical shells are formed by folding a two dimensional shape up into a three dimensional structure. However, the mechanics of the bending process are more complex than might be assumed. The most critical aspect to understand is that once bending is initiated along one axis, the structure becomes more rigid in the other axes. As a result, most of the bending observed in a released structure is only along one axis. For example, releasing a rectangular shaped pattern will result in the formation of a tube, not a spherical structure. Similarly, releasing a pattern such as a Mercator projection will also result in a tubular structure, not a spherical one. Considering this effect, the first pattern used in this work was the ribbon pattern, shown in Fig. 2a, composed of long arms radiating from a central point. This pattern has the advantage that the individual arms only need a length that is half the circumference of the sphere. Unfortunately, this arrangement has several drawbacks. First, all of the ribbons meet at a central point. If the length to width ratio of the ribbon is not large, the central core becomes very large relative to the size of the sphere. This core does not bend the same as the ribbon, creating a non-spherical base to the shell. In contrast, having long thin ribbons results in a spherical structure where a large percentage of the spherical surface is not covered. Therefore, we investigated the two alternative patterns. The petal pattern (Fig. 2b) is a direct projection of a slice of a three dimensional sphere onto a two dimensional surface. As can be seen, this projection narrows at both ends so that the problem with a large central core is largely mitigated. However, bending of the petal does not have the same radius of curvature in both directions. This problem is overcome by the final pattern of ridged ribbons. Bending on the main ribbons is primarily along the desired axis, but bending of the small side ribbons is on a second axis.

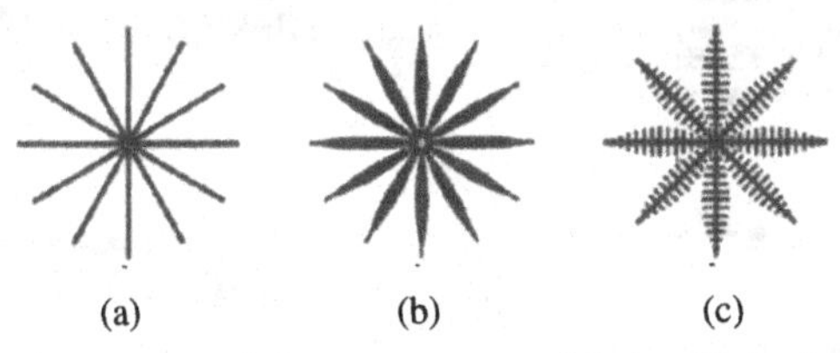

Figure 2. Images showing the three different patterns used during fabrication: (a) ribbon pattern, (b) petal pattern, and (c) ridged ribbon pattern.

Fabrication process

Spherical shells are fabricated through one of two similar processes. Both processes begin with a silicon-on-insulator (SOI) wafer. SOI wafers consist of three layers: a relatively thin (typically 0.1-2.0 μm) layer of silicon called the device layer; a thin layer (typically 0.1-1.5 μm) of silicon dioxide called the buried oxide, or BOX layer; and a thick layer of silicon (typically 0.5-1.0 mm) called the handle layer. In the first process, a bi-layer formed from the device silicon and the BOX layer is released from the handle layer to form the shell. In the second process, a tri-layer formed from a deposited silicon dioxide layer, the device silicon, and the BOX layer is released from the handle silicon to form the shell.

A schematic view showing the steps of the both processes is provided in Fig. 3. Before processing begins, the thicknesses of the device Si and BOX layers are measured by *Filmetrics* F20 system. A common series of steps is then used to etch the top two layers of the SOI wafer. First, a mask layer of SiO_2 is sputtered on the device Si layer. Photo-resist (Shipley 1813) is spin-coated onto the wafer, exposed with the desired pattern, and then developed. The SiO_2 layer is then etched using a buffered oxide etch (BOE 10:1, J.T.Baker). Subsequently, the exposed device Si layer is etched using a solution of 25% tetramethyl-ammonium hydroxide (TMAH) in

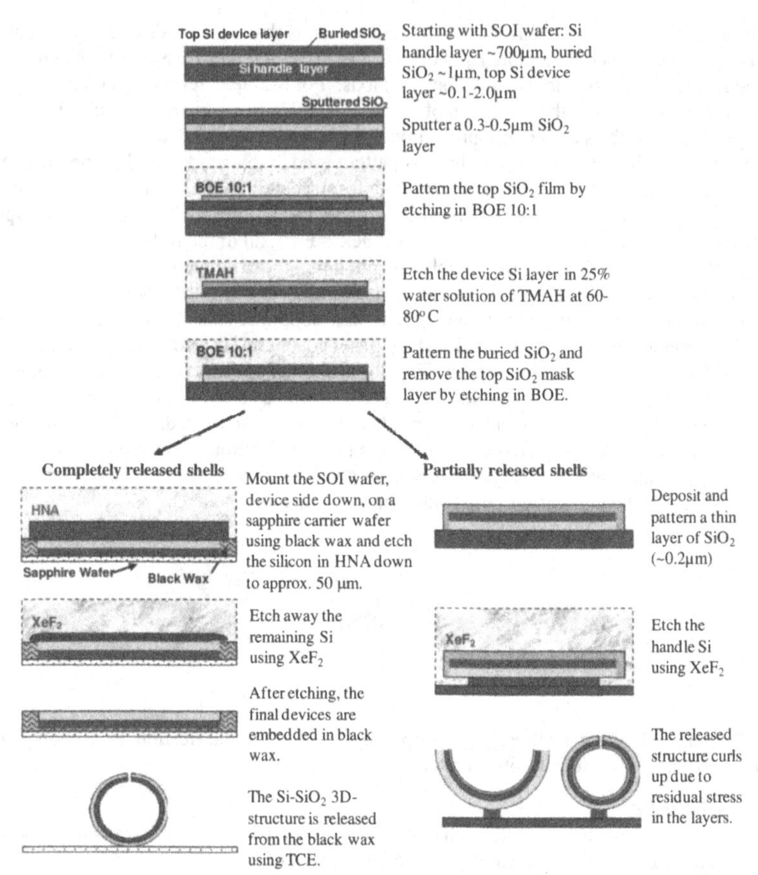

Figure 3. Schematic view of fabrication process. All fabrication begins with a silicon-on-insulator wafer. The top two layers of this wafer are then patterned to form the structure of the final shells. The shells are then be fully released using an intermediate handle wafer, or partially released using a front side silicon etch.

water at 60-80° C [3]. This etch stops on the BOX layer. Finally, the top SiO_2 mask layer and exposed BOX regions are removed by again etching in the 10:1 BOE.

At this point, the top two layers have been patterned; further processing is done using one of two separate release processes. The first process results in fully released spheres but requires the use of an intermediate handle wafer, while the second process only results in partially released spheres. Fully released spheres are formed by mounting the SOI wafer to a polished sapphire wafer. This is done by coating a polished sapphire wafer with melted black wax (Apiezon W100) at 100-120° C. The SOI wafer is then mounted by pushing the device side into the wax and slowly cooling. The wafer stack is then immersed into an HF, 25%:HNO_3, 35%:HAc, 40%

(a) (b)

Figure 4. Optical photograph of a sphere embedded in black wax (a) and scanning electron micrograph of a similar sphere after being fully released (b).

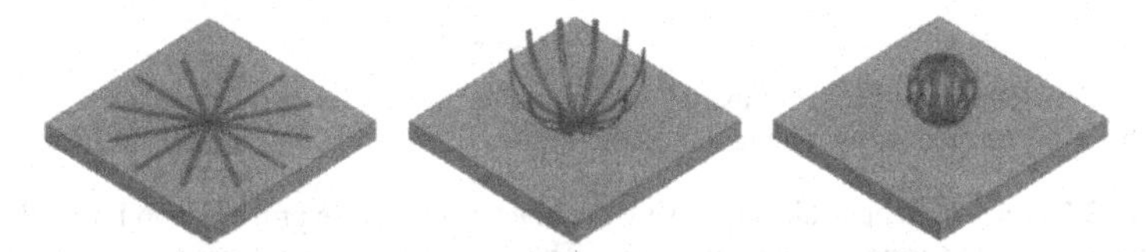

Figure 5. Selectively etching the silicon from underneath the oxide enclosed structure results in the edges of the structure being released first so that the gradually roll up into a sphere.

(HNA) solution [4] at room temperature, etching the silicon until it reaches a thickness of approximately 50 μm. Next, the wafers are etched in the XeF_2 system [5] to remove the remaining silicon of the handle layer, leaving the individual shells embedded in the black wax. Finally, the spheres are released by removing the black wax using trichloroethylene (TCE). Fig. 4 provides an optical image of a shell before it is released from the black wax and a scanning electron micrograph of the sphere after it is released from the black wax.

A second release process that does not require an intermediate handle wafer, but results in only partially released shells has also been explored. Since the device is not mounted in the black wax, it is necessary to protect the device silicon with an additional thin (approx. 0.2 μm) silicon dioxide layer. This layer is deposited through sputtering or chemical vapor deposition. Alternate materials such as silicon nitride could also be used but have not yet been explored. The SiO_2 protective layer is patterned using a swelled version of the original etch mask so that the remaining material covers all of the edges of the devices. The final release is then done by etching the handle silicon in XeF_2. As the XeF_2 undercuts the device layers, the shell curls up as illustrated in Figure 5.

RESULTS AND DISCUSSION

The fabrication process has been successfully used to fabricate spherical shells based on all three of the patterns and examples of these are shown in Fig. 1. These shells have diameters ranging from 0.5-3.0 mm, roughly in line with calculated values. Once the fabrication process was matured to the point where a reasonable yield (>50% of devices on each substrate) was achieved, an experiment to verify the bending calculations was performed. Three identical SOI wafers with a device layer of 3.0 μm and BOX layer thickness of 1.1 μm were plasma etched to thin the device layer of the wafers to the range of 0.55-3.0 μm so that shells with diameters from

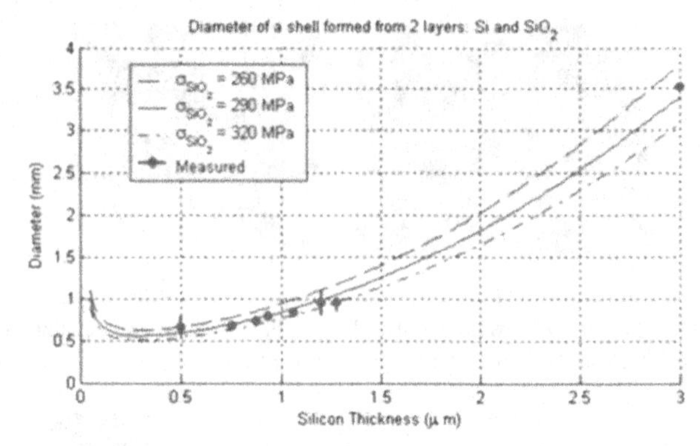

Figure 6. Calculated diameters compared with measured values for two layer shells. Calculations performed using Young's modulus values: E_{Si} = 150 GPa, E_{SiO2} = 39 GPa.

0.6-1.2 to 3.5 mm would be achieved. Using the black wax release process, two layer silicon shells were fabricated from each of the wafers. The diameters of the released shells were measured optically using a travelling microscope. Fig. 6 compares the results from this experiment with diameters predicted using the model described above. For the calculations, the Si and SiO_2 layers are assumed to have residual stress values of 0 and 290 MPa, respectively. As can be seen, the model is reasonably accurate for a Young's Modulus of silicon dioxide at 39 GPa. This value is significantly lower than typically reported values. This low value probably results from the fact that our model of the bending process does not yet account for elongation of the released films that occurs. In addition to this, the model does not take into account the effects of two dimensional folding. These effects will be modeled using finite element simulations.

ACKNOWLEDGMENTS

The authors gratefully acknowledge the support of Dr. Alvin Drehman (also from the Sensors Directorate) for providing the silicon dioxide depositions and measurements of the layer thicknesses. This work was supported by the Air Force Office of Scientific Research (PM: Kitt Reinhardt, AFOSR/NE) under LRIR 07SN06COR. The views expressed in this article are those of the authors and do not reflect the official policy or position of the United States Air Force, Department of Defense, or the U.S. Government.

REFERENCES

1. S. Hollar, A. Flynn, C. Bellow, and K. Pister, *IEEE International Conference on Micro Electro Mechanical Systems*, January 2003, pp. 706–711.
2. J.R. Reid, V. Vasilyev, R.T. Webster, *Proc. of Nanotech 2008,* Boston, MA, June 2008.
3. O. Tabata, R. Asahi, H. Funabashi, K. Shimaoka, and S. Sugiyama, *Sensors and Actuators.* **34**, 51 (1992).
4. B. Schwartz and H. Robbins, *J. Electrochem. Soc.* **123**, 1903 (1976).
5. P. B. Chu, J. T. Chen, R. Yeh, G. Lin, J. C. P. Huang, B. A. Warneke, and K. S. J. Pister, *Digest of the International Conference on Solid-State Sensors and Actuators.* **1,** 665 (1997).

Mater. Res. Soc. Symp. Proc. Vol. 1139 © 2009 Materials Research Society 1139-GG03-44

Transport of Charged Species across Solid-State Nanopores

Daisy Fung [1], E. Akdemir [1], M.J. Vitarelli [2], Eugene Sosnov [1], Shaurya Prakash [1,3]
[1]Department of Mechanical and Aerospace Engineering, [2]Department of Chemistry, [3]Institute of Advanced Materials, Devices, and Nanotechnology, Rutgers, The State University of New Jersey, Piscataway, NJ 08854 U.S.A.

ABSTRACT

Nanofluidic devices are finding growing interest for a variety of applications. An initial report is presented here on a wide range of parameters influencing transport of ionic species as they translocate across solid-state nanopores. AC electrical bias at low ionic concentration with overlapping electric double layers provides an enhancement of ionic flux over pure DC bias. Furthermore, results also indicate that concentration and pH gradients can be maintained across solid-state nanopores for extended periods of time that can last for several hours in the absence of driving forces such as electric fields.

INTRODUCTION

Nanofluidic systems have generated tremendous interest in recent times due to the promise of revolutionary new fluidic technologies for separations, proteomics, genomics, next generation lab-on-chip devices, and water purification. Given this broad scientific and technological interest, many new nanofluidic devices are being developed for a myriad of applications [1, 2]. Therefore, there is a need to enhance fundamental understanding of ionic transport in confined nanoscale systems to better engineer devices to meet the diverse array of challenges presented by nanofluidic devices including interfacial properties that influence confined fluid transport [1, 3], and changes in transport regimes as function of electric double layer (EDL) thickness [4].

In this paper, we present an initial report on a wide range of parameters influencing transport of ionic species as they translocate across solid-state nanopores. The nanopores used in this work are monodisperse, nuclear track etched membranes that have been classified as nanocapillary array membranes (NCAMs). The transport measurements are characterized by UV/VIS spectroscopy, bulk electrical conductivity measurements, and electrochemical impedance spectroscopy (EIS). This multi-technique characterization provides information on the influence of EDLs and applied electrical bias across NCAMs.

EXPERIMENT

In this work, nanocapillary array membranes (NCAMs) were used [4]. The nominal pore sizes vary from 10 to 800 nm; however, only 10 nm diameter pores were used in this report. The NCAM thickness is approximately 6 μm. The NCAMs are commercially available from GE Osmonics (MN, USA). All chemicals used are purchased from Sigma-Aldrich (MO, USA) and used without further purification. The experimental set-up consists of a traditional permeation cell (Fig. 1) with the NCAM separating two reservoirs [2]. The volume in each reservoir is 200 ml. In a typical experiment the electrolyte concentration, pH, and electrical bias across the nanopores are varied in a systematic manner. The electrolyte solution is potassium phosphate buffer, PB, an important salt for biological applications. The bulk conductivity of the source and permeate (or receiver) reservoirs is measured as a function of time to track the transport of salt ions. For certain experiments, the source is spiked with a solution of methylene blue (MB, 0.03 M). In time, MB diffuses to the permeate side. Periodically, samples are collected from the permeate reservoir. UV-VIS spectroscopy tracks the change in MB concentration of the permeate reservoir, and EIS is used to obtain a measure of the NCAM impedance.

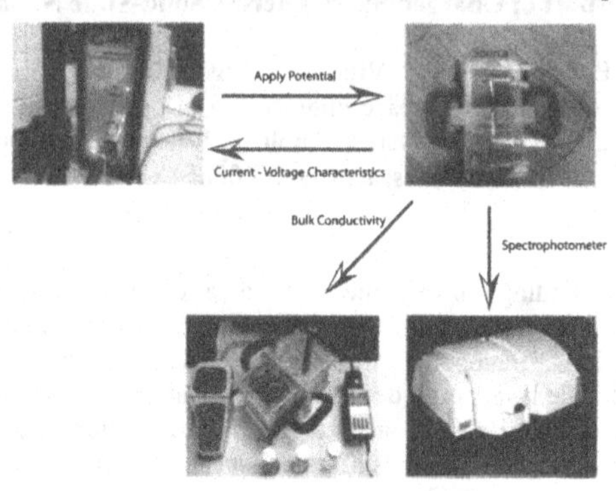

Figure 1: Digital photographs showing the experimental set-up and the multi-technique characterization methods.

RESULTS AND DISCUSSION

Figure 2 shows the concentration gradients that can be maintained across the NCAMs for extended periods of time.

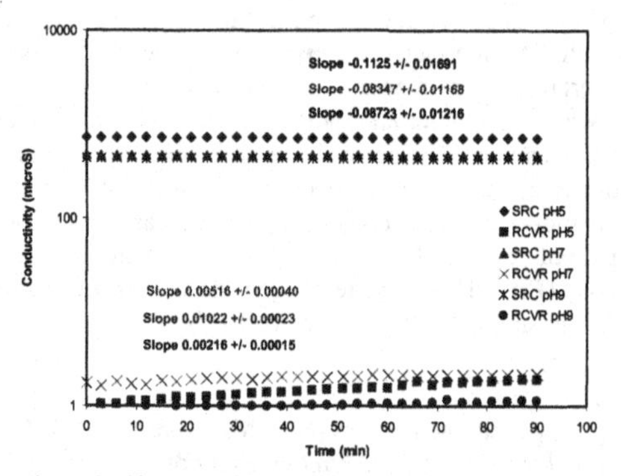

Figure 2: Change in bulk conductivity of source and permeate chambers as function of electrolyte pH. The pH can be used as a tool to alter the effective surface charge as surface dissociation of ionizable groups is a function of pH. The only driving force for ionic transport is a concentration gradient.

Figures 3 and 4 show results of applying an electrical bias across the NCAMs. It can be seen from Figure 3 that the AC bias is more effective at transporting ions across the NCAMs at low ionic concentrations. At such concentrations the EDLs within the NCAMs are interacting.

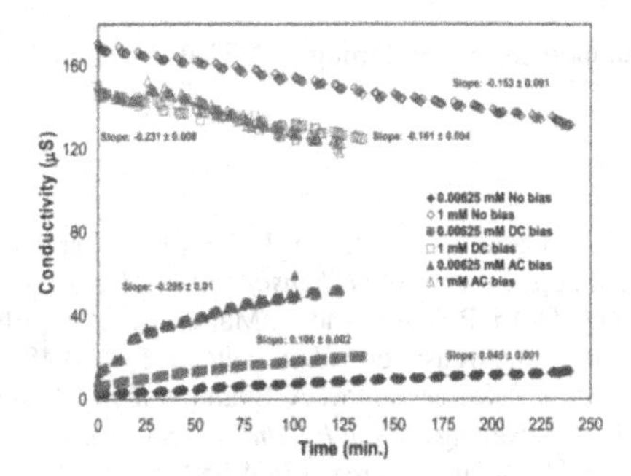

Figure 3: Change in bulk conductivity of source and permeate chambers as function of applied electrical bias for various electrolyte concentrations. Electrolyte concentrations can be used as a tool to change the EDL thickness. With an AC bias a higher flux can be achieved in contrast to a DC bias.

Figure 4 shows the EIS modulus data for impedance of the NCAMs as function of electrical bias. Low electrical biases (< 500 mV) are used to minimize the influence of Faradaic reactions. The EIS data shows that impedance of the NCAMs decreases with increasing electrical bias. The decrease in impedance of the NCAM to ionic flow can be expected as with increasing driving force it is likely that electromigration has a larger influence on ionic transport. These results are in agreement with the results reported previously [4, 5].

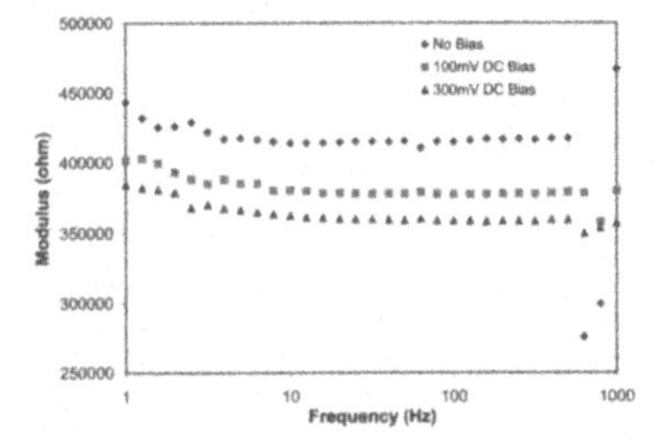

Figure 4: Evaluating influence of applied membrane bias as function of AC frequency. Impedance to ionic flow decreases with increasing AC amplitude.

CONCLUSIONS

AC electrical bias at low ionic concentration with overlapping electric double layers provides an enhancement of ionic flux over pure DC bias. Further, results indicate that concentration and pH gradients can be maintained across solid-state nanopores for extended periods of time that can last for several hours in the absence of driving forces such as electric fields.

ACKNOWLEDGMENTS

The authors acknowledge support from NSF through CBET 0813944 and the NSF-STC under agreement number CTS 0120978.

REFERENCES

1. Prakash, S., A. Piruska, E.N. Gatimu, P.W. Bohn, J.V. Sweedler, and M.A. Shannon, 2008, Nanofluidics: systems and applications. *IEEE Sensors Journal*. **8** 441-450.
2. Karnik, R., R. Fan, M. Yue, D. Li, P. Yang, and A. Majumdar, 2005, Electrostatic control of ions and molecules in nanofluidic transistors. *Nano Letters*. **5** 943-948.
3. Prakash, S., J. Yeom, N. Jin, I. Adesida, and M.A. Shannon, 2007, Characterization of ionic transport at the nanoscale. *Proceedings of the Institution of Mechanical Engineers, Part N: Journal of Nanosystems and Nanoengineering*. **220** 45-52.
4. Kemery, P.J., J.K. Steehler, and P.W. Bohn, 1998, Electric field mediated transport in nanometer diameter channels. *Langmuir*. **14** 2884-2889.
5. Lakshminarayanaiah, N., *Transport phenomena in membranes*. 1969, New York: Academic Press.

Mater. Res. Soc. Symp. Proc. Vol. 1139 © 2009 Materials Research Society 1139-GG03-45

Fabrication and Piezoelectric Characterization of AlN Mesa Structures

R. Farrell[1], A. Kabulski[1], V. R. Pagán[1], S. Yeldandi[1], X. A. Cao[1], P. Famouri[1], J. Peter Hensel[2] and D. Korakakis[1,2]
[1]Lane Department of Computer Science and Electrical Engineering, West Virginia University, PO Box 6109, Morgantown, WV 26506, U.S.A.
National Energy Technology Laboratories,
3610 Collins Ferry Road, Morgantown, WV 26507, U.S.A.

ABSTRACT

Circular AlN mesa structures, consisting of a 500 nm thick AlN layer between two ~200 nm metal layers, have been created for this study. By decreasing the contact size and limiting the surface area of the AlN film, clamping effects are reduced. This configuration also minimizes non-normal electric field lines between the top and bottom contacts and further isolates the d_{33} piezoelectric coefficient. The aluminum nitride mesa structures exhibit a piezoelectric coefficient (d_{33}) typical for aluminum nitride (4-6 pm/V) but can be further controlled (and even increased above 6 pm/V) depending on the thickness of the film, the diameter of the mesa, and the types of metals used for the electrical contacts. The methods used to fabricate these mesas and the piezoelectric properties of the structures are studied. The effect of different contact metals as top and bottom electrodes on the displacement of the AlN mesas is also reported.

INTRODUCTION

Aluminum nitride (AlN) films have been explored for piezoelectric and high temperature microelectromechanical systems (MEMS) applications but few device geometries have been explored in depth. One of the main advantages of using AlN as an active material is its ability to retain piezoelectric properties at high temperatures (>300°C). Typically, a metal layer covering the entire AlN film is used as a top electrode in d_{33} measurements [1]. In this work, the effect of different contact metals and contact geometries on the measured piezoelectric response of the film is studied and the fabrication techniques used to make these structures are described.

EXPERIMENT

AlN mesa structures are fabricated to isolate the d_{33} piezoelectric coefficient by reducing the effect of non-normal electric field lines on the piezoelectric response of the film. *Figure 1a* illustrates the emergence of non-normal field lines in AlN thin films when a voltage is applied to the top and bottom electrodes. *Figure 1b* shows how the AlN mesa structure reduces the non-normal field lines reducing the contribution of the lateral displacement in the film caused by the d_{31} piezoelectric coefficient. This configuration also allows for better confinement of the electric field and therefore more accurate d_{33} measurements.

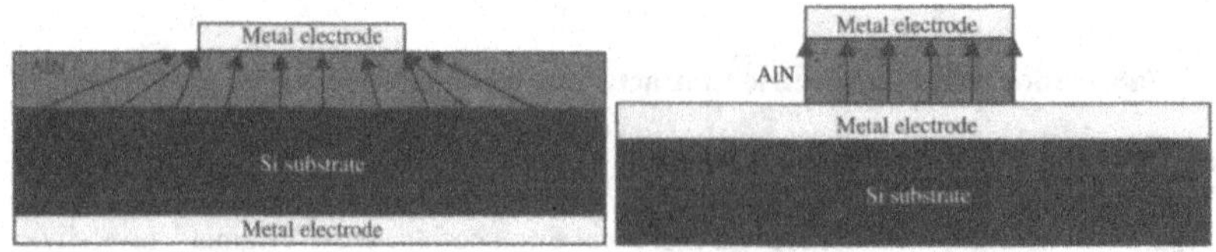

Figure 1 – (*a*) Non-normal field lines in AlN film, (*b*) Isolation of field lines in mesa structure

Two types of AlN films were investigated as discussed below; low power density sputtered and high power density sputtered. DC reactive sputtering was used to deposit polycrystalline AlN using an Al target with an Ar/N flow of 3/27 sccm. The quality of the material can be changed by adjusting the power density of the sputter deposition. Low power density depositions produce AlN films with a refractive index of ~1.9. AlN mesas sputtered with low power density were fabricated using a bottom up approach because the heat from the sputter deposition is relatively low at low power densities. A thin layer of metal is sputtered onto a clean Si wafer then a standard negative photolithography process is performed to create a photoresist mask with openings down to the metal layer. AlN is deposited with an 8" Al target at 500W under a pressure of 30 mTorr then a metal layer is deposited on top. After a standard liftoff technique, what remains are AlN mesa structures with metal contacts on top, standing on a thin metal layer below.

The bottom-up fabrication approach cannot be used to make higher quality AlN mesa structures (refractive index ~2.0-2.05). The higher sputtering power density generates too much heat causing the photoresist mask to burn so a top-down approach to the mesa fabrication must be implemented. A thin metal is first sputtered onto a Si wafer followed by a high quality AlN film sputtered with a 2" Al target at 200W under a pressure of 45 mTorr. The same negative photolithography technique is used to create a photoresist mask with openings down to the AlN layer then metal is deposited. After liftoff, what remains are metal contacts on an AlN film on top of a metal layer. A positive photolithography technique is used to create a reversal of the photoresist mask used to deposit metal contacts, so there is a thick layer (~5 μm) of photoresist covering the metal contacts on the AlN film, leaving the rest of the film exposed. Finally, a dry etch of AlN is performed using an Inductively Coupled Plasma (ICP) system with Cl_2/BCl_3 chemistry. The exposed AlN film is etched down to the bottom metal layer and an Asher process using an O_2/CF_4 chemistry is used to rid the sample of the photoresist mask.

DISCUSSION AND CONCLUSIONS

Combinations of Pt and Al were used as the top and bottom electrodes and the apparent effect of contact material on the piezoelectric response is investigated [2]. The top electrode is referring to the metal deposited on top of the AlN mesa and the bottom electrode refers to the contact metal deposited underneath the AlN mesa. An MSV-100 Polytec scanning head laser Doppler vibrometer (LDV) was used to measure the picometer displacement of the AlN mesas at different AC voltages at a frequency of 3.5 kHz. The results of the sputtered low quality AlN mesas and higher quality AlN mesas can be seen in *Figure 2a* and *Figure 2b* respectively.

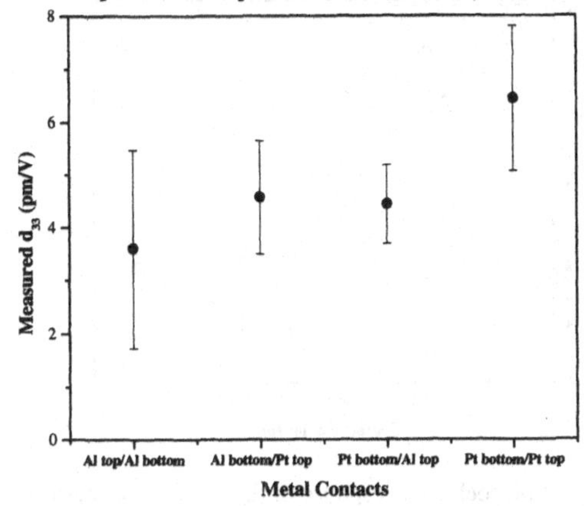

Figure 2a – Measured piezoelectric response of low quality AlN mesas

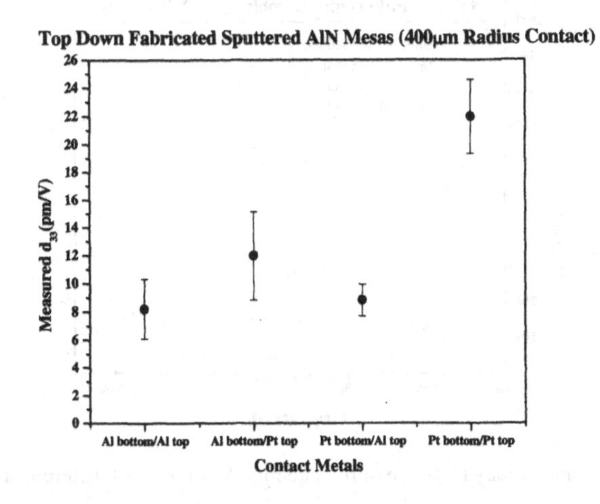

Figure 2b – Measured piezoelectric response of high quality AlN mesas

The measured piezoelectric response of the mesa structures shows a similar trend with Pt/Pt contact combination exhibiting the highest measured d_{33} value on average. The results in *Figures 2a* and *2b* were all measured from 400 μm radius mesas. To determine whether contact size has a relationship with the piezoelectric response, *Figure 3* plots responses of different metal contacts and different contact sizes to show that the measured piezoelectric response of the AlN mesas is dependent on both contact material and contact geometry.

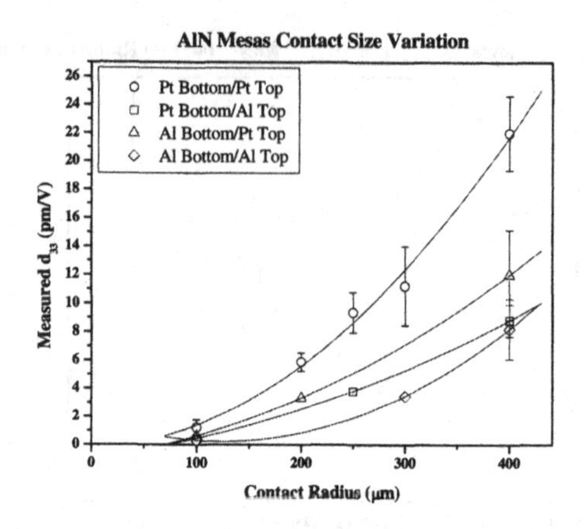

Figure 3 – Measured piezoelectric response of different sized mesa structures

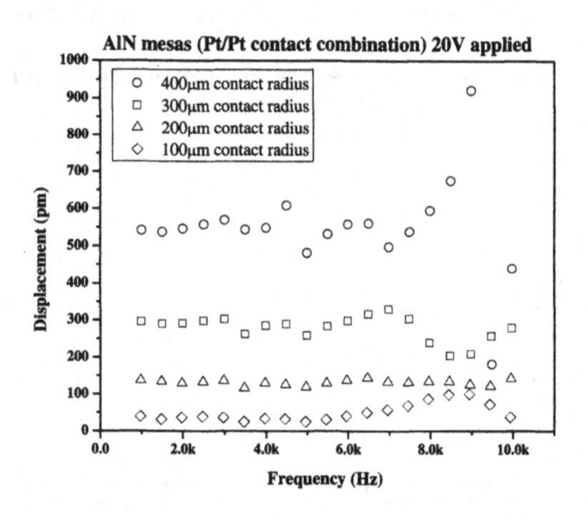

Figure 4 – Frequency response of high quality AlN mesas of different sizes

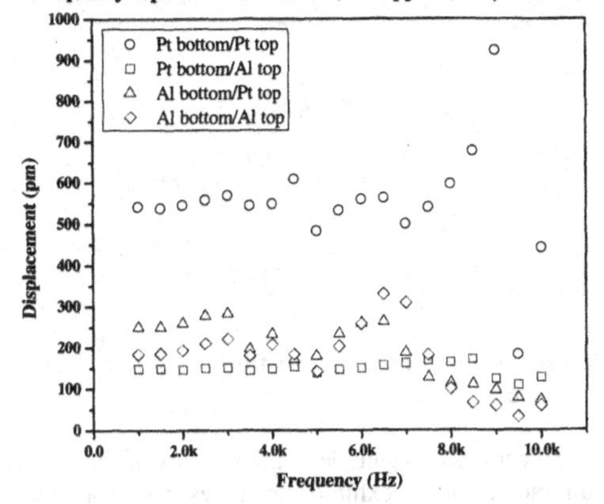

Figure 5 – Frequency response of high quality AlN mesas with different metal contacts

The frequency dependence (*Figures 4* and *5*) shows that there is no resonance at the probing frequency (3.5 kHz). The MSV-100 Polytec (LDV) is capable of scanning over a large area and mapping the piezoelectric response at different points on a sample. These measurements show how AlN responds at different points on different sized mesas. The Pt/Pt contact combination exhibited the largest displacement (as high as 22 pm/V in *Figure 3*) in the high quality sputtered AlN mesas so the rest of this study is focused on these mesas. *Figures 6a, 6b, 6c,* and *6d* show the LDV surface mapping results of 400 μm, 300 μm, 200 μm, and 100 μm radius contacts, respectively.

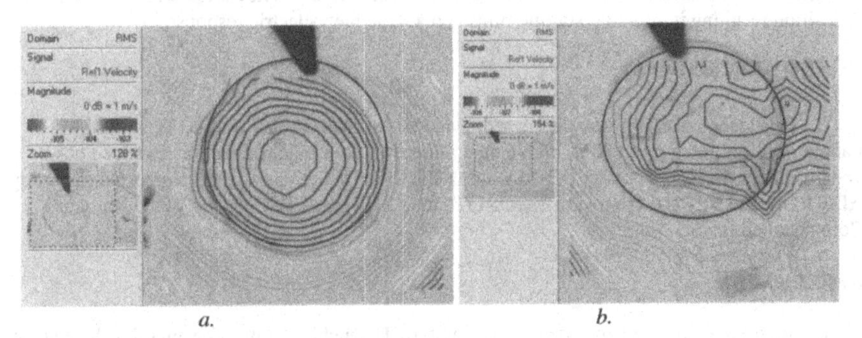

a. *b.*

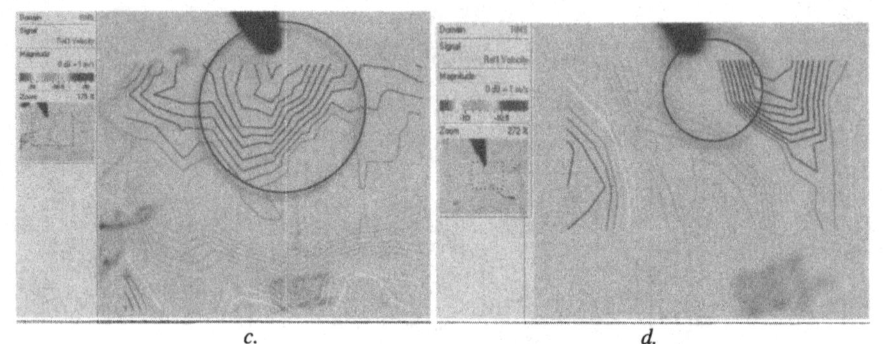

c. d.

Figure 6 – Distribution of piezoelectric response across film. The edge of the mesas is highlighted with a darkened circle.

The surface scans display the dB magnitude of the velocity of the AlN mesas over a designated area. The 400 μm radius contact exhibits the highest response in the center of the contact with a gradual decrease in response extending out to the area of the sample where no AlN is present. The smaller sized contacts exhibit a different trend, where the highest response is not in the center of the mesa and the isolines are not evenly distributed. This result is not intrinsically understood and is still under investigation.

AlN mesa structures have been fabricated using both a bottom-up and top-down approach with different metals used for top and bottom contacts. The Pt/Pt contact combination exhibits the highest measured piezoelectric response in mesa structures fabricated using both techniques which is consistent with literature [2]. The residual stress in the metallic thin films was not directly measured in this study; however there is reportedly no correlation between the stress caused by the AlN/electrode mismatch (10.9% for Pt) and the measured d_{33} value. Pt exhibits properties of symmetry similar to sapphire providing an optimal nucleation surface for AlN deposition resulting in more crystalline films and a large piezoelectric response [3].

ACKNOWLEDGMENTS

This technical effort was performed in support of the National Energy Technology Laboratory's on-going research in high temperature flow control hardware for advanced power systems under the RDS contract DE-AC26-04NT41817. This work was also supported in part by AIXTRON and NSF RII contract EPS 0554328 for which WV EPSCoR and WVU Research Corp matched funds.

REFERENCES

1. Kazuhiko Kano, Kazuki Arakawa, Yukihiro Takeuchi, Morito Akiyama, Naohiro Ueno, Nobuaki Kawahara. Sensors and Actuators A 130-131 (2006) 397-402
2. J. Harman, A. Kabulski, V.R. Pagan, P. Famouri, K.R. Kasarla, L.E. Rodak, J. Peter Hensel, and D. Korakakis. J. Vac. Sci. Technol. B 26(4), Jul/Aug 2008
3. Marc-Alexandre Dubois and Paul Muralt. J. App. Phys. V. 89(11) 2001

Mater. Res. Soc. Symp. Proc. Vol. 1139 © 2009 Materials Research Society 1139-GG03-46

Investigating the stress and crystal quality of AlN Air-Bridges through micro-Raman scattering

Sridhar Kuchibhatla[1], L.E. Rodak[1], D. Korakakis[1,2]
[1]*Lane Department of Computer Science and Electrical Engineering, West Virginia University, Morgantown ,WV 26506*
[2]*National Energy Technology Laboratory, Morgantown, WV 26507-0880*

ABSTRACT

In this work, we report the post-growth investigation of the microstructure and stress in the AlN films grown on patterned amorphous dielectrics through micro-Raman spectroscopy. The surface texture of AlN/SiO_2 structures was characterized by randomly oriented crystallites typical of polycrystalline films. Post growth analysis of the AlN/SiO_2 structures using micro Raman spectroscopy did not reveal phonon modes corresponding to wurtzite AlN. The presence of randomly oriented crystallites with a possibility of oxidized Al phase in the AlN film could have suppressed the appearance of wurtzite AlN phonon peaks in the Raman spectrum. Profiling the stress and the microstructure of AlN/SiO_2 structures across the width of the bridges is thus limited by these factors. AlN structures on SiO_2 when subjected to wet etching in buffered HF (10:1) showed a clear change in texture. Micro-raman spectroscopy on the etched areas revealed wurtzite AlN like phonon modes. The appearance of wurtzite AlN modes can be attributed to the removal of oxidized Al phase in the AlN film after wet etching.

INTRODUCTION

Wide bandgap group III-nitride materials such as GaN, AlN and its alloys have attracted considerable attention for the fabrication of MEMS/NEMS devices due to their excellent bio compatibility, well developed growth techniques for high quality thin films and chemical stability at high temperatures in comparison to other common materials such as quartz, polyvinylidelflouride (PVDF), and metal oxides (ZnO, PZT, BaZiO2, BaPbTiO3) [1]. Prior work has demonstrated that AlN thin film suspended structures can be fabricated through the process of lateral overgrowth on substrates such as Si, sapphire, and sapphire/GaN patterned with SiO_2 [2] using metalorganic vapor phase epitaxy (MOVPE) techniques also called metalorganic chemical vapor deposition (MOCVD). SiO_2 films were deposited by plasma enhanced chemical vapor deposition (PECVD) techniques. Given the amorphous microstructure of the sacrificial layer it is expected that the growth of AlN films using MOVPE techniques does not result in crystalline films on the amorphous dielectrics. However, the material properties of the structures have not been thoroughly studied to evaluate potential MEMS/NEMS applications such as piezoelectric micro and nanofluidic channels. Also the mismatch in the thermal expansion coefficients between AlN and SiO_2 can induce significant residual stresses in the AlN film. Residual stress can promote cracking, which is detrimental to device fabrication. This study deals with investigating the microstructure of AlN films on SiO_2. Si (111) and GaN (thin film)/sapphire substrates were used in this work. The carrier concentration of GaN was around 10^{18} cm^{-3}. In order to determine the local variation in the microstructure quality and the residual stress, across the width of the AlN/SiO_2 structures, micro-Raman scattering was used. In addition, SEM and optical images were used to investigate the micro-bridges.

EXPERIMENTAL DETAILS

SiO_2 films of thickness 100nm were deposited on Si (111) and GaN/Sapphire substrates using PECVD techniques. The films were patterned into stripes, having widths ranging from 5μm to 110μm, using standard photolithography and wet etching process. Using MOVPE techniques AlN film of thickness 300nm was grown on the patterned substrates as shown in Figure 1.

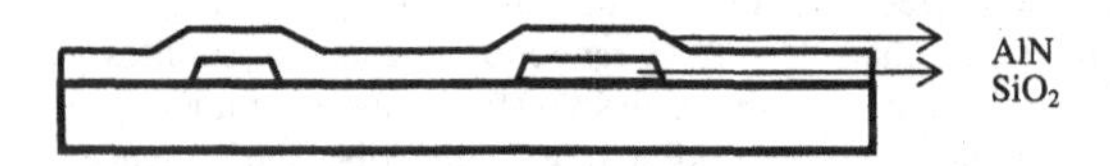

Figure 1. Process utilizing PECVD and wet etching techniques to fabricate SiO_2 stripes on Si(111) and GaN/sapphire substrates. MOVPE technique used to grown AlN films on the patterned structure.

In this study Renishaw In Via Raman spectrometer was used to conduct micro-Raman analysis at room temperature to determine the local variation in the microstructure quality and stress, across the width of the AlN/SiO_2 structures. Micro Raman spectroscopy offers high lateral spatial resolution. The laser excitation wavelength was at 514nm and was focused to a spot having a diameter between 2-3μm at a power of 40mW. Micro-Raman measurements were taken on AlN/SiO_2 structures after the growth. To determine the crystal orientation of the AlN films on the sacrificial layer, measurements were taken in the $z(y, y)\bar{z}$ backscattering geometry. The samples were subjected to wet etching in buffered HF (10:1), typically used to etch SiO_2 films consisting of NH_4F added to aqueous HF (10:1) in order to maintain the pH value of the solution, to determine the chemical stability of the AlN films on SiO_2. To observe this effect the edges of samples were masked with photo-resist to prevent lateral etching of SiO_2. Raman measurements were again taken across the width of the AlN/SiO_2 structures on the post-etched samples.

RESULTS AND DISCUSSION

The Raman active modes that are observed in the backscattering geometry $z(y,y)\bar{z}$ and $z(y,x)\bar{z}$ for wurtzite crystals, assuming that the c-axis is oriented parallel to the z axis direction with y and x axis perpendicular to the z axis, are the $E_2^H, E_2^L, A_1(LO)$ modes and E_2^H, E_2^L modes respectively [3]. With high sensitivity to stress the peak position of the E_2^H phonon mode is typically used to characterize the biaxial stress in group III nitrides such as AlN. In addition, the full width at half maximum (FWHM) of the phonon mode reflects the crystal quality [4].

In order to establish a baseline for the measurement of Raman spectra on AlN/SiO_2 for the current growth conditions, the spectrum of wurtzite AlN film on the crystalline substrates i.e. GaN/sapphire and Si (111) was taken in the backscattering configuration $z(y,y)\bar{z}$. In addition to the peaks from substrate material the other observable peaks were at 648 cm^{-1} and 652 cm^{-1} for the AlN films grown on GaN/sapphire and Si(111) substrates respectively as shown in Figure 2. These peaks correspond to the E_2^H phonon mode of wurtzite AlN. The red shift of the peaks relatively to the stress free value of 655 cm^{-1} for unstrained AlN [5-7] is an indication of the tensile stress in the AlN films relative to the underlying material. The FWHM of the phonon mode of AlN films on GaN and Si (111) was equal to 12.97 ± 0.3 cm^{-1} and 13.66 ± 1 cm^{-1} respectively. These FWHM values lie within the range of 3 cm^{-1} and 50 cm^{-1} for the reported value of highest quality AlN single crystal [8] and highly defective AlN crystal [9] respectively. This implies that even though the grown AlN films are of high quality they still contain significant defect concentrations. The AlN films grown on GaN have a lower FWHM value when compared to those grown on Si (111) implying that the AlN films grown on GaN have a low defect concentration and hence better quality. In addition the estimated FWHM of the lorentzian fitted phonon modes shows a variation of 0.3 cm^{-1} for AlN grown on GaN and 1 cm^{-1} for AlN grown on Si (111) further implying that the AlN films grown on GaN are of better quality when compared to the AlN films grown on Si(111). The apparent absence of the relatively weak $E_2^L, A_1(LO)$ could be due to the thickness of the film and low integration time.

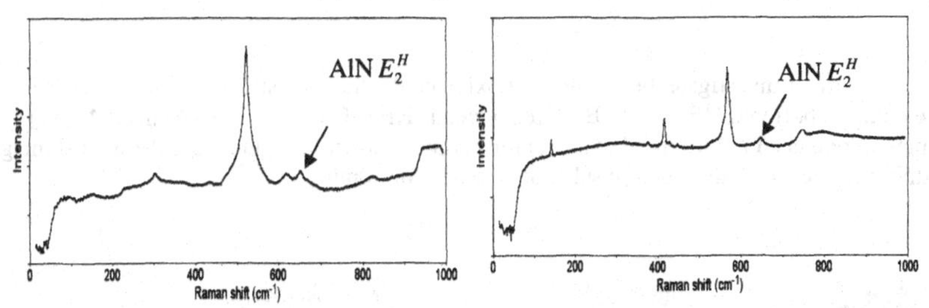

Figure 2. Left: Raman spectra of AlN film on Si (111), Right: Raman spectra of AlN on highly doped GaN/sapphire taken in the backscattering configuration $z(y,y)\bar{z}$ with the intensity in logarithmic scale. The observable peak of the AlN film is the E_2^H.

The SEM image analysis of the AlN films grown on SiO_2 showed the formation of randomly oriented crystallites on SiO_2 typical of polycrystalline films with an apparently rough texture when compared to the AlN films on the substrates. It was also observed that the size of the randomly oriented crystallites was smaller for AlN/SiO_2 structures grown on GaN when compared to the structures on Si (111). The AlN films grown on SiO_2 on GaN and Si (111) substrates did not exhibit pin holes. The AlN films in this work were grown at a temperature of 1000°C using the MOVPE process. Such high temperature CVD processes usually lead to strong oxidation of Al and the oxygen incorporation also varies with the thickness of the AlN film grown [10]. The AlN/SiO_2 interface is characterized by Al-O, Si-O and Si-N bonds [10]. The Raman spectrum of the AlN/SiO_2 structures on the Si (111) and GaN/sapphire did not yield any observable peaks related to c-axis oriented AlN. The presence of an oxidized Al phase existing in the AlN film and formation of randomly oriented crystallites could be attributed to this effect. Shown in Figure 3 is the Raman spectrum of AlN/SiO_2 structure on GaN/sapphire taken in the backscattering geometry $z(y,y)\bar{z}$. The raised baseline in the spectrum shown in Figure 3 is due to the background noise the origins of which can be traced to sources such as the underlying SiO_2 and possibly from the overgrown AlN film. The observable peaks are those corresponding to wurtizte GaN film and sapphire substrate. A similar trend was observed for the AlN/SiO_2 structure on Si (111) substrate with only the Si peaks observable.

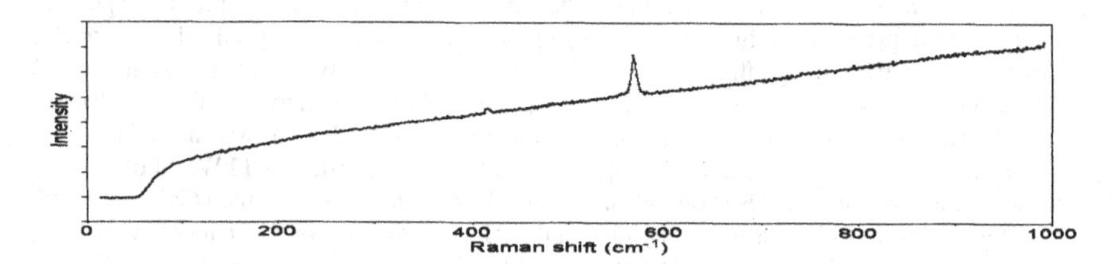

Figure 3. Raman spectrum of AlN_SiO_2 structure grown on highly doped GaN/sapphire taken in the backscattering configuration $z(y,y)\bar{z}$. The dominant observable peak corresponds to that of the wurtzite GaN film E_2^H.

To further investigate the presence of oxidized Al, the AlN structures were subjected to wet etching in buffered HF (10:1). Buffered aqueous HF is known to etch oxidized Al [11]. The samples were etched in buffered HF (10:1) for 30min. Shown in Figure 4 are the optical images of AlN structures on GaN/sapphire substrates etched for 30min.

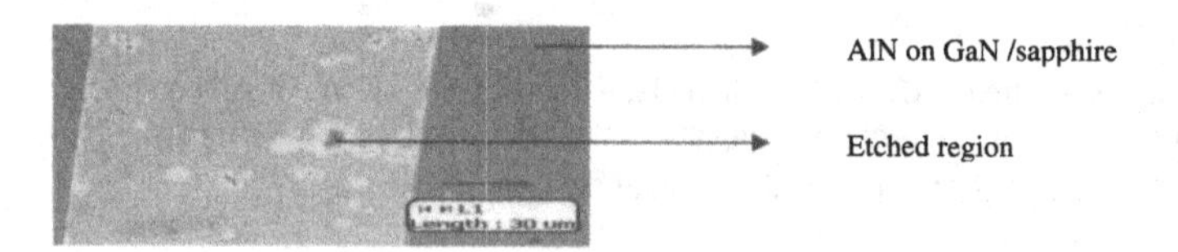

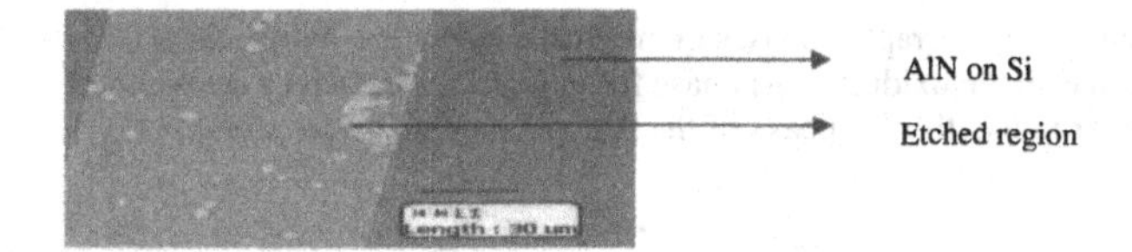

Figure 4. Optical images of AlN_SiO_2 structures on GaN/sapphire and Si (111) substrates etched for 30min BOE (10:1).

The fact that areas on the AlN structure have been etched lends credence to the argument that an oxidized Al phase exists in the AlN film. A few studies show that buffered HF etches polycrystalline AlN deposited by sputtering [12] but the effect of oxygen incorporation in the films on wet etching has not been addressed. Micro Raman analysis was conducted on the etched regions in the backscattering geometry $z(y,y)\bar{z}$. Show in Figure 5 is the Raman spectrum of AlN/SiO_2 structures on GaN/sapphire substrate corresponding to etched regions revealing the AlN E_2^H phonon mode, tentatively assigned based on its position in the spectrum. The baseline of the spectrum is lowered compared to Figure 3 indicating a reduction in the background noise. Similar results were observed from etched region for the AlN/SiO_2 structures on Si (111).

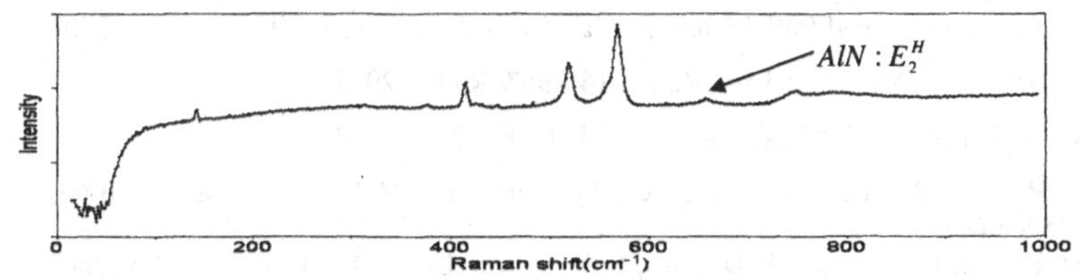

Figure 5. Raman spectrum of etched region 1 of AlN structures on GaN/sapphire substrate, taken in the backscattering geometry $z(y,y)\bar{z}$ with the intensity in logarithmic scale. The observable peak of the AlN film is the E_2^H.

CONCLUSIONS

This study shows the effects of growing AlN films grown on amorphous sacrificial layers such as SiO_2. The presence of a matrix in the AlN film grown on SiO_2 consisting of randomly oriented crystallites and with the possible formation of oxidized Al phase resulted in the suppression of wurtzite AlN phonon peaks in the Raman spectra during post growth measurements. Therefore the information regarding the built in residual stress and microstructure quality could not be extracted. Wet etching of the AlN/SiO_2 structures yielded wurtzite AlN like phonon modes in the AlN Raman spectra, pointing to the presence of oxidized Al phase in the AlN film. Further study using micro Raman spectroscopy needs to be undertaken to identify various crystallographic phases in the etched AlN film. Wet etching needs to be further pursued to etch the oxidized Al phase from the AlN film across the width of the AlN/SiO_2 structure in order to identify the peaks of the AlN phonon modes for stress analysis.

ACKNOWLEDGEMENTS

This technical effort was performed in support of the National Energy Technology Laboratory's on-going research in high temperature flow control hardware for advanced power systems under the RDS contract DE-AC26-04NT41817. This work was also supported in part by, AIXTRON and NSF RII contract EPS 0554328 for which WV EPSCoR and WVU Research Corp matched funds. The authors would like to thank Dr.Kolin Brown, Eric Schires, and Dr.Ting Liu for their guidance and material growth and characterization lab members for their technical support.

REFERENCES

1. V Cimalla, J Pezoldt and O Ambacher, J. Phys. D: Appl. Phys. **40**, 6386–6434 (2007).
2. L. E. Rodak, Sridhar Kuchibhatla, Ting Liu, P. Famouri and D. Korakakis; SympQ: Nitrides and Related Bulk Materials, Mater. Res. Soc. Proc. Vol.**1040E** (Dec –2007).
3. H. Harima ,J. Phys.: Condens. Matter. **14** R967–R993 (2002) .
4. V. Lughia and D. R. Clarke, App. Phys. Lett. **89**, 241911 (2006).
5. A. R. Goni, H. Siegle, K. Syassen, C. Thomsen, and J. M. Wagner, Phys.Rev. B **64**, 035205 (2001).
6. M. Kuball, J. M. Hayes, A. D. Prins, N. W. A. van Uden, D. J. Dunstan, Y.Shi, and J. H. Edgar, Appl. Phys. Lett. **78**, 724 (2001).
7. C. Bungaro, K. Rapcewicz, and J. Bernholc, Phys. Rev. B **61**, 6720 (2000).
8. K. Kuball, J. M. Hayes, Ying Shi, and J. H. Edgar, Appl. Phys. Lett. **77**, 1958 (2000).
9. P. Perlin, A. Polian, and T. Suski, Phys. Rev. B. **47**, 2874 (1993).
10. R.Berjoan, B. Armas, D. Perarnau, J. Phy IV (Colloque), **3**, n C3, 171-6 (Aug-1993). .
11. W.S.Yang, S.W.Kang, Thin.Solid.Film. **500**, 1-2, 231-6 (2006).
12. L.La Spina, H.Schellevis,N.Nenadovic, L.K.Nanver, 25th International Conference on Microelectronics (IEEE Cat. No. 06TH8868), 4pp (2006).

Mater. Res. Soc. Symp. Proc. Vol. 1139 © 2009 Materials Research Society 1139-GG03-47

Aluminum Nitride Thin Film Based Surface Acoustic Wave Sensors

A. Kabulski[1], V. R. Pagán[1], D. Cortes[2], R. Burda[2], O.M. Mukdadi[2], and D. Korakakis[1,3]
[1]Lane Department of Computer Science and Electrical Engineering, West Virginia University, Morgantown, WV, U.S.A.
[2]Department of Mechanical and Aerospace Engineering, West Virginia University, Morgantown, WV, U.S.A.
[3]National Energy Technology Laboratory, 3610 Collins Ferry Road, Morgantown, WV, U.S.A.

ABSTRACT

Surface Acoustic Wave (SAW) devices have drawn increasing interest in the field of sensor devices for the accurate detection of gases and liquids. In order to detect a range of gases and liquids, of which some will be capable of attacking the active material of the SAW device, a robust material is desired for fabrication. Aluminum Nitride (AlN) acts as a piezoelectric SAW material that can withstand both high temperatures and harsh environments. SAW devices have been fabricated by reactively sputtering an AlN film directly on silicon as well as on a freestanding SiO_2/Si diaphragm. The selective absorption of the gases or liquids modifies the mechanical properties of the surface of the device which can be detected as a change in resonance and phase velocity of the surface waves. Experimental analysis has been conducted to observe dispersion curves, Rayleigh velocities, and resonance frequencies and the results were found to qualitatively agree with those modeled by theoretical simulation. These results indicate the ability of further analyses to optimize the design and sensitivity of SAW sensors for medicine, engineering, and basic biology and chemistry sciences.

INTRODUCTION

Surface acoustic wave sensors operate on the principle that a change in the environment surrounding a SAW device will produce a measurable variation of one or more parameters of the wave propagation. Many types of SAW sensors have been designed around this concept and some of these designs include a special layer which absorbs or attracts a desired substance, causing a chance in the mechanical properties of the layer and consequently a change in the characteristics of the wave propagation which therefore allows for the detection of the substance [1]. Other sensors are based on the propagation of "leaky" surface acoustic waves and change the interaction between the sensor and the surrounding fluid rather than changing the properties of the sensor. These sensors allow for the detection of changes in the properties such as density, viscosity, etc. of the surrounding fluid [2].

Several types of surface waves have been applied to the design of ultrasonic SAW sensors. These surface waves include Rayleigh waves, which are characterized by an elliptical displacement of surface particles, shear-horizontal surface waves, which are characterized by an oscillatory displacement of surface particles parallel to the surface and perpendicular to the propagation, and Lamb waves, which are similar to Rayleigh waves but with a higher amplitude [3,4]. Typically, the displacements of surface waves are confined to the surface of the sensor with the penetration depth of the saves equal to about one wavelength. When the thickness of the substrate is finite (comparable to the wavelength), the propagated wave is a Lamb wave. For very high frequencies the wave velocity of these modes tends to the Rayleigh wave velocity. Sensors based on Lamb waves usually have a strong coupling with the surrounding fluid and a high sensitivity which allow for improved devices as compared to Rayleigh wave sensors [3]. In this work, the sensor was designed to monitor variations in the density of the surrounding fluid by detecting changes in the attenuation of the phase velocity of the wave [5,6].

EXPERIMENT

Lamb wave SAW devices are fabricated by depositing a thin layer of silicon dioxide (SiO_2) onto both sides of (100) double-side polished p-type silicon wafers using e-beam evaporation. Photolithography and chemical etching are used to create square "windows" in the SiO_2 layer on the backside of the wafer. The silicon substrate is chemically etched through the "windows" using a Preferential Silicon Etchant (PSE) leaving freestanding SiO_2 diaphragms.

Fig.1 SEM image showing backside of freestanding diaphragm

A thin, ~200 nm platinum layer is reactively sputtered using a CVC 610 DC magnetron sputter deposition system. The target used for Pt deposition is a 50 mm-diam., 99.999% Pt disk. The system base pressure of $5x10^{-6}$ mbar is routinely achieved before Ar gas is introduced into the chamber at a flow rate of 30 sccm. The Pt film is deposited at room temperature in a chamber pressure of ~60 mT using a power of 100 W applied to the Pt. target. After Pt deposition, an AlN layer, ~1.5 μm thickness, is sputtered on top of the Pt/SiO_2 layer comprising the diaphragms. For AlN deposition, a 200 mm-diam., 99.999% Al disk is used with N and Ar gases introduced into the chamber at flow rates of 27 and 3 s ccm respectively. The AlN film is also deposited at room temperature, but in a chamber pressure of ~30 mT using a power of 500 W applied to the Al target. Rayleigh wave SAW devices are fabricated by depositing AlN thin films directly on a silicon substrate using the same AlN deposition process described above.

In order to achieve the ~2μm feature sizes typical of the Pt IDTs, an extensive photolithograph process is needed. First, the samples are repeatedly cleaned using acetone and methanol and dried using nitrogen gas. A dehydration bake is then performed by heating samples to 110°C for at least 15 minutes using a hotplate. A hexamethyldisilazane (HMDS) adhesion layer is spun onto the sample at 4000 rpm for 40 seconds before spinning on AZ5214E image reversal photoresist at 4000 rpm for 40 seconds. The photoresist-covered sample is then placed on a glass slide and heated at 95°C for 1 minute using a hotplate. Before exposure is performed, the mask is cleaned using methanol and blow dried with nitrogen. A Karl Suss MA6 aligner is used to align the IDTs on top of the diaphragms and then expose the samples for ~25 seconds. After exposure, image reversal is achieved by heating the samples at 120°C for 2 minutes and then using a flood exposure over the samples for ~80 seconds. After reversal, development is performed by constantly agitating the samples in AZ 300MIF. The samples are then rinsed in DI water for one minute and the features can be inspected using an optical microscope. Once the development is verified, a thin, ~200nm, Pt layer is deposited on the surface of the sample using the sputtering process listed previously. The photoresist is then lifted off in order to leave behind the Pt IDTs as shown in Fig. 2.

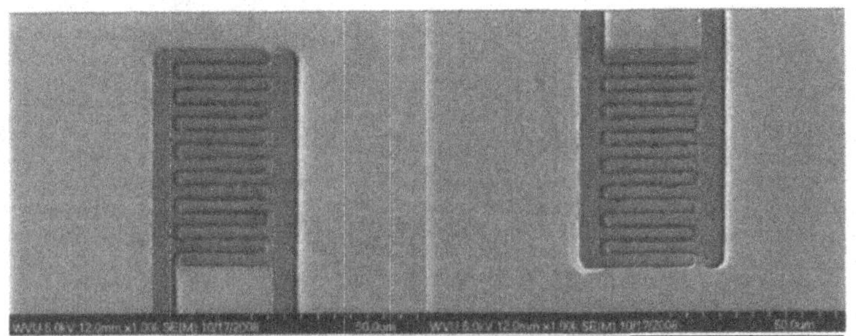

Fig. 2 SEM images showing pair of Pt IDT contacts deposited on AlN thin film

Characterization of the IDTs was performed using a four probe setup along with a Tektronix AFG3102 Arbitrary Function Generator to generate the input signal on one IDT and a Tektronix TDS2024B Digital Storage Oscilloscope to detect the output signal created on the other IDT. Continuous sine and pulse signals ranging from 100 kHz to 10 MHz were used as input signals. All measurements were performed in a controlled ambient environment.

DISCUSSION

Rayleigh wave SAW devices were fabricated by depositing an AlN thin film directly on a cleaned Si wafer. Due to the ease of fabrication, these samples were considered for initial characterization. Lamb wave SAW devices were also successfully fabricated with a significantly more complex fabrication process along with the difficult IDT processing. The Rayleigh wave SAW devices presented here consist of an AlN layer with a thickness of ~300 nm and Pt IDT contacts with thicknesses of ~200 nm.

A 10 V (peak to peak) continuous sine waveform was applied across one set up IDTs and the measured output signal is shown in Fig. 3. For these measurements, the frequency of the input signal was varied but a 100 kHz input signal is shown in these results.

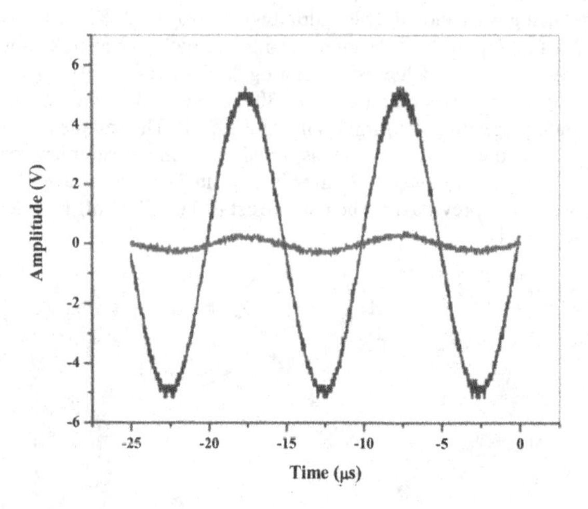

Fig. 3 Input/output response using continuous sine wave

As seen in the figure, there is a considerable amount of attenuation between the input and output signal though it is clear that the output does follow the input when using a continuous sine wave. This confirms the potential of the SAW device to function as a sensor for low frequencies. The attenuation of the input signal is influenced by the distance between the IDTs and the properties of the material. A higher quality (more crystalline) AlN film with less than 200 μm separating the IDTs could lead to an output signal with larger amplitude. Utilizing Lamb waves rather than Rayleigh waves can also contribute to a larger output signal.

A pulse input was used to determine the acoustic velocity of the device. A 10 V input pulse was generated across one IDT and the resulting output was measured on the other IDT contact. Fig. 4 shows that there is some delay between the input and output signals. This delay, also called the arrival time, is determined by the acoustic velocity of the material. A shorter delay corresponds to a greater velocity.

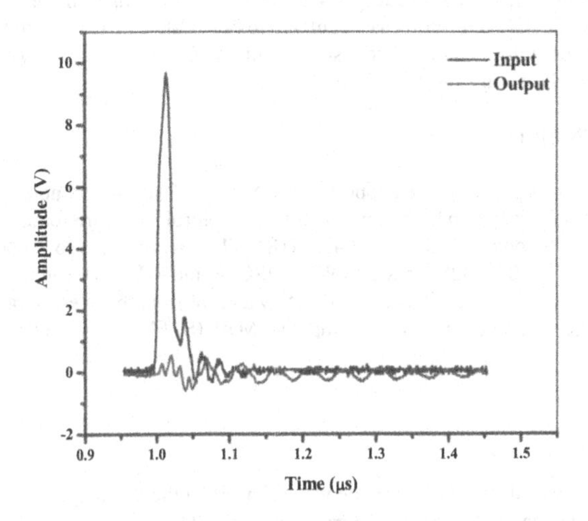

Fig. 4 Input/output response using single pulse

The experimental measurement of the Rayleigh wave velocity can be obtained by measuring the time required for the wave to travel the distance between IDTs. This time can be calculated from the wavelet transformation of the signal acquired from the receiving IDT. The delay measured in Fig. 4 is approximately 75 ns, which corresponds to an acoustic velocity of about 3300 m/s. The theoretical velocity of AlN has been calculated as anywhere between 5000 and 5600 m/s depending on the wave number and thickness of the AlN film [7]. While the experimental results are lower than the theoretical velocity, the decrease is due to the fact that the deposited AlN is polycrystalline rather than a perfect crystal as is considered in the theoretical prediction. The measured velocity, however, is consistent with theoretical predictions for a polycrystalline AlN film, which can be obtained by solving the motion equation with the proper electrical and mechanical boundary conditions and were estimated at approximately 3300 m/s [8].

CONCLUSIONS

AlN thin film SAW devices designed to operate as either Rayleigh or Lamb wave mode sensors were successfully fabricated. Characterization of the Rayleigh wave SAW devices indicated that an input signal can be detected across a distance of 180-240 μm with some amount of attenuation though the attenuation can be decreased by increasing the crystallinity of the AlN film or by switching to a Lamb wave device. The Lamb wave devices were fabricated by patterning and etching a Si substrate to create a freestanding diaphragm on which an AlN thin film is deposited. An alignment has been performed to place an IDT contact on each edge of the diaphragm though these structures have yet to be characterized. Literature indicates that the

Lamb wave devices will be exhibit larger amplitude and greater sensitivity. By using a single pulse input and calculating the delay in response measured on the output of the signal, the acoustic velocity of the AlN thin film was calculated to be ~3300 m/s, consistent with theoretical predictions for AlN of similar quality. The results in this work indicate the ability of further analyses to optimize the design and sensitivity of SAW sensors.

ACKNOWLEDGMENTS

This technical effort was performed in support of the National Energy Technology Laboratory's on-going research in high temperature flow control hardware for advanced power systems under the RDS contract DE-AC26-04NT41817. This work was also supported in part by NSF RII contract EPS 0554328 for which WV EPSCoR and WVU Research Corp matched funds. V.R.P. was supported in part by a grant from the West Virginia Graduate Fellowship in Science, Technology, Engineering, and Math (STEM) program and the WVNano Bridge Award.

REFERENCES

1. T. M. A. Gronwold, "Surface acoustic wave sensors in the bioanalytical field: Recent trends and challenges," *Analytica Chimica Acta*, vol. 603, pp. 119-128, 2007.
2. G. Lindner, "Sensors and actuators based on surface acoustic waves propagating along solid-liquid interfaces," *Journal of Physics D: Applied Physics*, vol. 41, pp. 123002, 2008.
3. D. S. Ballentine, R. M. White, S. J. Martin, A. J. Ricco, G. C. Frye, E. T. Zellers, and H. Wohltjen, *Acoustic Wave Sensors: Theory, Design and Physico-Chemical Applications*, vol. Academic Press. San Diego, 1997.
4. B. A. Auld, *Acoustic Fields and Waves in Solids*, vol. II, 2nd ed. Malabar, FL: Krieger, 1990.
5. S. W. Wenzel and R. M. White, "A multisensor employing an ultrasonics Lamb-wave oscillator," *IEEE Transactions on Electron Devices*, vol. 35, pp. 735-743, 1988.
6. R. M. White and S. W. Wenzel, "Fluid loading of a Lamb-wave sensor," *Applied Physics Letters*, vol. 52, pp. 1653-1655, 1988.
7. K. Tsubouchi, K. Sugai, and N. Mikoshiba, "AlN Material Constants Evaluation and SAW Properties on AlN/Al_2O_3 and AlN/Si," *IEEE Ultrasonics Symposium*, pp. 375-380, 1981.
8. A.N. Nayfeh, *Wave propagation in layered anisotropic media*. Amsterdam: Elsevier. 1995.

Mater. Res. Soc. Symp. Proc. Vol. 1139 © 2009 Materials Research Society 1139-GG03-48

Active Field Effect Capacitive Sensors for High-Throughput, Label-Free Nucleic Acid Analysis

Manu Sebastian Mannoor[12], Teena James[12] Dentcho V. Ivanov[12], Bill Braunlin[3] and Les Beadling[3]
[1]Microelectronics Research Center, New Jersey Institute of Technology, Newark, NJ, USA.
[2]Department of Biomedical Engineering, New Jersey Institute of Technology, Newark, NJ, USA
[3]Rational Affinity Devices LLC, Newark, NJ-07103, USA

ABSTRACT

We report a highly selective technique for rapid and label-free analysis of nucleic acid sample using Metal Oxide Semiconductor (MOS) capacitive sensors. The binding of charged macromolecules such as DNA on the surface of these Field Effect Devices modifies the charge distribution in the Semiconductor (Si) region of the sensor. These changes are manifested as a significant shift in the Capacitance-Voltage (C-V) characteristics measured across the device. The speed and selectivity of the detection process is enhanced by the use of external electric field of controlled intensity. This simple and high-throughput sensing technique holds promises for future electronic DNA arrays and Lab-on-a Chip devices.

INTRODUCTION

The monitoring and detection of Deoxyribonucleic acid (DNA) re-association by base pairing (hybridization) at solid-liquid interfaces is of utmost importance for the development of high-throughput sequencing, genetic screening and biomolecular sensing with great potential application in future drug and diagnostic development. For instance, an increasingly large number of infectious diseases are diagnosable by molecular analysis of nucleic acids[1]. The hybridization event can be used as an analytical technique when the sequence of one of the members in the double stranded complex is known, as it infers the presence of a sequence complementary to the known sequence, in a sample of unknown nucleotide sequences. Several methods have been developed for the label-free electrical detection of oligonucleotide hybridization, among which Field-Effect devices utilizing the intrinsic molecular charge of the nucleotide sequences for the generation of the sensing signal, receives special attention. The compatibility of these devices with the standard micro and nano fabrication techniques makes them cost-effective. Most of the successful research work in this direction have made use of a Field Effect Transistor (FET) structure[2].

In this work, we demonstrate the potential of a simple and easy to fabricate Metal Oxide Semiconductor (MOS) capacitive structure in attaining selective detection of oligonucleotide hybridization. Our experiments show that there is almost no advantage in using a MOSFET (Metal Oxide Semiconductor Field Effect Transistor) structure to obtain the information equally accessible with a MOS capacitor, which is much easier to fabricate. In particular, the simultaneous shift in the C-V characteristics of a MOS capacitive sensor along both the capacitance and voltage axis makes the C-V method more informative than a static DC measurement on the transistor structure.

We also demonstrate the possibility of enhancing the speed and selectivity of the molecular detection process electronically, by the application of an external electric field of appropriate bias and intensity. Selective melting or dehybridization of the target-probe pair, depending on the degree of their complementarity has been achieved by controlling the electric field to the appropriate level. This active electronic bio-sensing technique is capable of providing massively parallel, highly selective, monitoring of a wide variety of biomolecular interactions.

THEORY

Biomolecular Sensing Principle

During the immobilization of DNA molecules, the intrinsic negative charge due to its phosphate backbone will effectively alter the surface potential at the gate metal (Au) which induces a change in charge distribution in the silicon underneath as described above. The presence of an additional charged molecular layer due to hybridization enhances this effect. This charge redistribution in silicon layer due to the presence of biomolecules modifies the depletion capacitance ($C_{depletion}$) and is reflected as a shift along the capacitance axis of the C-V characteristics. The immobilization and hybridization event can be electrically modeled as the transfer of a certain quantity of charge from the solution to the gate metal (Au).The use of a thin film of gold as the gate material in our sensing structure offers several advantages. The chemical inertness of gold protects it from getting oxidized by the electrolytic environment. Moreover, the chemistry of covalent immobilization of probe sequences using alkyl thiol linkage has been well studied and the immobilization is easily achievable.

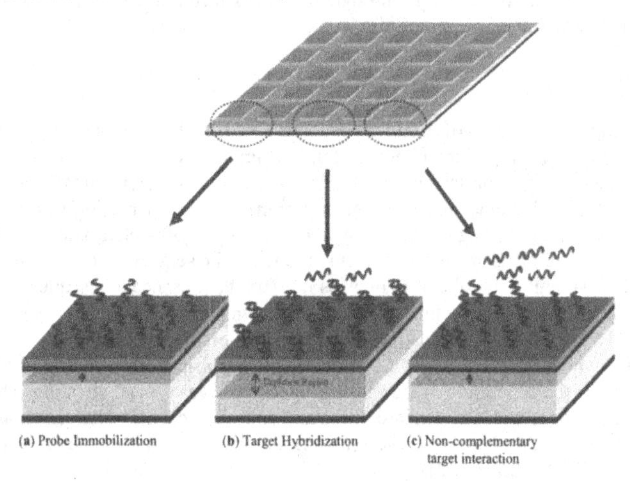

Figure 1. Schematic showing the biosensing principle of the Field Effect Sensing devices. The immobilization of the negatively charged single stranded probe oligomers (green) on to the gate surface (Au) extends the depletion region in the Silicon (a). Hybridization of the complementary target sequences (red) to the immobilized probe sequences further increases the depth of the depletion region in Si (b). Exposure of the bio-functionalized sensor to non-complemetary target sequences (blue) produces negligible change in the depth of the depletion region (c).

Electric field assisted rapid and highly selective detection technique

Our experiments demonstrate that the rate and selectivity of hybridization reaction can drastically be improved by the application of an external electric field. Since oligomers in solution carry a net negative charge, they can be transported towards the probe molecules immobilized on the sensor surface (Au) by the application of a positive bias on the Au surface with respect to an electrode in solution. The movement of the target nucleotide molecules toward the immobilized probe sequences, facilitated by electric field can result in a concentrating effect of the target molecules near the surface enabling the binding of probe and target sequences at a much higher rate.

The electric field induced transport of nucleotide molecules is also used for enhancing the selectivity of the sensing process. The sensor selectivity depends on the specificity of the binding between the target and probe sequences. The unhybridized target molecules which stay non-specifically bound on the sensor surface (Au electrode) will also contribute to the sensing signal. By applying an appropriate negative bias on the Gold (Au) surface these nonspecific target nucleotide molecules is released and repelled away from the sensing area, thereby eliminating their effect on the sensing signal.

By adjusting the electric field to the appropriate level, selective dehybridization or melting of the target-probe pair is also demonstrated, which is promising for single nucleotide polymorphism (SNP) analysis. Similar technique has been developed by Nanogen Inc. for the optical detection of hybridization [3-5]. Fixe et.al has demonstrated the use of electric field pulses to attain rapid hybridization of probe-target sequences [6]

EXPERIMENTAL

Materials

DNA oligonucleotides including the thiol labeled probe sequences used for the experiments were purchased from IDT (Integrated DNA Technologies). 20xSSC buffer solution (3.0M Sodium Chloride + 0.3M Sodium Citrate) used for the hybridization assay was purchased from Sigma-Aldrich. The chemicals and other materials used for microfabrication of the device were obtained from Microfabrication Center at New Jersey Institute of Technology. Single stranded DNA sequences pre-modified by thio linker (5'-CACGTAGCAG/3 - $(CH_2)_3SH$-3') were immobilized on the gold metal gate using a concentration of 10µM in 0.05xSSC buffer. By taking advantage the high affinity of sulphur atoms to gold substrate the DNA molecules with thiol end groups are chemically assembled onto the gold surface from the solution[7, 8].

Sensor Fabrication and Measurement Setup

The MOS capacitive sensors used in this work were fabricated on n-type Silicon wafer with resistivity 10 ohm-cm. A thermal oxide layer of <10 nm was grown as the gate insulator. The sensitivity towards gate surface potential changes was enhanced by the use of this extremely thin layer of gate insulator[9]. Circular regions of 150nm thick, 2mm diameter gold regions were deposited as the gate electrode. A 10nm Titanium layer was used to promote the adhesion of gold to SiO_2. Back side ohmic contact was provided by a 150nm layer of Al on silicon. The performance of the MOS capacitive sensors in monitoring biomolecular interactions were studied by analyzing Capacitance-Voltage (C-V) characteristics across the device. The C-V characteristics were obtained using HP4145B Semiconductor Parameter Analyzer at frequencies of 100 Hz and 1 kHz frequencies. Capacitance-Voltage (C-V) curves were obtained by varying the gate voltage from 1.5V (accumulation) to -1.5V (inversion) through the depletion region.

RESULTS

The Capacitance-Voltage characteristics for the bio-functionalization procedure were monitored in real-time. Figure 2a shows the C-V characteristics at various time intervals of the MOS capacitive sensor during the immobilization procedure. The constant shift of the characteristics to the side of positive gate voltage confirms the presence of more negative charge on the gate surface. A shift of 140 mV is observed due to immobilization after the exposure of the gold electrode to the thiol labeled probe sequences for 25 mins. The substrates with immobilized oligomers were then allowed to interact with 1 μM concentration of complementary oligomers (5'-CTG CTA CGT G-3') over a short period of time. Figure 2b shows the C-V characteristics of the MOS structure during hybridization event at various time intervals. Due to the hybridization of complementary sequences, a further shift of 73 mV towards the positive side is observed after 20 mins.

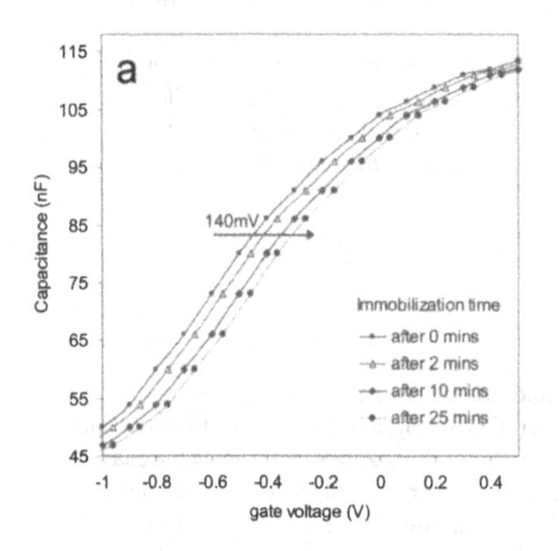

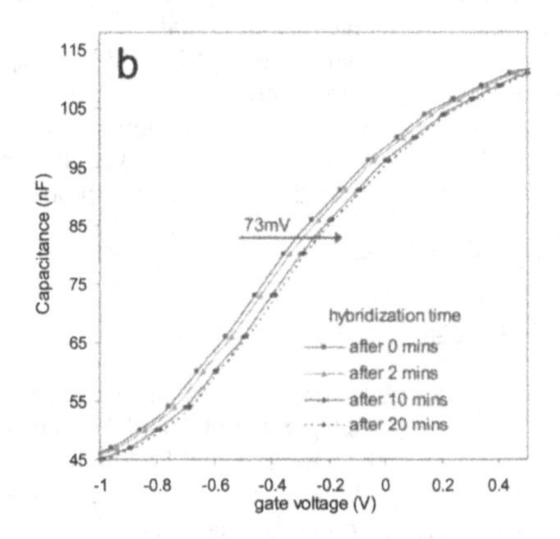

Figure 2a. Capacitance-Voltage characteristics measured at various time intervals during the immobilization of single stranded probe DNA sequences. **2b**. Capacitance-Voltage characteristics measured in real-time during the hybridization of immobilized probe sequences with complementary target sequences.

This agrees with the presence of an additional layer of negatively charged DNA molecules. As a control experiment, non-complementary target sequences (5'-ATG GCC CTG T-3') of same concentration as the complementary target solution is allowed to interact with the immobilized probe layer. Fig. 3a shows negligible change in the dielectric property upon exposure to the non complementary sequence for a time period of 25mins. This supports the relationship between capacitance change and specific nucleotide interaction. The immobilization and hybridization experiments has been repeated with the same concentration of probe and complementary target oligomer sequences under the effect of 0.3 to 0.7 V applied between the Al back contact and a Pt electrode immersed in the sample solution.

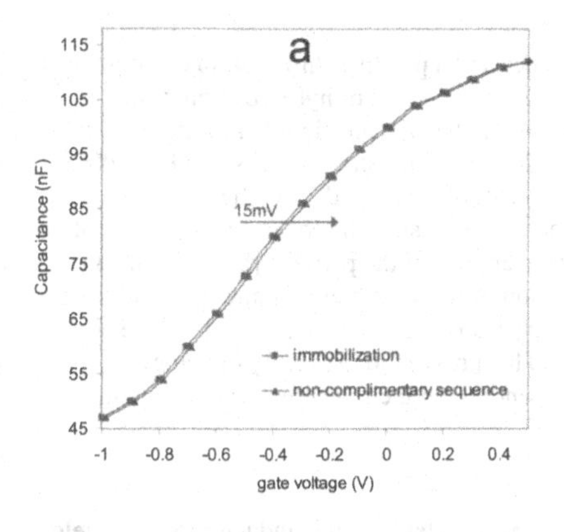

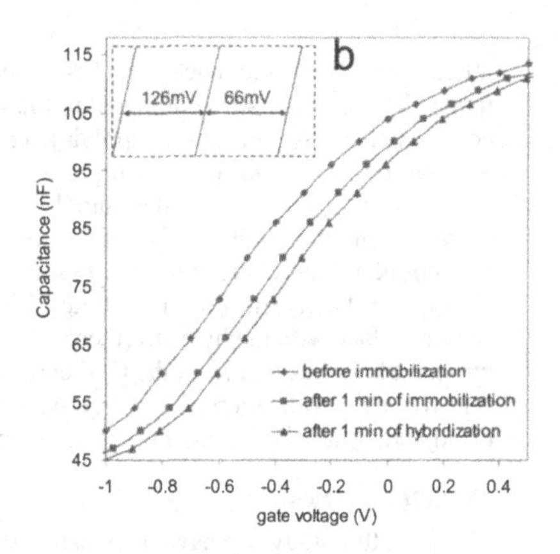

Figure 3a. C-V characteristics as a result of the interaction of immobilized probe sequences with non-complementary targets. **3b**. C-V characteristics of the Field Effect Capacitive sensor after 60s of electric Field assisted immobilization and hybridization procedures.

Fig. 3b shows the results for the field assisted immobilization and hybridization procedures. The electric field assisted transport of thiol-labeled probe oligomers towards the gold surface resulted in the attainment of immobilization at a much faster rate. A positive shift of 126 mV is observed after the exposure of the sensor surface to the probe molecules for 1 min, under the presence of electric field. The localized concentrating effect of the target molecules near the immobilized probe surface resulted from the field assisted transport enabled the binding of probe and target sequences at a much higher rate. From Fig. 3b, a positive shift of about 66 mV is observed after interaction of probe molecules with the complementary target sequences for 1 minute.

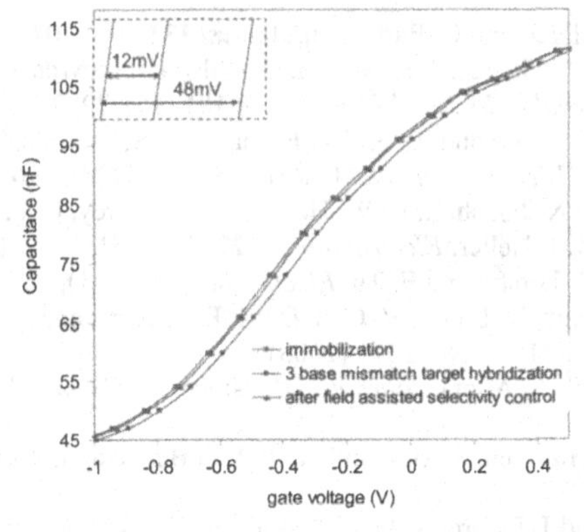

Figure 4. C-V characteristics showing the Field assisted selectivity improvement using 3 base pair mismatch target sequences.

The application of electric field is also used to enhance the selectivity of the molecular detection process. The sensor surface with the immobilized probe sequences was allowed to interact with target sequences of 3 base pair mismatches and a positive shift of 48mV is observed in the C-V characteristics compared to the shift of 73 mV as a result of hybridization of the complementary sequences. By applying a negative bias on the gate metal with a potential varying from -0.3 V to -0.5 V a negative shift in the C-V characteristics was observed, resulting in a net shift of 12 mV from the immobilization characteristics. This indicates the dehybridization or melting of the partially hybridized 3 base mismatch sequences as a result of the applied negative bias. Fig. 4 shows the C-V characteristics of the partial hybridization and field induced dehybridization (melting). The application of negative bias of same potential to the sensor surface with the hybridized complementary target-probe sequences did not produce any measurable negative shift in the C-V characteristics. This proves the feasibility of achieving selective dehybridization of the target-probe pair, depending on their degree of complementarity by adjusting the electric field to the appropriate level.

CONCLUSIONS

In this study, we have demonstrated the use of Metal-Oxide-Semiconductor (MOS) Field Effect Capacitive sensors in providing label-free, real-time monitoring of oligonucleotide hybridization. The immobilization of probe oligomers on the sensor surface and their hybridization with the target oligomers of complementary sequences has produced a significant shift (140mV and 73mV respectively) in the Capacitance-Voltage characteristics measured across the device. The improvement in the detection speed and selectivity by the use of external electric field has also been demonstrated. The experiments with 3 base pair mismatches proved the possibility of obtaining selective melting of the hybridized nucleotide pairs depending up on their degree of complementarity. Future work will be concentrated on optimizing this technique to be able to discriminate single base pair mismatches and to improve the concentration sensitivity of the sensor. The active area of the sensor can be further miniaturized using standard photolithographic techniques promoting massively parallel screening of nucleic acid samples in array formats.

REFERENCES

1. B. Pejcic, R. D. Marco and G. Parkinson, *Analyst* **131** (10), 1079-1090 (2006).
2. E. Souteyrand, J. P. Cloarec, J. R. Martin, C. Wilson, I. Lawrence, S. Mikkelsen and M. F. Lawrence, *Journal of Physical Chemistry B* **101** (15), 2980-2985 (1997).
3. C. F. Edman, D. E. Raymond, D. J. Wu, E. Tu, R. G. Sosnowski, W. F. Butler, M. Nerenberg and M. J. Heller, *Nucleic Acids Research* **25** (24), 4907-4914 (1997).
4. C. Gurtner, E. Tu, N. Jamshidi, R. W. Haigis, T. J. Onofrey, C. F. Edman, R. Sosnowski, B. Wallace and M. J. Heller, *Electrophoresis* **23** (10), 1543-1550 (2002).
5. M. J. Heller, A. H. Forster and E. Tu, *Electrophoresis* **21** (1), 157-164 (2000).
6. F. Fixe, H. M. Branz, N. Louro, V. Chu, D. M. F. Prazeres and J. P. Conde, *Nanotechnology* **16** (10), 2061-2071 (2005).
7. R. G. Nuzzo and D. L. Allara, *Journal of the American Chemical Society* **105** (13), 4481-4483 (1983).
8. T. Wink, S. J. Van Zuilen, A. Bult and W. P. Van Bennekom, *Analyst* **122** (4), 43R-50R (1997).
9. E. H. Nicollian and J. R. Brews, *MOS Physics and Technology* Wiley, New York (2007).

Mater. Res. Soc. Symp. Proc. Vol. 1139 © 2009 Materials Research Society 1139-GG03-52

Growth of Epitaxial Potassium Niobate Film on (100)$SrRuO_3$/(100)$SrTiO_3$ by Hydrothermal Method and their Electromechanical Properties

Mutsuo Ishikawa[1], Shintaro Yasui[1], Satoru Utsugi[1], Takashi Fujisawa[1], Tomoaki Yamada[1], Takeshi Morita[2], Minoru Kurosawa[1] and Hiroshi Funakubo[1]

[1] Department of Innovative and Engineered Materials, Tokyo Institute of Technology, J2-1508, 4259 Nagatsuda-cho, Midori-ku, Yokohama, Kanagawa 226-8502, Japan

[2] Graduate School of Frontier Sciences, The University of Tokyo, 5-1-5 Kashiwanoha, Kashiwa, Chiba 277-8563, Japan

ABSTRACT

Epitaxially-grown $KNbO_3$ thick films over 8 μm in thickness were successfully obtained at 220 °C for 6 h on $(100)_c SrRuO_3//SrTiO_3$ substrates by a hydrothermal method. Epitaxial $SrRuO_3$ layers grown on $(100)_c SrTiO_3$ substrates by sputter method were used as bottom electrode layers. Relative dielectric constant and the dielectric loss were 530 and 0.11, respectively. Clear hysteresis loops originated from the ferroelectricity were observed and a remanent polarization was 25 μC/cm^2 at a maximum applied electric field of 540 kV/cm. In addition, the hydrothermal $KNbO_3$ thick film was able to transmit and receive of the ultrasonic waves over 50MHz.

INTRODUCTION

Lead-free piezoelectric materials have been widely investigated for the requirement of the exclusion of the toxic element such as lead, which is essential one for the present high performance piezoelectric materials, for example PZT. $KNbO_3$ [1], $KNbO_3$-based and related materials, such as (K, Na)NbO_3 [2] and $(1-x)$ $(K_{0.5}Na_{0.5})NbO_3 - xLiNbO_3$ [3], have been widely investigated. Films of $KNbO_3$ or $KNbO_3$-based materials have been reported by several methods [4-6], however the difficulty of the K/Nb ratio control was pointed out due to the high volatility of potassium, especially for the deposition methods operated under vacuum condition.

Hydrothermal method is a unique method for growing the ferroelectric materials because the films were grown under high pressure instead of the widely investigated low pressure and crystalline films were grown at relatively low temperature [7]. In addition, excellent conformability is expected due to their inhomogeneous nucleation on the substrate surface, which results in the conformal growth even on the surface with complex shape. However, the electrical property of the hydrothermally grown $KNbO_3$ films has been hardly reported, even though their epitaxial films were reported [8, 9]. In the present study, epitaxial $KNbO_3$ films were grown on (100)$SrRuO_3$//(100)$SrTiO_3$ substrates and the ferroelectric properties were ascertained for the first time. Moreover, the transmitting and receiving of ultrasonic waves over 50MHz using epitaxial $KNbO_3$ films were demonstrated.

EXPERIMENTAL PROCEDURE

The $KNbO_3$ thick films were grown on the $(100)_c$ $SrRrO_3$// (100)$SrTiO_3$ substrates at the deposition temperature of 220 °C by the hydrothermal method. Epitaxial $(100)_c$-oriented $SrRuO_3$ layers were grown on (100)$SrTiO_3$ substrates by sputtering methods and were used for the bottom electrodes, $(100)_c$ $SrRrO_3$// (100)$SrTiO_3$ [10] .

The 30 *ml* solution of potassium hydroxide (KOH, Kantoukagaku Co., Ltd.) and 1.6 g powders of niobium oxide (Nb_2O_5, purity 99.9%, Kantounkagaku Co., Ltd.) were put in an autoclave (PARR, 4748) as the source materials together with the $(100)_c$ $SrRrO_3$//$SrTiO_3$ substrates. The autoclave was closed to retain water vapor and was placed in a constant-temperature oven for the hydrothermally chemical reaction for 6 h.

The film composition was measured with X-ray fluorescence spectroscopy (XRF, HORIBA, 7593H). The thickness of the obtained film was determined by a scanning electron microscopy (SEM, HITACH S-4800) and their crystal structure was characterized with high resolution X-ray diffraction (HRXRD, Philips X'Pert MRD system) analysis using a four-axis diffractometer with Cu $K\alpha_1$ radiation. The electrical properties were measured at room temperature as Pt/$KNbO_3$/$SrRuO_3$ capacitors after making platinum top electrodes with 100 μm in diameter, which were deposited by the electric beam evaporation.

The characteristics of the transmitting and the receiving of the ultrasonic wave were investigated for the hydrothermally-grown $KNbO_3$ thick films grown on the $(100)_c$ $SrRrO_3$//$SrTiO_3$ substrates. The $SrRuO_3$ layer was used as the bottom electrode, while Au layer with a 1×6 mm^2 rectangular shape was vacuum-evaporated on the surface of the $KNbO_3$ thick films as the counter electrode. An Au/ $KNbO_3$/$SrRrO_3$/$SrTiO_3$ transducer was clamped to a test-fixture in brass and both electrodes were connected to Palser-Receiver (PANAMETRICS, 5900PR, 200MHz bandwidth) for the high frequency transmitting and receiving of the ultrasonic wave. The received signal was amplified at 26 dB and was filtered with high-pass filter at 10 MHz because the fundamental resonance frequency of the expected thickness vibration mode, which could be calculated from $SrTiO_3$ thickness, was estimated to be higher than this frequency.

RESULTS AND DISCUSSIONS

The K/Nb ratio of the obtained films was evaluated to be 51/49, suggesting that the films with almost stoichiometric composition were obtained in the present condition. X-ray θ - 2θ pattern is shown in Fig. 1 for the $KNbO_3$ film grown on $(100)_c$ $SrRrO_3$//$SrTiO_3$ substrate. Two sprit peaks were observed at around 22 ° and 45 ° together with those from the substrate. The lattice constants calculated from each sprit peaks were found to be approximately 4.06 Å and 4.00 Å. The reported data [11 ,12] suggest that the lattice parameters of (100)/(001) tetragonal $KNbO_3$ are 4.062 Å/3.993 Å, while those of (100)/(110) orthorhombic $KNbO_3$ with pseudo cubic unit cell are 4.036 Å/3.974 Å, respectively. The detail crystal structure analysis of the obtained film is under investigated because the lattice parameters of the obtained films are possible to be strained by the clamping force from the substrate.

X-ray pole figure measured at a fixed 2θ angle at 31.5° is inserted in Fig.1, corresponding to the tetragonal {110} or orthorhombic {111}. It had a four fold symmetry at inclination angle of about 45°. This suggests the cube-on-cube epitaxial growth of this film.

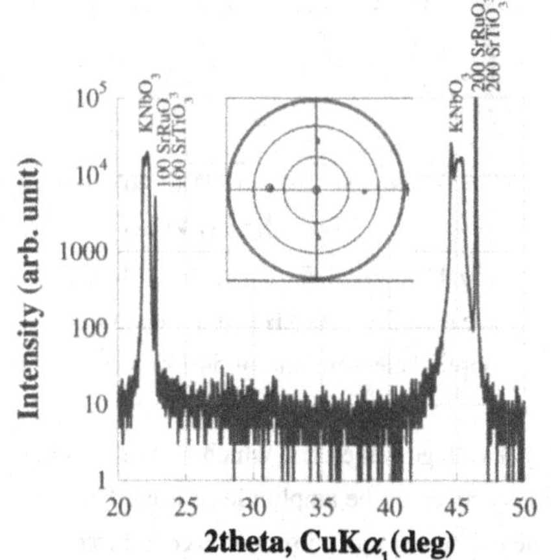

Fig.1 XRD pattern of $KNbO_3$ film deposited on $(100)_c SrRuO_3$//$SrTiO_3$ substrate together with the X-ray pole figure measurement fixed at 2θ of 31.5°.

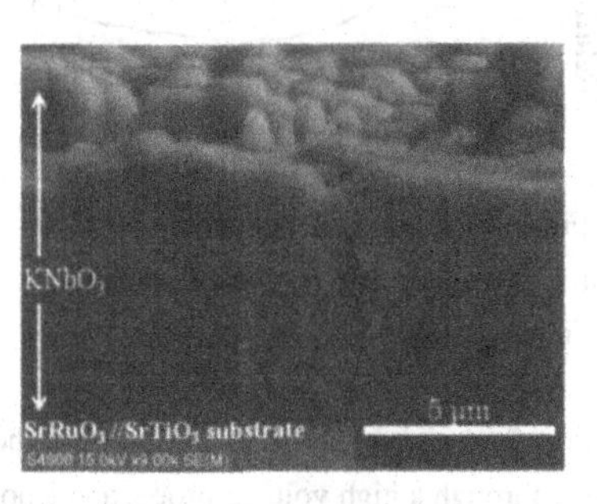

Fig.2 Cross-sectional SEM image of 8 μm-thick epitaxial $KNbO_3$ film deposited on $(100)_c SrRuO_3$//$SrTiO_3$ substrate.

Figure 2 shows a cross-sectional SEM image of the same $KNbO_3$ film whose x-ray profile is shown in Fig.1. Relatively dense film with large thickness over 8 μm was ascertained to be grown on $(100)_c$ $SrRrO_3$//$SrTiO_3$ substrate. Figure 3 shows the frequency dependency of the relative dielectric constant, ε_r, and the dielectric loss of the Pt/(8μm-$KNbO_3$)/$SrRrO_3$/ capacitor. The ε_r and the dielectric loss at 100 kHz were 530 and 0.11, respectively. The ε_r was about 50% and the dielectric loss value was three times larger compared to the reported ones for the sintered

$KNbO_3$ [13]. The $P-E$ relationships measured at 100 kHz at room temperature is shown in Fig. 4 for the Pt/(8μm-$KNbO_3$)/$SrRrO_3$/ capacitor. Clear hysteresis loops originated from their ferroelectricity were observed for the first time as the hydrothermally - grown epitaxial thick film, even though the contribution from the leakage was not perfectly negligible. Observed remanent polarization, P_r, was 25μC/cm^2 at the maximum electric field of 540 kV/cm and this P_r value was almost the similar to the reported one for the sintered body [13]. The results shown in Fig.4 indicated that the present $KNbO_3$ film could drive for the ultrasonic transmitting over 100 kV/cm, so that the characteristics of transmitting and receiving of ultrasonic wave of the hydrothermal oriented $KNbO_3$ film were investigated as the next step.

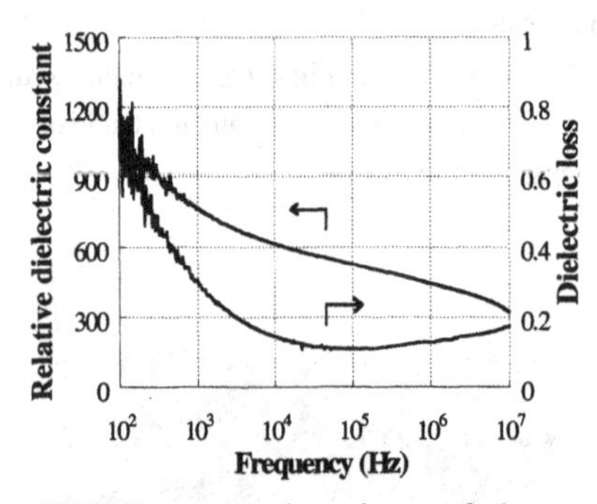

Fig.3 Frequency dependence of the relative dielectric constant and the dielectric loss.

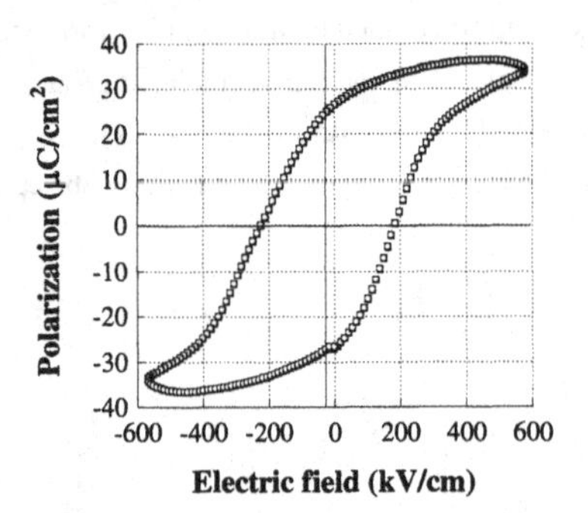

Fig.4 $P-E$ hysteresis loops measured at 100kHz and maximum applied electric field of 540 kV/cm.

Figures 5 (a) and (b) show an amplitude of the input voltage waveform, which was measured directly through a high voltage-probe, and a power spectrum of the amplitude measured by fast Fourier transform (FFT). The signal was confirmed to be a single pulse with a center frequency of 25MHz.

The transducer was driven with the aforementioned single pulse. The generated ultrasonic signal, which was excited by the piezoelectric and the electrostrictive effects at the epitaxial $KNbO_3$ film, was propagated into the $SrTiO_3$ substrate. After this process, the ultrasonic signal was reflected at the boundary between the $SrTiO_3$ substrate-edge face and air. This reflected ultrasonic signal was received by the hydrothermal epitaxial $KNbO_3$ film.

Figures 6 (a) and (b) show the amplitudes of the receiving voltage waveform and the power spectrum of the obtained waveform by FFT, respectively. As shown in Fig.6(a), the propagation time of the reflected wave into the $SrTiO_3$ substrate was observed at 143 ns after the driving signal was applied. This period time, 143 ns, is in agreement with the propagation time into the $SrTiO_3$ substrate because the half of the travel distance within the $SrTiO_3$ substrate is estimated to be 7900 m/s × 143 ns × 1/2 = 0.556 mm, which was almost the same with the thickness of the $SrTiO_3$ substrate. The power spectrum shown in Fig. 6(b) consisted of several peaks at several frequencies; the peak at approximately 25MHz is the feed through noise as the transmitting, and the components distributed from 50 MHz to 200MHz represent the acoustic signals. This result clearly demonstrates the transmitting and the receiving of the ultrasonic wave.

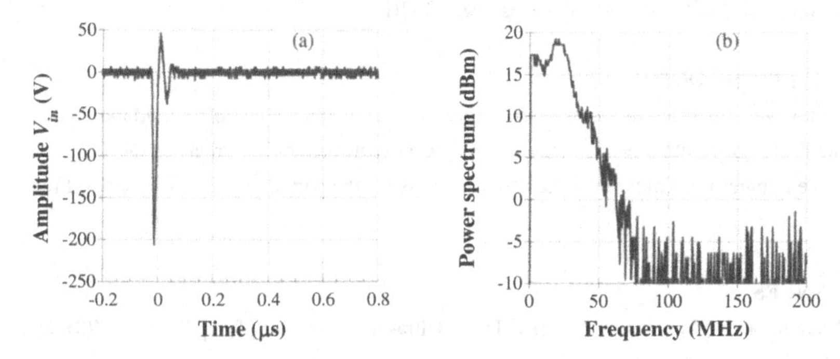

Fig.5 Amplitude of the input pulse voltage (a) and the power spectrum of the pulse signal (b).

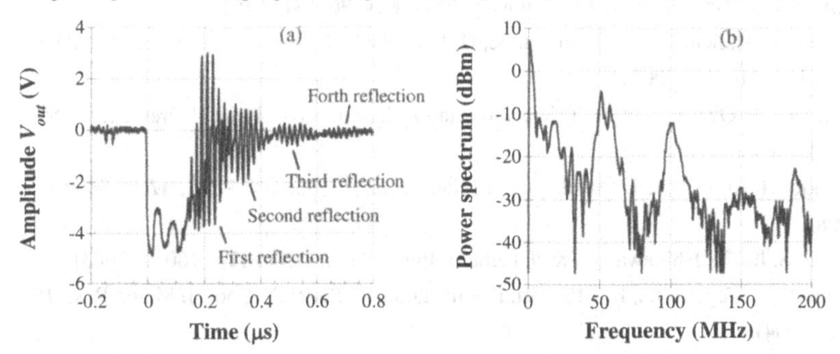

Fig.6 Amplitudes of the received waveform (a) and the power spectrum of the ultrasonic wave form (b).

These results demonstrate that the epitaxial grown 8 μm-thick $KNbO_3$ film at the low process temperature of 220 °C, although there are no reports on $KNbO_3$ thick films over 5μm in thickness by the PLD method or the sputtering method. Furthermore, the clear hysteresis loops and the characteristics of transmitting and receiving of ultrasonic wave over 50MHz were obtained for the first time. This performance indicates the excellent performance of the growth of $KNbO_3$ films even at low temperature process of the hydrothermal methods.

CONCLUSIONS

The epitaxially-grown $KNbO_3$ thick films over 8 μm in thickness were successfully obtained on the $(100)_c SrRuO_3//SrTiO_3$ substrates at 220 °C for 6 h by the hydrothermal method. The dielectric constant ε_r and dielectric loss were 530 and 0.11, respectively. The clear hysteresis loops originated from the ferroelectricity was observed and the P_r was 25 $\mu C/cm^2$ at the maximum applied electric field of 540 kV/cm. The hydrothermal $KNbO_3$ thick film was able to transmit and receive of the ultrasonic waves over 50 MHz.

ACKNOWLEDGEMENTS

This study was supported by Inoue Foundation for Science and the authors would like to thank Associate Professor Kentaro Nakamura and Technical department of the Precision and Intelligence Laboratory, Tokyo Institute of Technology for the partial measurement support for this work.

REFERENCES

1) K. Nakamura, and Y. Kawamura, IEEE Trans. Ultrason., Ferroelec., Freq., Contr., **47**(3), 750 (2000).
2) G. Shirane, R. Newnham, and R. Pepinsky, Phys. Rev. **96**, 581 (1954).
3) N. Klein, E. Hollenstein, D. Damjanovic, H. J. Trodahl, N. Setter, and M. Kuball, J. Appl. Phys., **102**, 014112 (2007).
4) K. Shibata, F. Oka, A. Ohishi, T. Mishima, and I. Kanno., Appl. Phys. Express, **1**, 011501 (2008).
5) S. Kakio, H. Kurosawa, T. Suzuki, and Y. Nakagawa, Jpn. J. Appl. Phys., **47**(5), 3802 (2008).
6) T. Arai, S. Ito, K. Ishikawa and K.Nakamura, Jpn. J. Appl. Phys., **42**(9), 6019 (2003).
7) T. Morita, Y. Wagatsuma, H. Morioka, H. Funakubo, N. Setter, Y. Cho., J. Mater. Res., **19**, 1862 (2004).
8) Z. B. Wu, T. Tsukada, M. Yoshimura, J. Mater. Sci., **35**, 2833 (2000).
9) W. L. Suchanek, Chem. Mater., **16**, 1083 (2004).
10) T. Kamo, K. Nishida, K. Akiyama, J. Sakai, T. Katoda, H. Funakubo, Jpn. J. Appl. Phys., **46**(10), 6987 (2007).
11) E. A. Wood, Acta Cryst., **4**, 353 (1951).
12) N. Kumada, T. Kyoda, Y. Yonesaki, T. Takei, N. Kinomura, Jpn. J. Appl. Phys., **47**(5), 3802 (2008).
13) H. Birol, D. Damjanovic, and N. Setter, J. Am. Ceram. Soc., **88** (7), 1754 (2005).

Mater. Res. Soc. Symp. Proc. Vol. 1139 © 2009 Materials Research Society 1139-GG03-53

Bio-compatible micro-sensor for blood pressure measurement using SiC technology

Nupur Basak, G.L. Harris, James Griffin, K.D. Wise, (Howard Nanoscale Science and Engineering Facility, Department of Electrical and Computer Engineering, Howard University, 2300 Sixth Street, NW, Washington DC 20059, USA, K.D. Wise (Center for Wireless Integrated Microsystem, 2114-E, EECS,1301 Beal Avenue, Ann Arbor, MI 48109-2122, USA)

ABSTRACT

A monolithic SiC pressure sensor for blood pressure measurement has been investigated. For the implantable blood pressure measurement sensor, the sensor is wireless, biocompatible and linear. The SiC is used for its superior physical, chemical and mechanical characteristics. The sensor uses a wireless telemetry system that can be implanted in patient's body and coupled with an external inductor. The resonant peak method is used for sensing the changes in pressure. The sensor is made with planar inductor and a variable capacitance sensor. Together they make a resonant circuit, which loosely couples with an external inductor. This sensor gives the continuous, real-time measurement of the blood pressure.

There is strong evidence that SiC material is biocompatible when used as a coating on vascular stents [1]. The sensor must function optimally in the pressure range between 50-250 mmHg. The paper discusses the design, layout, modeling, fabrication and characterization of the blood pressure measurement sensor. The layout and modeling of the sensor is done using FEA (finite element analysis) software Coventor and Tanners L-Edit. The simulation for membrane deflection, stress analysis and electro-mechanical analysis were performed. Various fabrication techniques for fabricating the planar inductor and the capacitor are detailed. The fabricated planar inductor gives a value of 1.43μH at 13MHz. 3C-SiC has a Young's Modulus of about twice that of silicon (SiC =488-700 GPa) and is chemical inert. A larger Young's Modulus improves the sensitivity of the pressure sensor in this case. The real-time analysis for electromechanical characteristics of this sensor is done in a simulated laboratory set up. The changes in resonant peak frequency w.r.t. pressure are recorded.

INTRODUCTION

High blood pressure is one of the biggest killers in the United States and around the world. Though there are reliable ways of measuring blood pressure, there is a need for better and comfortable measurement. Patients whose blood pressure changes rapidly and critical patients who have undergone stoke, needs continuous diagnosis or measurement of the blood pressure An implantable device for continuous measurement of the pressure has been the topic of many researches in the area of microelectronics and Micro-electro-mechanical systems (MEMS) [1]. This devices could be made biocompatible and the size small enough to be inserted in the wall of the blood vessel without any complications. This paper reports on the fabrication and characterization of an implantable blood pressure sensor made with Silicon Carbide (SiC).

The blood pressure can be measured in millimeter of mercury (mmHg). The highest pressure at the beginning of the cardiac cycle is called the systolic pressure and the lowest pressure at the end of cardiac cycle is called diastolic pressure. For a normal healthy person the blood pressure should be 120 mmHg (16 kPa) systolic and 80 mmHg (11kPa) diastolic [2] but My Vary during the day. When the blood pressure is abnormally high (above 140/90) it is referred to as hypertension; and when it's abnormally low its called hypotension [2].

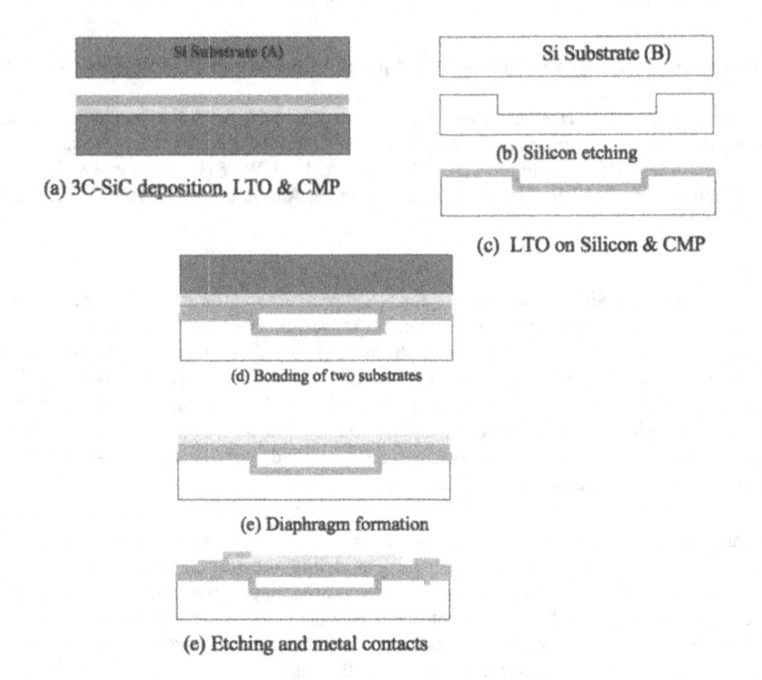

Figure 1a. Capacitor Fabrication Procedure

EXPERIMENT

The proposed pressure sensor in the study is wireless and made with a Silicon Carbide (SiC) epilayer structure. The sensor is capacitive with one electrode flexible for pressure measurement. A planar inductor is made with gold. This inductor acts as an antenna. The inductor and capacitor in parallel makes a wireless resonant circuit. SiC is considered the hardest material after diamond. Other than being the hardest material, it is also the least chemically reactive material with superior mechanical properties. For these reasons, it is a more reliable material for implantable biomedical devices. Because of the physical and chemical properties

SiC possesses, it is difficult to fabricate SiC MEMs or other electronic devices. Various studies have been done before to make sensors for measurement of the pressure [3]. Some uses piezorestive sensors and some capacitive sensor. The capacitive sensor can be connected in parallel to an inductor for data telemetry and hence the sensor becomes wireless [5]. The wireless telemetry system can be implanted in a patient's body and coupled with an external inductor. Together this system acts as a loosely coupled transformer consisting of an external antenna and the primary planar inductor implanted in the body. One of the main concerns in making a capacitive sensor is the output linearity so the capacitance is linear with blood pressure. [6]

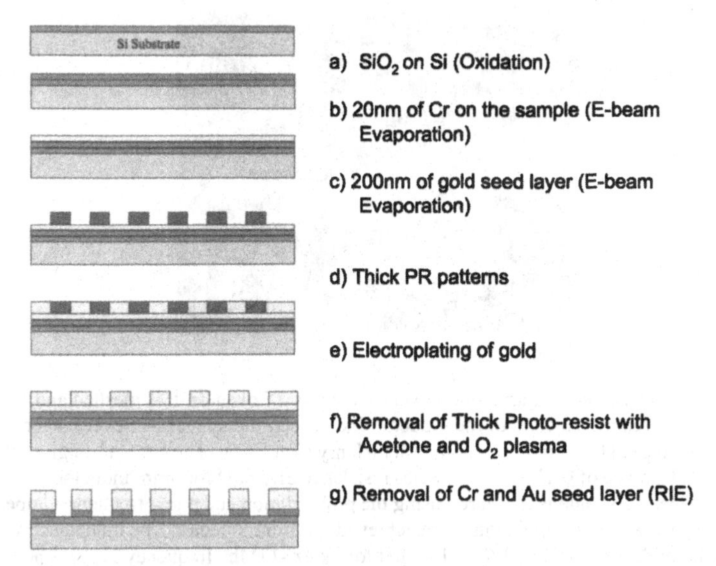

Figure 1b. Inductor Fabrication procedure

The Layout for first generation device was done using L-Edit. The mask consists of the coupling inductor, capacitor, MESFET, circular MESFET, diode and resistor. A separate mask is made again for the final sensor device once all the parameters were extracted from the first generation mask set. The sensor fabrication involved many standard fabrication steps like lithography, e-beam deposition, electroplating, chemical vapor deposition, wafer bonding, oxidation, wet etching and reactive-ion etching. A detail processing procedure is shown in Figure#1a and 1b for the capacitor and inductor. The 20nm thick Chromium (Cr) layer is deposited on the sample followed by a 200nm of Gold (Au) seed layer using e-beam evaporation. BPR-100 (Rohm-Haas) was used as the ideal photoresist because the Au plating of the inductor required a thick photoresist with near vertical sidewalls. The SiC layer was growth on the Si using a horizontal CVD process. Photoelectrochemical etching (PEC), was employed using UV light from a 550 W light source. The etch bath was a 10 % HF solution. The chemical mechanical planarization (polishing) was used to aid in the bonding of the SiO_2 layers.

DISCUSSION

The oxidation thickness of 1725Å was achieved with 2 hours of dry oxidation. The surface of the SiO_2 is considered very smooth with average roughness of 0.415 nm after polishing. Electroplating resulted in a deposition rate greater than 0.5 µm/min which gave a non-uniform porous surface whereas a deposition rate between 0.2-0.3 µm/min gave a more uniform and dense surface. After bonding the Si, the sample surface is etched away using TMAH at 85°C to release the membrane and this process worked well. Figure #2 shows the thin membrane of SiO_2.

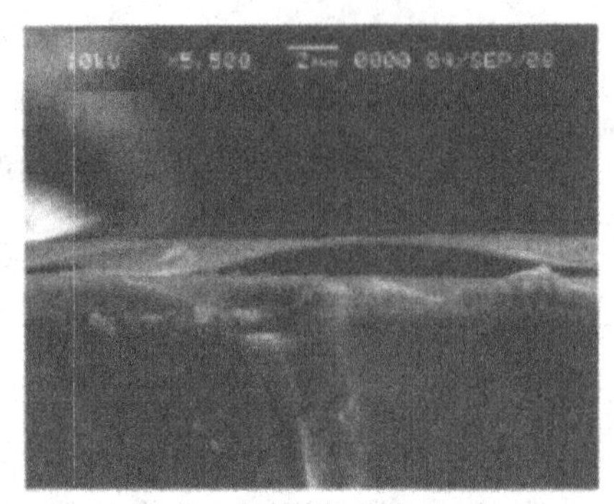

Figure#2 Bonded SiO_2/SiO_2 surfaces

A solid model of the planar inductor was made using Coventor. It is meshed and simulated for different parameters of the inductor like the self inductance, series resistance and parasitic capacitance. This is done using the MemHenry simulation. The simulation gives the value of self inductance of 0.413μH with series resistance of 1.68Ω for 3mm inductor.

The inductance value is measured using the probe station and a low frequency impedance analyzer. The following measurements were achieved for 3mm inductor. The inductance value for 3mm inductor decreases from 3.6μH-1.43 μH for 1kHz -13Mhz frequency range. The reactance value increases in the same frequency range. The parasitic capacitance value is 0.22 pF. This includes the lead capacitance and inductance.

The capacitive sensor is simulated in Coventor to give the simulated result which is the reflection of the actual measurement. The test sensor is 100 μm^2 with membrane thickness t_m= 0.4 μm, the gap (d) between the electrodes is 2 μm. First, different load is applied in the dynamic range of the sensor which is between 50 -350mmHg for blood pressure measurement. The capacitance varies between 2.313 pF to 2.319 pF. The sensitivity of the sensor in this range is 0.108 fF/ kPa or 0.014fF / mmHg. The maximum nonlinearity is approximately 0.013 % with average non-linearity of 0.008%. The sensitivity of the capacitive sensor depends on the size and the thickness of the sensor membrane. As the area of the sensor increases and the thickness of the membrane decreases, greater sensitivity is achieved.

CONCLUSIONS

The planar inductors were fabricated with different outer diameters (O.D.), from 2mm to 5mm giving inductances of 0.11μH to 4.1 μH. The inductor with O.D. 3mm with inductance of 0.385 μH was chosen for the sensor's resonant circuit. These inductors were also simulated for inductance and series resistances. The simulation is performed using Coventor's MemHenry and measured values are obtained using an impedance analyzer. The measured inductance and resistance values are compensated for connecting wire and probes inductance and resistance. The calculated, simulated and the measured values closely match.

A variable capacitor with sizes of $100\mu m^2$ and 400 μm^2 was simulated for membrane deflection, stress analysis and electro-mechanical analysis. These simulations were completed using FEA software Coventor's MemMech and MemElectro. The results are promising for pressure measurement in the range of 50-350mmHg. The capacitance variation for the 100 μm^2

sensor is between 2.313 pF to 2.319 pF with sensitivity of 0.0154 fF/mmHg and average non-linearity of 0.008%. An improved model was simulated with sensor size of 400 μm^2 and gives better sensitivity of 0.21fF/mmHg with average non-linearity of 0.232%. The capacitance changes from 0.934-0.997pF.

The resonant circuit was made with the variable capacitor and sensor and gives a sensitivity of 27.462 kHz/mmHg with a resonant frequency changes from 248MHz to 256 MHz. A similar resonant pressure sensor was fabricated on glass and Si substrate, which required front to back interconnect and an EDP etch [6].

In conclusion, a SiC sensor system was fabricated using a planar inductor and a SiC capacitive sensor. The simulation results show promising results for blood pressure measurement application with sensitivity of 0.21 fF/mmHg. The sensor itself was 400μm^2 in size; the sensor with a smaller area 100 μm^2 gives less sensitivity

ACKNOWLEDGMENTS

This research was supported by NSF/WIMS Center at University of Michigan G. Harris acknowledges the support of the AMSRD-ARL-RO-SI Proposal Number: 50845-RT-ISP. Part of the research was done at the Howard Nanoscale Science and Engineering Facility a node of the National Nanofabrication Infrastructure Network. The authors gratefully acknowledge C. Taylor and T. Gomez for technical support.

REFERENCES

1. A.J. Rosenbloom, Y. Shishkin, D.M. Sipe, Y. Ke , R.P. Devaty, and W.J. Choyke, Porous Silicon Carbide as a Membrane for Implantable Biosensors Mater. Sci.Forum Vols. 457-458 (2004), p. 1463
2.] S. Chatzandroulis,D. Tsoukalas and P.A. Neukomm, "A miniature pressure system with a capacitive sensor and a passive telemetry link for use in implantable application", *J. Microelectromechanical systems, Vol9, No.1,* March 2000
3. americanheart.org
4. Y.S. Lee and K.D. Wise, "A batch-fabricated silicon capacitive pressure transducer with low temperature sensitivity", *IEEE ED vol. 29 Issue 1,* Jan 1982
5. A. DeHennis and K.D. Wise, "A double-sided single –chip wireless pressure sensor", *International conference on MicroElectroMechanical Systems, Las Vegas*, Jan2002
6. Q.Wang, W.H. Ko, "Modeling of touch mode capacitive sensor and diaphragms," *Sensors and Actuators,* vol. 75, pp. 230-241, Feb 1999
7. O. Akar, T. Akin, T. Harpster, K. Najafi, " A wireless batch sealed absolute capacitive pressure sensor", *Eurosensors XIV*, pp. 323-324, 2000

Micro/Nanomechanics

Mater. Res. Soc. Symp. Proc. Vol. 1139 © 2009 Materials Research Society 1139-GG04-01

CMOS-Integrated Stress Sensor Systems for Mechanical Sensing and Packaging Reliability Testing

Oliver Paul, Pascal Gieschke and Benjamin Lemke
Department of Microsystems Engineering (IMTEK), University of Freiburg,
Georges-Koehler-Allee 103, D-79110 Freiburg, Germany

ABSTRACT

This paper presents a selection of microsensors and microsystems based on complementary metal oxide semiconductor technology for measuring mechanical stress and tries to identify some of the next challenges. As the stress acting on the sensors may originate from effects internal or external to the package, the sensors lend themselves for the measurement of (1) time-dependent thermomechanical forces acting on a packaged microchip and (2) external mechanical constraints, i.e., forces and moments. Recently developed applications include tactile sensors, smart brackets, and a solid-state-only joystick with four degrees of freedom, among others. The systems implemented in these applications rely on sensors for in-plane stresses. For packaging testing applications, in addition, sensors for out-of-plane stress components are beneficial. First designs of such new sensors and corresponding results are presented. The calibration of piezoresistive elements is performed on dedicated experimental setups, such as optimized four-point bending and torsional bridges for the application of well-controlled in-plane stresses, and setups in the first stage of their development for the exertion of vertical stress components.

INTRODUCTION

Stress-sensitive elements have been a central component of mechanical microsensors since the inception of the first silicon-based microelectromechanical systems (MEMS) [1]. They have been instrumental in the development of the first membrane-based pressure sensors [2]. Even accelerometers, where nowadays electrostatic effects are preferred, have relied on stress measurements. The reason for this early success resides in the high piezoresistive gauge factors offered by silicon [2,3] and the convenient realization of such structures using established silicon technologies.

Generally speaking, stress sensors make it possible to assess the mechanical constraints imposed on the sensor chip. These constraints may originate from within the encapsulated system or from external loads to which the system is exposed. As a consequence, two categories of applications can be envisaged. In the first scenario, the stress sensors are exploited to sense the stress state just below the interface between the chip and the packaging materials caused, e.g., by intrinsic and thermal stresses of the involved materials. One may thereby hope to identify critical locations of a package, to confront and support finite element simulations with measured data, to assess drift phenomena caused by fatigue or creep, and even to monitor failure phenomena in-situ.

In the second category of applications, external forces and moments are transmitted via the encapsulation to the sensor chip and translate there into an on-chip stress signature that is probed by a sufficient number of sensors at dedicated locations on the chip. Thus, after appropriate calibration of the system, the external force and moment components are extracted from the measured signature.

Besides the longitudinal and transverse piezoresistance effects, advantage has been taken of the shear piezoresistance effect, also termed pseudo-Hall effect. In the first two effects, resistance changes along a resistor due to normal stresses parallel and perpendicular to the resistor are measured [2,3]. In the shear piezoresistance effect a potential difference arises and is measured in the direction perpendicular to the current flow, when the conductive structure is exposed to a shear stress [3]. It is illustrated in Fig. 1, where the current is rotated in an octagonal device with eight contacts [4]. As the bias voltage V_{in} and the associated current are rotated, the perpendicular component of the electrical field and therefore the magnitude of the voltage V_{pH} across the contact pair perpendicular to the current direction switches twice from positive to negative and back. Depending on the stress, i.e., σ_{xy} or $(\sigma_{xx} - \sigma_{yy})$, the maxima occur at current directions $n\pi/2$ and $(2n + 1)\pi/4$, respectively. This angular evolution can be taken advantage of to clean the stress signals from signal components of other angular orders than the mechanical contribution, such as those due to the Hall effect or thermoelectric voltages [4]. When the number of contacts are restricted to four per device, either σ_{xy} or $(\sigma_{xx} - \sigma_{yy})$ can be determined, depending on the orientation of the structures.

Sensors can be realized as standard planar diffusions [4,5] with standard contacts or as field-effet transistors (FET) with multiple source/drain contacts [6]. They are then termed piezo-FETs. Their gate makes it possible to conveniently arrange the structures into arrays with interconnection overhead scaling linearly with the linear dimension of the array. In contrast, diffused structures require interconnections proportional to the number of sensors in the array, i.e., to its area.

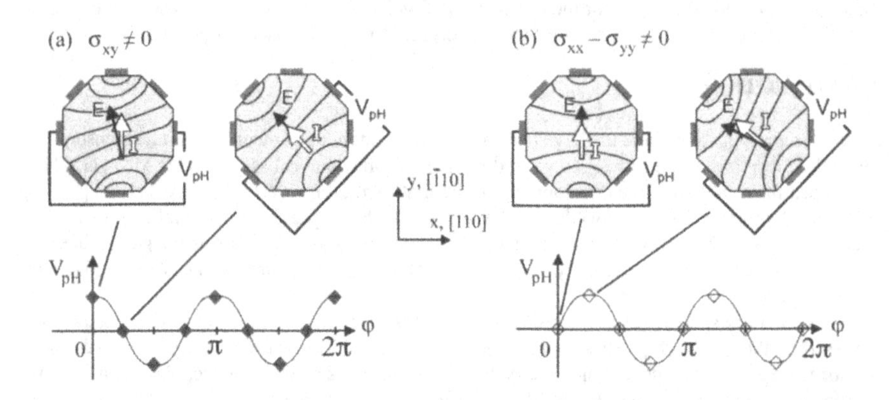

Figure 1: Illustration of current rotation in eight-contact piezoresistors: a current ***I*** (white arrow) is injected, driven by a field ***E*** (black arrow), in the eight directions $\varphi = 0, \pi/4, \pi/2, ..., 7\pi/4$: (a) A stress σ_{xy} causes a pseudo-Hall voltage V_{pH} varying as $\cos(2\varphi)$; sampling the signal in a four-contact device in the directions $\varphi = 0, \pi/2, \pi$, and $3\pi/2$ provides a measure of σ_{xy}; (b) With $(\sigma_{xx} - \sigma_{yy})$, V_{pH} varies as $\sin(2\varphi)$; sampling at uneven multiples of $\pi/4$ thus gives a signal proportional to $(\sigma_{xx} - \sigma_{yy})$.

STRESS SENSORS AND SYSTEMS

Piezo-FETs are shown in Figs. 2 (a) to (d) [7]. They consist of FETs of symmetric design with overall dimensions of about 40×40 μm^2 including the guardring and four peripheral source/drain and the gate contacts. Average sensitivities $S = V_{in}^{-1}\, dV/d\sigma_{xy}$ of -3.26×10^{-10} Pa^{-1} and -4.48×10^{-10} Pa^{-1} for the sensors in Figs. 2 (a) and (b) and $S = V_{in}^{-1}\, dV/d(\sigma_{xx}-\sigma_{yy})$ of 3.61×10^{-10} Pa^{-1} and 4.77×10^{-10} Pa^{-1} for those in Figs. 2 (c) and (d), respectively, were determined for these structures [7]. The devices were integrated in systems such as that shown in Fig. 2 (e). The chips include 32 piezo-FETs, an instrumentation amplifier with gain selectable between 100 and 700, an 8-bit analog-to-digital converter (ADC), and digital logic for signal conditioning and communication via an I^2C bus [7].

Calibration of the structures has shown the high reproducibility of their mechanical response, with sensitivities varying by $\pm1.22\times10^{-12}$ Pa^{-1} over the 32 sensors of individual chips with n-channel and p-channel piezo-FETs. Zero stress offset signals of 0.63±0.45 mV and 0.45±2.21 mV are observed for n-channel and p-channel sensors, likely due to small deviations

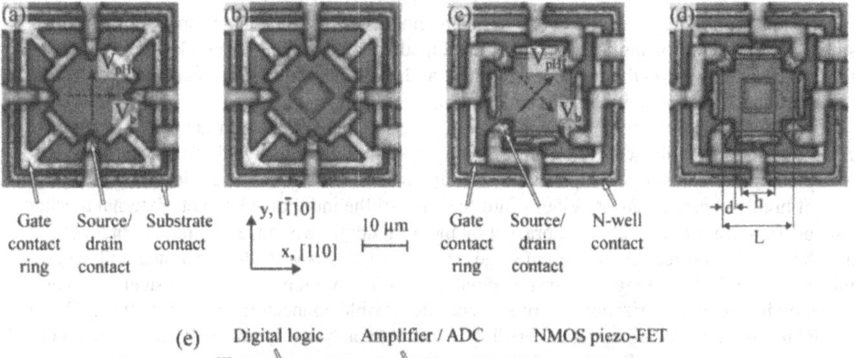

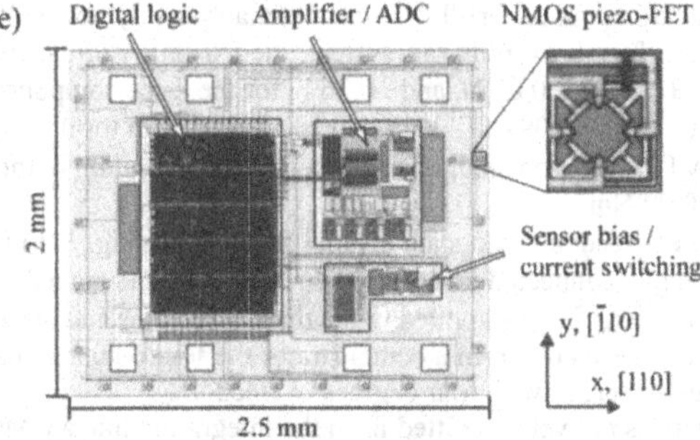

Figure 2: Four-contact FET-based piezoresistive sensors and CMOS system incorporating these devices. (a), (b): NMOS sensors without and with central hole; (c), (d): PMOS sensors without and with central hole; (e) stress sensor chip with 32 NMOS piezo-FETs.

from ideal symmetry due to technology and residual stresses in the chip [7]. The resolution of the piezo-FET sensors is limited by $1/f$ and thermal noise. For the sensors in Figs. 2 (a) and (c), the noise voltage spectral densities show corner frequencies of 1 kHz and 400 Hz, respectively. The corresponding thermal noise floor lies at 10 $nV/Hz^{1/2}$ and 18 $nV/Hz^{1/2}$.

Sensor systems similar to that in Fig. 2 (e) with sizes of 3×3 mm^2, 3×6 mm^2, and 1.8×2.5 mm^2 with 32, 32, and 10 sensors, respectively, were reported previously [8]. These were based on piezoresistive diffusions with eight contacts each. As a consequence, maps of both σ_{xy} and $(\sigma_{xx} - \sigma_{yy})$ were extracted, typically at a frame rate of a few 10 Hz.

A chip with an array of 32×32 stress sensors in an area of 1 mm^2 was built on piezo-FETs similar in design to those in Figs. 2 (a) and (c). It alternates sensors of p-type and n-type, thus making it possible again to determine distributions of both σ_{xy} and $(\sigma_{xx} - \sigma_{yy})$ stresses. A spatial resolution of about 45 μm was demonstrated in tactile experiments [9,10]. The framerate of the system was about 1 Hz.

APPLICATIONS

Force and moment sensing

Force and torque sensors are important elements of instrumentation in manufacturing, safety systems, control applications, and the biomedical area, among others. Figure 3 shows three applications currently under development at the authors´ laboratory based on CMOS systems like those described in the previous section.

The first is a smart bracket aiming at the measurement of three-dimensional forces in the fixed orthodontic appliances. Such information constitutes an objective feedback for orthodontists planning or adjusting therapies with such appliances. This goal was achieved by the integration of highly compact sensor systems into the body of the individual bracket. External mechanical constraints are imposed on the bracket by the orthodontic wire inserted into the bracket slot attached to the bracket. At the present stage, 1:1 scale brackets were demonstrated with a system incorporating the following stack of components [11]: a tooth represented by a steel pin to be mounted in the characterization setup; a polyimide flexible connector; a sensor chip flip-chip attached to the cable, including an underfill; a bracket slot adhesively bonded to the chip. A complete system is shown in Fig. 3 (a). With this system, the feasibility of the concept was demonstrated. Accuracies of ±0.07 N, ±0.07 N, and ±0.26 N for the force components in x, y, and z directions and $\pm 0.76\times10^{-3}$ Nm, $\pm 1.09\times10^{-3}$ Nm, and $\pm 0.22\times10^{-3}$ Nm for the three moments were achieved, respectively [11]. These values enable a sufficient resolution in the clinically relevant ranges of ±2 N and ±0.02 Nm.

A four-degree-of-freedom solid-state joystick is shown in Fig. 3 (b) [12]. It consists of a sensor system attached and connected to a printed circuit board and embedded in a molded epoxy pillar topped by a sphere. Forces applied in the three directions and out-of-plane moments are clearly separated from each other. In a demonstrator the four output signals allow the user to control a digital object on-screen with four degrees of freedom.

A structure that has not yet benefitted from the integration into a CMOS system is shown in Fig. 3 (c). It is a tactile element for the characterization of steep surfaces of microstructures, such as deep, narrow holes, where reflection-based optical techniques reach their limit. The structure consists of a slender beam with an overall length of 7 mm, a thickness of 380 μm, and a width of 150 μm [13]. Close to the suspension, the structure was thinned into a 100-μm-thick

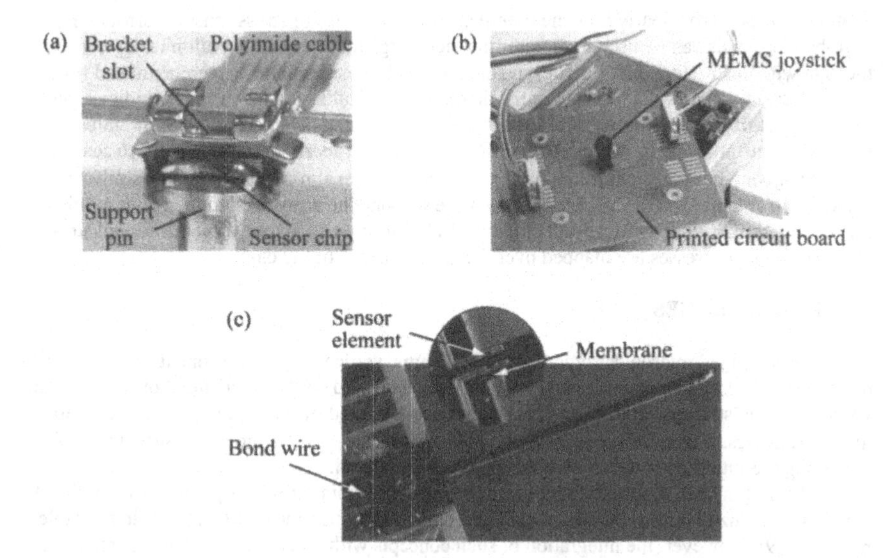

Figure 3: Three applications of stress sensors for mechanical sensing: (a) 1:1-scale MEMS-based smart bracket; (b) four-degree-of-freedom miniaturized MEMS joystick; (c) tactile sensor for the geometric characterization of deep narrow holes.

membrane hinge on which a diffused p-type four-contact stress sensor is placed. The structure shows a mechanical resilience of 3.4 μm/mN at a sensitivity up to 10.9 mV/(V mN) for contact at its tip. The sensitivity is increased by concentrating the current flow to the periphery of the sensor by implementing a square hole [13,14], similar to those shown in Figs. 2 (b) and (d).

Packaging characterization

Packaging stress test chips have been in use now for many years. Research groups at Sandia [15], University College Cork [16], Philips [17], Auburn University [18,19], ETH Zurich [20,21], University of Freiburg [22], and Bosch/Infineon [23] have reported work in this direction. However, to this day the use of stress sensor test chips has not become a standard method of packaging characterization, despite the evident appeal to observe the mechanical and thermomechanical evolution of packaged microelectronic and MEMS systems in a non-destructive way. We hypothesize that this is partly due to the fact that the early chips required a considerable level of technical proficiency for their operation and the interpretation of their output. Most of the early test chips did not present signal conditioning circuitry or even multiplexers to address the multitude of sensors covering the dies.

One case where dedicated stress sensors have greatly helped to understand and optimize a packaging technology is the characterization of wire and flip-chip bonding. A variety of CMOS force sensors have been developed and utilized for this purpose [20-22]. Reference 21 has demonstrated particularly clearly how the time-dependent response of stress sensors surrounding wire-bonding pads can be used to monitor the metallurgical processes in real time. By correlating the data with destructive reliability test results, optimal process windows were extracted in-situ.

Recently, a new CMOS-integrated thermomechanical test chip was produced in a joint development of Robert-Bosch GmbH, Stuttgart, with Infineon, Munich [23]. Sensor chips with sizes between 1 and 7 mm^2 were reported. One of them carries an array of up to 10×6 sensor fields each containing (1) a current mirror with two orthogonal n-MOSFETs, (2) a similar PMOS current mirror, and (3) a diode-based temperature sensor. The array is combined with multiplexing and on-chip driving and read-out circuitry. In the absence of an out-of-plane normal stress, all three in-plane stresses are mapped over the entire surface of the chip.

NEXT CHALLENGES

In packaged microelectronic and MEMS chips, vertical stress components definitely cannot be neglected. As an example, in the vicinity of a ball grid contact to a chip, both out-of-plane vertical normal stress σ_{zz} and shear stress components σ_{xz} and σ_{yz} may be expected in addition to in-plane stresses. Vertical stresses have the potential to cause the failure of the structure by fracture along the interface between chip and packaging material.

Little has been reported in the past about stress sensors for out-of-plane stress. With planar sensors realized on wafers of non-standard orientation, such measurements are in principle possible [19]. However, the integration of such concepts with CMOS circuitry is incompatible with standard CMOS technology with its customary (100) substrate orientation.

A first step at the authors′ laboratory towards the measurement of out-of-plane normal stress has been made with the structure schematically shown in Fig. 4 [24]. The sensor was produced in a 0.6 μm CMOS technology of X-FAB, Erfurt, Germany. It is designed to incorporate vertical and horizontal current flows. Its resistance variation between a central contact and a peripheral contact is due to the changes of vertical resistive elements, R_{V1} and R_{V2} and a horizontal element R_H under stress and temperature. The relative resistance change is

$$\Delta R/R = [\pi_{11} + \pi_{12} - \Gamma(\pi_{11} - \pi_{12})](\sigma_{xx} + \sigma_{yy})/2 + [\pi_{12} + \Gamma(\pi_{11} - \pi_{12})]\sigma_{zz} + \alpha\Delta T, \qquad (1)$$

where the thermal sensitivity of the device is mediated by the coefficient α. The factor $\Gamma = (R_{V1} + R_{V2})/(R_{V1} + R_{V2} + R_H)$ denotes a resistive mixture factor tending towards 0 for sensors with purely horizontal current flow and towards 1 for exclusively vertical current flow. When measurements are performed with two structures designed with different Γ values Γ_1 and Γ_2, or when one structure is operated under different bias condition leading to different values Γ_1 and Γ_2, access is provided to the temperature-compensated stress combination $(\sigma_{xx} + \sigma_{yy})/2 - \sigma_{zz}$, according to

$$\Delta R_1/R_1 - \Delta R_2/R_2 = (\Gamma_2 - \Gamma_1)(\pi_{11} - \pi_{12})[(\sigma_{xx} + \sigma_{yy})/2 - \sigma_{zz}], \qquad (2)$$

and to a second combination not compensated for temperature variations and given by

$$\Delta R_1/R_1 - (\Gamma_1/\Gamma_2)\Delta R_2/R_2 = (1 - \Gamma_1/\Gamma_2)\,[(\pi_{11} + \pi_{12})(\sigma_{xx} + \sigma_{yy})/2 + \pi_{12}\sigma_{zz} + \alpha\Delta T]. \qquad (3)$$

Therefore, using an independent temperature sensor, the two components $(\sigma_{xx} + \sigma_{yy})/2$ and σ_{zz} can in principle be separated from each other in a temperature compensated way. Unfortunately however, in reality the system of equations is rather badly conditioned for silicon, since $\pi_{11} + \pi_{12} \approx -\pi_{12}$. Nevertheless, if a well-defined relationship between $(\sigma_{xx} + \sigma_{yy})/2$ and σ_{zz} is provided in a context by different methods, e.g., finite element simulations, σ_{zz} and thus $(\sigma_{xx} + \sigma_{yy})$ can still be extracted, namely using Eq. (2) alone, even without the help of a temperature sensor. This was demonstrated for the case of forces applied to bonding balls, using a sensor operated under different biasing conditions modulating the width w_{eff} of the central channel by expanding the depletion zone surrounding the central vertical resistance path [24]. Sensors for out-of-plane shear stress components realized in standard CMOS technology are yet to be demonstrated.

The calibration of such out-of-plane sensors also constitutes a challenge. Whereas well-controlled in-plane normal and shear stresses can be applied with bending bridges [25-27] and torsional bridges [28] suitable for the characterization of stripes of silicon wafers, well-controlled out-of-plane normal and shear stresses seem much harder to achieve. Experimental techniques aiming in this direction are currently under development at the authors' laboratory.

Other extensions of the current work may address the ultimate resolution of the sensors and operation modes to push these limits, especially in monolithically integrated systems. For some applications it appears interesting to transmit the energy and data by different methods than through a printed circuit board or a flexible cable, as in the smart bracket of Fig. 2. A viable, yet challenging alternative may be wireless energy and data transmission.

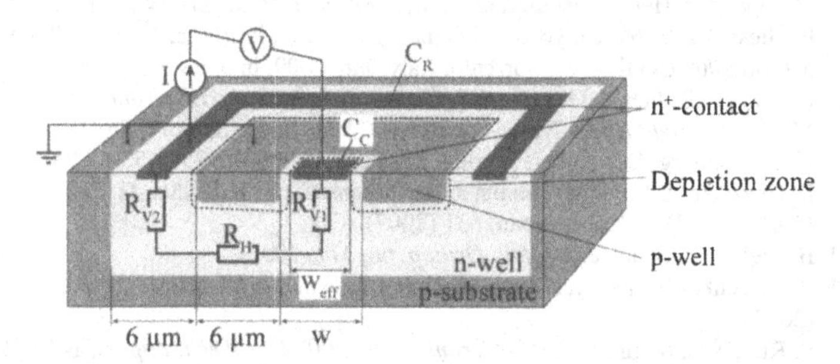

Figure 4: CMOS sensor for the measurement of stress distributions including the vertical normal stress component σ_{zz}. Signal contributions, i.e. resistance changes are due to the piezoresistance effect on horizontal and vertical resistances characterized by their mixture coefficient $\Gamma = (R_{V1} + R_{V2})/(R_{V1} + R_{V2} + R_H)$. Γ can be varied by changing w_{eff} using different p-well/deep n-well biasing voltages.

CONCLUSIONS

Significant progress has been made in recent years in the understanding, design, operation, and application of CMOS based stress sensors. Higher spatial resolutions, integration into large arrays, and the combination of sensors with integrated analog and digital circuitry makes it possible to realize user-friendly compact systems-on-a-chip that will certainly find their way into other applications. However it is concluded by the authors that the out-of-plane components of the stress tensor will have to become accessible to measurement before stress sensor chips will be able to also make significant contributions to the advancement of packaging technologies and to the assessment of their reliability. First steps in this direction were described in this paper. Further devices suitable, e.g., for the measurement of out-of-plane shear stress are under development.

ACKNOWLEDGMENTS

Financial support by the German Research Foundation DFG through grant PA792/5-1 and by INTEL, Micronas GmbH, and Zeiss IMT GmbH through Ph.D. grants is gratefully acknowledged.

REFERENCES

[1] K. E. Peterson, Proc. IEEE **70**, 420 (1982).
[2] H. J. Timme, *CMOS-Based Pressure Sensors*, Ch. 6 in *CMOS-MEMS*, O. Brand and G. Fedder Eds., Wiley-VCH, Weinheim, 2005.
[3] Y. Kanda, Sens. Actuators **28**, 83 (1982).
[4] J. Bartholomeyczik, P. Ruther, O. Paul, *IEEE Sensors J.* **5**, 872 (2005).
[5] J. Bartholomeyczik, M. Doelle, P. Ruther, O. Paul, *Sens. Actuators* **A127**, 255 (2006).
[6] M. Doelle, J. Held, P. Ruther, O. Paul, *J. Microelectromech. Syst.* **16**, 1232 (2007).
[7] P. Gieschke, Y. Nurcahyo, M. Herrmann, M. Kuhl, P. Ruther, O. Paul, *Tech. Digest IEEE MEMS 2009 Conference*, Sorrento, Italy, Jan. 2009, in press.
[8] O. Paul, P. Ruther, J. Gaspar, *IEEJ Trans. Electrical and Electronic Eng.* **2** (3), 199 (2007).
[9] M. Doelle, *Field Effect Transistor Based CMOS Stress Sensors*, Ph.D. Thesis, IMTEK, Univ. Freiburg; Der Andere Verlag, Marburg, ISBN 3-89959-458-4 (2006).
[10] P. Gieschke, J. Held, M. Doelle, J. Bartholomeyczik, P. Ruther, O. Paul, *IEEE MEMS 2007 Conf. Tech. Dig.*, Kobe, Japan, 631 (2007).
[11] B. Lapatki, O. Paul, *J. Orofac. Orthop.* **68**, 377 (2007).
[12] P. Gieschke, J. Richter, J. Joos, P. Ruther, O. Paul, *IEEE MEMS 2008 Conf. Tech. Dig.*, 86 (2007).
[13] P. Ruther, S. Spinner, O. Paul, *Transducers 2007 Dig. Tech. Papers*, 1469 (2007).
[14] M. Doelle, D. Mager, P. Ruther, O. Paul, *Sens. Actuators* **A127**, 261 (2006).
[15] J. N. Sweet, D. W. Peterson, A. H. Hsia, *InterPACK '99 Proc.*, 1 (1999).
[16] J. Barrett, C. Cahill, T. Compagno, M. O. Flaherty, T. Hayes, W. Lawton, J. O. Donovan, C. Mathuna, G. McCarthy, O. Slattery, F. Waldron, A. C. Vera, M. Masgrangeas, P. Pipard, C. Val, I. Serthelon, *45th Electronic Comp. Technol. Conf.*, 656 (1995).
[17] S. A. Gee, W. F. van den Bogert, V. R. Akylas, *IEEE Trans. Compon. Hybr. Manuf.* Technol. **12**, 587 (1989).
[18] J. C. Suhling, R. C. Jaeger, *IEEE Sens. J.* **1**, 14 (2001).
[19] A. Mian, J. C. Suhling, R. C. Jaeger, *IEEE Sens. J.* **6**, 340 (2006).
[20] M. Mayer, J. Schwizer, O. Paul, H. Baltes, *Proc. InterPACK 99* **EEP-Vol. 26-1**, 973 (1999).
[21] J. Schwizer, CMOS Force Sensors for Wire Bonding and Flip Chip Process Characterization, Ph.D. Thesis ETH Zürich, No. 15293 (2003).

[22] M. Doelle, C. Peters, P. Ruther, O. Paul, *J. Microelectromech. Syst.* **15**, 120 (2006).
[23] H. Kittel, *Proc. ZVEI Expert Meeting "Stressarme MST Packages"*, (2008).
[24] B. Lemke, K. Kratt, R. Baskaran, O. Paul, *Tech. Digest IEEE MEMS 2009 Conference*, Sorrento, Italy, Jan. 2009, in press.
[25] S. A. Gee, V. R. Akyla, W. F. van den Bogert, IEEE Proc. Microelectronic Test Structures **1**, 185 (1988).
[26] J. Richter, M. B. Arnoldus, O. Hansen, E. V. Thomsen, Rev. Sci. Instrum. **79**, 044703 (2008).
[27] J. Bartholomeyczik, S. Brugger, S. Kibbel, P. Ruther, O. Paul, *Eurosensors XIX Proc.*, TB23 (2005).
[28] M. Herrmann, P. Gieschke, Z. Liu, J. Korvink, P. Ruther, O. Paul, *IEEE Sensors Conf. Proc.*,1528 (2008).

Mater. Res. Soc. Symp. Proc. Vol. 1139 © 2009 Materials Research Society 1139-GG04-02

Evaluation of the Mechanical Properties of Aluminum Thin Films as a Function of Strain Rate using the Wafer-Scale Microtensile Technique

Joao Gaspar, Marek E. Schmidt, Jochen Held, and Oliver Paul
Department of Microsystems Engineering – IMTEK, Microsystems Materials Laboratory, University of Freiburg, Georges-Koehler-Allee 103, 79110 Freiburg, Germany

ABSTRACT

This paper reports on the mechanical characterization of aluminum thin films as a function of strain rate using the wafer-scale microtensile technique. Multiple test structures are processed on the same silicon substrate and sequentially measured with an automated setup. Each structure is composed of an inner movable part connected to an outer fixed frame through micromachined parallel springs. Such design allows the in-plane, uniaxial elongation of a bridging aluminum microtensile specimen when a displacement is imposed to the inner frame. Engineering stress-strain curves of the Al specimens are acquired from the measured load-displacement data of the test structures, from which mechanical properties are obtained. Parameters extracted include the Young's modulus E, fracture strain or maximum elongation ε_{max}, ultimate tensile strength σ_{max}, 0.2% offset yield strength $\sigma_{0.2\%}$, strength coefficient K and strain-hardening exponent n, obtained for strain rates $d\varepsilon/dt$ between 2.5×10^{-4} to 2.5 s^{-1}.

INTRODUCTION

Among several mechanical characterization methods available, the tensile test is one of the most used and well established. Its main advantage resides in the fact that uniaxially loaded specimens experience a uniform stress distribution, allowing a straightforward analysis of load-displacement data and extraction of mechanical parameters [1]. Deviations in the mechanical material behavior from the macro to the micro-scale make it necessary to measure the mechanical properties of samples with similar dimensions to those employed in integrated circuits (ICs) and microelectromechanical systems (MEMS). This has been the driving force for the development of the microtensile technique.

Major difficulties in dealing with microtensile specimens are sample handling, alignment and attachment to actuators. Several configurations have been proposed and implemented to overcome those issues [2-6]. In fact, the microtensile test method has matured to a level where wafer-scale measurements are now possible [7,8]. The microtensile technique has been used to characterize a wide range of brittle and ductile thin-film materials. These include crystalline and polycrystalline silicon, passivation layers such as carbides, nitrides and oxides, metals like aluminum, copper, gold, nickel and titanium, and polymers [2-14].

In the case of brittle samples, these materials deform elastically until failure, independently of loading speed, thus simplifying the data analysis. In contrast, the mechanical properties of ductile materials are known to depend on the rate of deformation. These dependences have been extensively studied on macroscopic bulk samples, the strength of which tends to increase with increasing strain rate [1,15-19]. Few data exist however for the microscale case. Ductile materials such as patterned metal films used in ICs and MEMS devices may be subjected to a wide range of elongations and elongation rates and, therefore, apart from obvious scientific interest, their mechanical characterization under such operation conditions is of utmost importance. This motivates the work presented here, which reports on the characterization of the strain rate dependence of the mechanical properties of evaporated aluminum (Al) thin films using the recently developed wafer-scale microtensile test method [7,8].

TEST STRUCTURE DESIGN

The layout of the arrangement of microtensile samples over a 4-inch substrate and close-ups of a single test structure and respective microtensile specimen are schematically shown in Fig. 1. The individual test structure consists of a fixed crystalline silicon (c-Si) outer frame connected to an inner moving part through four parallel springs. All flexures are identical with length, width and thickness of 4 mm, 100 µm and 525 µm, respectively. Such design allows the in-plane motion of the inner frame along the x-direction when a force F is applied to it. Consequently, it leads to the uniaxial elongation of the bridging Al microtensile specimen with length $L = 400$ µm and width $L = 50$ µm.

The alignment of the specimen to the acting force is ensured by construction. A window in the movable frame is provided for the insertion of an external actuator. Both frames are rigid enough for the elongation of the tensile specimen to differ only negligibly from the displacement d imparted on the movable frame.

EXPERIMENTAL PART

Sample fabrication

The fabrication of micromachined test structures with thin-film tensile specimens is summarized in Fig. 2 (a). The process starts with the evaporation of an Al layer with a thickness t of 2 µm on a 525-µm-thick double-side polished silicon wafer, using a commercial system (Leybold Univex 500) with a deposition rate of 50 nm/min. This layer is then patterned into microtensile specimens using photolithography and wet etching. Then follows the deposition of a 5-µm-thick silicon oxide (SiO_2) film on the rear side of the substrate by plasma-enhanced chemical vapor deposition (PECVD) and its opening by reactive ion etching. This patterned layer serves as an etch mask for the subsequent and final c-Si deep reactive ion etching (DRIE) step, which is responsible for defining the gaps and parallel flexures in the micromachined test structure. The DRIE recipe is highly selective to Si with respect to the Al layer. This results in an

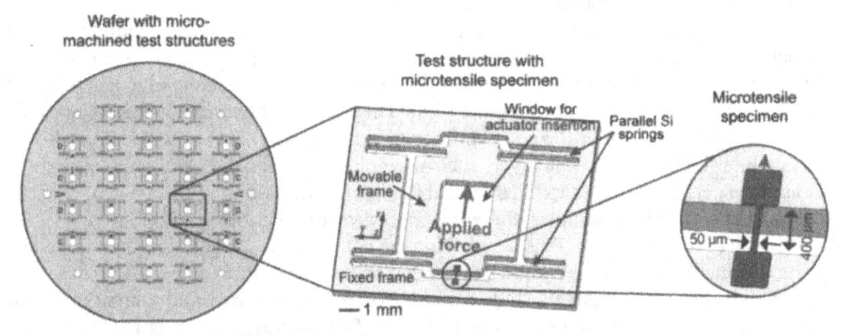

Figure 1. Arrangement of microtensile samples over a 4-inch silicon wafer and close-ups of one test structure and its bridging Al tensile specimen. The specimen is uniaxially elongated when an in-plane displacement is imposed to inner movable frame.

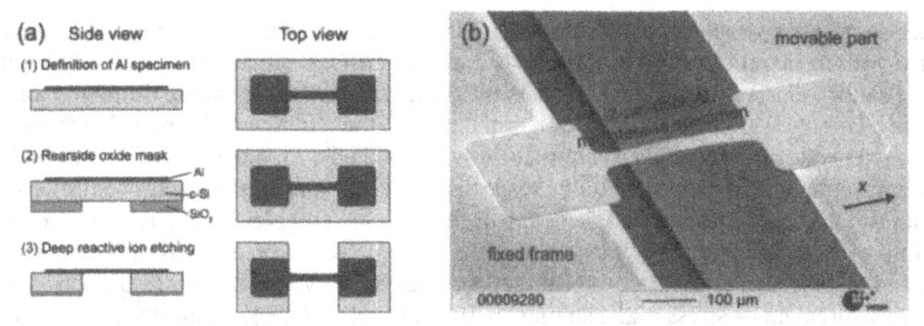

Figure 2. (a) Schematic overview of the fabrication process of microtensile test structures and (b) SEM micrograph of an Al microtensile specimen bridging the frames.

Al microtensile specimen suspended across the gap between the movable and fixed parts of the frame. Figure 2 (b) shows the scanning electron microscope (SEM) graph of an Al sample bridging the frames of the test structure. Prior to the tensile testing procedure, the exact thickness of the films is measured using a mechanical profilometer, and the planar dimensions of each individual specimen are measured using an optical profilometer.

Measurement setup

The schematic of the automated full-wafer tensile setup is presented in Fig. 3 (a). The wafer is placed and fixed on a vacuum chuck mounted on a motorized $xy\theta$ table. This table allows to step the position of the 26 test structures and to align them to the stainless steel loading tip with spherical elements. A close-up photograph of the tip, taken during a measurement, is shown in Fig. 3 (b). The tip is mounted on a force sensor with a resolution of 1 mN connected to a PZT actuator with a displacement range of 30 µm for actuation of the sample. An actuation head comprising programmable *xyz* stages enables one to insert the tip into the test structure window. This actuation head can be used as well to displace the inner frame of the test structures with larger travel range than the PZT actuator. The design-limited displacement range of 400 µm for the inner frame of the test structure is thus achievable, corresponding to specimen strain values of 100 %. The stages of the actuation head can be actuated with speeds ranging from 10^{-1} to 10^{3} µm/s, thus covering the range of strain rates $d\varepsilon/dt$ from 2.5×10^{-4} to 2.5 s^{-1}.

As schematically shown in Fig. 3 (a), displacements are measured with a compact laser-

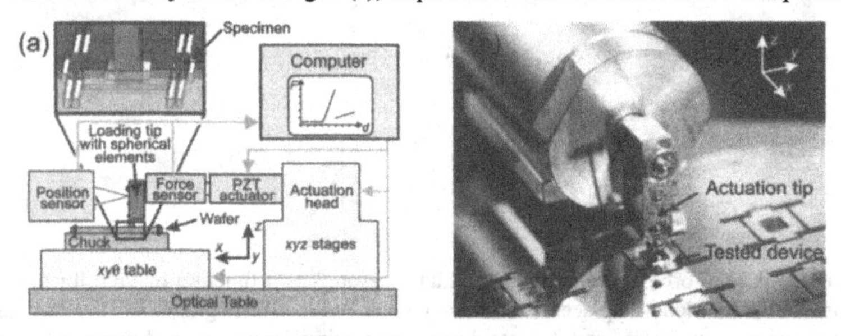

Figure 3. (a) Schematic of the automated wafer-scale microtensile setup and (b) close-up photograph of the loading tip during the measurement of one Al-based test structure.

deflection position sensor with a resolution of 10 nm focused on the top part of the loading tip, aligned with the axial direction of the load cell. Such non-ideal situation is taken into account using a calibration procedure that considers the finite stiffness values of the different parts of the setup [8]. Upon positioning and insertion of the tip into the window in the movable frame, a function generator triggers the actuation cycle synchronized with the measurement of both force and displacement data, connected to the respective load cell and laser-detector controllers. Both loading and unloading responses are recorded, enabling the complete characterization of the test structure, with and without specimen, in a single attempt. To the resolutions obtained using the force and displacement sensors, 1 mN and 10 nm, respectively, correspond resolutions achieved in the measurement of stress σ and strain ε of 10 MPa and 0.0025 %, respectively, for typical specimen dimensions.

RESULTS AND DISCUSSION

Force-displacement data

Figure 4 (a) shows typical measurement results of force and displacement data plotted as a function of time for one test structure. The same data are shown in Fig. 4 (b) in the form of a load-displacement diagram. In this case, both loading and unloading cycles are performed at a rate of 1.25 μm/s. After the initial contact, the force increases due to the elongation of the specimen and deformation of the 4 c-Si parallel flexures. When the specimen fails, only the parallel springs contribute to the force response during the rest of the loading cycle and entire unloading sequence of the test structure. No hysteresis effects are detected in the mechanical response of the flexures, as observed from measurements of test structures without specimens. A SEM graph of the fractured specimen is shown in the inset of Fig. 4 (b), evidencing that failure occurs in its gage section where the stress is uniform.

The load-displacement characteristic of the tensile specimen can thus be singled out by subtracting the measured contribution of the silicon flexures from the overall response of the test structure. Engineering stress-strain curves, such as the one depicted in Fig. 4 (c), are obtained by scaling the resulting difference by the geometrical dimensions of the specimen, yielding

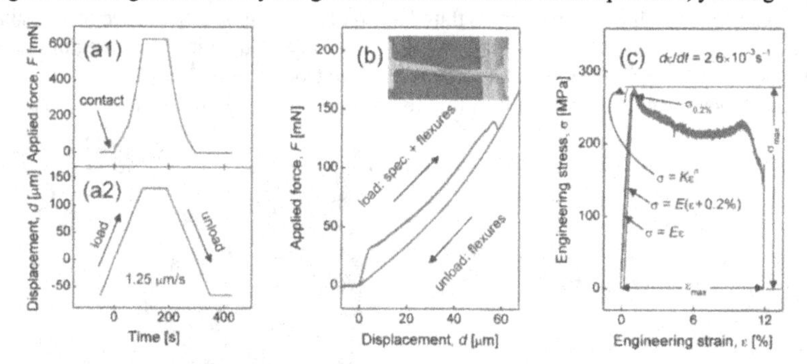

Figure 4. (a) Force and displacement data measured as a function of time for one test structure, (b) load-displacement characteristic with SEM graph of the fractured specimen as an inset, and (c) resulting engineering stress-strain diagram of an Al microtensile specimen loaded at a strain rate $d\varepsilon/dt$ of 2.6×10^{-3} s^{-1}.

$\sigma = F/wt$ and $\varepsilon = d/L_{eff}$ for the engineering stress and engineering strain, respectively, where L_{eff} is the effective specimen length [8]. Similar behaviors have been observed in Al engineering σ-ε diagrams by other authors [20]. As illustrated in Fig. 4 (c), mechanical parameters such as Young's modulus E, maximum elongation ε_{max}, maximum strength σ_{max}, 0.2% offset yield strength $\sigma_{0.2\%}$, strength coefficient K and strain-hardening exponent n can then be extracted [1].

Strain-rate dependences

Figures 5 (a), (b) and (c) show the experimentally obtained dependences of Young's modulus, maximum elongation and maximum stress, respectively, as a function of the applied strain rate. The Young's modulus, $E = 63\pm9$ GPa, appears to have no dependence on $d\varepsilon/dt$, within the span of strain rates tested. This value is well within the range of literature data reported for similar films [11-13].

Regarding the fracture strain, it varies between 1.27±0.19 and 33.2±3.9%: a power law decrease is observed for rates from 2.5×10^{-4} to 10^{-1} s^{-1}, $\varepsilon_{max} = 1.5\times10^{-3}(d\varepsilon/dt)^{-0.76}$, followed by a saturation at $\varepsilon_{max} = 1.27\%$ for $d\varepsilon/dt > 10^{-1}$ s^{-1}.

Opposite trends are obtained for the strength parameters σ_{max}, $\sigma_{0.2\%}$ and K. As shown in Fig. 5 (c), the maximum stress σ_{max} increases logarithmically with $d\varepsilon/dt$, i.e. $\sigma_{max} = 153\{1+0.197 \ln[(d\varepsilon/dt)/10^{-4}]\}$ MPa, for $2.5\times10^{-4} < d\varepsilon/dt < 10^{-1}$ s^{-1}, suggesting a deformation mechanism governed by Johnson-Cook thermally activated dislocations [16,18]. Beyond 10^{-1} s^{-1}, the material reaches its maximum strength at $\sigma_{max} = 380\pm6$ MPa with no dependence on strain rate. Such change in mechanical behavior suggests a different deformation mechanism, which is tentatively explained here by a reduction of the ductility to the point of brittle fracture, sometimes observed in macroscopic metal samples when actuated at sufficiently low temperatures or equivalent high deformation rates. The 0.2 % offset yield strength $\sigma_{0.2\%}$ is about 30 MPa below σ_{max} and the strength coefficient K is approximately 2.5 GPa. A strain-rate independent strain-hardening exponent of $n = 0.48\pm0.08$ is extracted.

CONCLUSIONS

The mechanical characterization of thin-film evaporated Al layers is performed using the

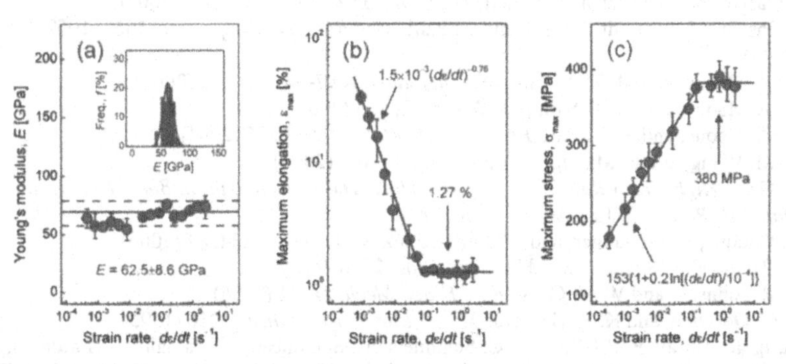

Figure 5. (a) Young's modulus, (b) maximum elongation, and (c) maximum stress obtained as a function of strain rate for evaporated thin-film Al microtensile specimens.

wafer-scale microtensile test for strain rates ranging between 2.5×10^{-4} to $2.5\ s^{-1}$. The value extracted for Young's modulus, 63 GPa, appears to have no dependence on deformation rate, within the experimental uncertainty. The maximum elongation and fracture strength of these films are however strongly influenced by the strain rate. The fracture strain decreases by almost 2 orders of magnitude with increasing deformation speed, reaching a plateau of 1.27 % for $d\varepsilon/dt > 10^{-1}\ s^{-1}$. The maximum stress experienced by the Al layers shows an opposite trend. It increases logarithmically for strain rates up to $10^{-1}\ s^{-1}$, compatible with Johnson-Cook thermally activated dislocations mechanisms, and saturates at 380 MPa for higher speeds. The latter could be an evidence of the transition from ductile to semi-brittle behavior of these films at room temperature. Natural expansions and improvements of this work are its extension to temperature series studies for the identification of plastic deformation mechanisms in MEMS ductile materials, online monitoring of the local cross section reduction (necking) via resistance measurements and localized strain measurements in order to obtain more accurate stress-strain relations.

REFERENCES

1. See, for example, J. R. Davis, in *Tensile Testing*. Ohio: ASM International, 2004.
2. W. N. Sharpe, K. M. Jackson, K. J. Hemker, and Z. Xie, *IEEE J. Microelectromech. Syst.* **10**, 317 (2001).
3. S. Greek, F. Ericson, S. Johansson, M. Furtsch, and A. Rump, *J. Micromech. Microeng.* **9**, 245 (1999).
4. T. Tsuchiya, O. Tabata, J. Sakata, and Y. Taga, *IEEE J. Microelectromech. Syst.* **7**, 106 (1998).
5. K. Sato, T. Yoshioka, T. Ando, M. Shikida, and T. Kawabata, *Sens. Actuators* **A70**, 148 (1998).
6. S. Kamiya, J. Kuypers, A. Trautmann, P. Ruther, and O. Paul, *IEEE J. Microelectromech. Syst.* **16**, 202 (2007).
7. J. Gaspar, M. Schmidt, J. Held, and O. Paul, *Tech. Dig. 21st IEEE MEMS Conf.*, 439 (2008).
8. J. Gaspar, M. E. Schmidt, J. Held, and O. Paul, "Wafer-scale microtensile testing of thin films," *IEEE J. Microelectromech. Syst*, in press (2008).
9. J. Gaspar, M. Schmidt, and O. Paul, *Tech. Dig. Transducers'07*, 575 (2007).
10. J. Gaspar, M. Schmidt, J. Held, and O. Paul, *Mat. Res. Soc. Symp. Proc.* **1052**, 1052-D01-02 (2008).
11. M. A. Haque, and M. T. A. Saif, *Sens. Actuators A* **97-98**, 239-245 (2002).
12. M. A. Haque, and M. T. Saif, *Exp. Mech.* **42**, 123 (2002).
13. M. A. Haque, and M. T. Saif, *Proc. Nat. Acad. Sci.* **101**, 6335 (2004).
14. Y. M. Wang, and E. Ma, *Appl. Phys. Lett.* **83**, 3165 (2003).
15. G. Gray, *High Strain-Rate Testing of Materials: The Split Hopkinson Bar. Methods in Materials Research*. John Wiley Press, 1997.
16. H. Couque, R. Boulanger, and F. Bornet, *J. Phys. IV France* **134**, 87 (2006).
17. T. Özel, and Y. Karpat, *Mat. Manufact. Proc.* **22**, 659 (2007).
18. G. R. Johnson, and W. H. Cook, *Eng. Fract. Mech.* **21**, 31 (1985).
19. R. C. Dorward, and K. R. Hasse, *J. Mater. Eng. Performance* **4**, 216 (1995).
20. M. Ignat, S. Lay, F. R.-Dherbey, C. Seguineau, C. Seguineau, C. Malhaire, X. Lafontan, J. M. Desmarres, and S. Brida, "Micro tensile tests on Aluminum thin films: tensile device and in-situ observations," *Mater. Res. Symp. Soc. Proc.*, in press (2008).

Mater. Res. Soc. Symp. Proc. Vol. 1139 © 2009 Materials Research Society 1139-GG04-04

Micro Tensile Tests on Aluminium Thin Films: Tensile Device and In-Situ Observations

M. Ignat[1], S.Lay[1], F. Roussel-Dherbey[1], C. Seguineau[1,3] , C. Malhaire[2], Xavier Lafontan[3], J.M. Desmarres[4], S.Brida[5]

[1]ENSEEG INPG, SIMAP UMR 5266, Université Joseph Fourier, F-38402, France.
[2]Université de Lyon, INSA-Lyon, INL, CNRS UMR 5270, Villeurbanne, F-69621France.
[3]Nova MEMS, 10 av. de l'Europe, Ramonville, F-31520, France.
[4]CNES, DCT/AQ/LE, bpi 1414, 18 Avenue Edouard Belin, Toulouse F-31401, France.
[5]AUXITROL S.A. , Esterline Sensors Group, Bourges, F-18941, France.

ABSTRACT

The results of micromechanical tensile experiments performed on thin aluminum samples are presented and discussed. The micro tensile test system and the design of the samples, based on finite element modeling (FEM), and their production by micromachining are briefly described. Some examples of the stress strain curves are presented. The Young's modulus and critical parameters (flow and rupture stress and strains) are reported. The micro structural changes induced by the tensile experiment were observed during and after the testing by scanning electron microscopy (SEM) and transmission electron microscopy (TEM).

INTRODUCTION

To understand the reliability issues that limit MEMS design and application, it is essential to have adequate understanding of the mechanical stability of the materials to be used; cconsequently, determining the mechanical properties for MEMS materials continues to be a fundamental research objective. Material parameters, such as Young's modulus and the flow and failure stress, which depend both on microstructure and applied loading modes, need to be measured; see, for example, references[1-3] . In contrast with the case of bulk materials, for thin films a lack of information on their mechanical properties will be a continuing problem. Since their mechanical properties are strongly dependent on the microstructure and, as the microstructure will change with the deposition process, for the same thin material, different processes will produce different microstructures, and different mechanical behavior. Generally, micro tensile experiments on thin self standing samples, differ either in the way of attaching the samples to the tensile device (detached or semidetached with respect to the substrate), or the sample geometry, which often determines how the displacement and force are applied and measured during a tensile experiment [1-3]. The new equipment is based on a device previously developed to perform in-situ tests in a SEM, on film on substrate systems (for details of that previous device and some results obtained with it see references [4-6]).

Even though with that device it was not possible to obtain significant results on self standing thin films, the original concept: to induce a symmetrical displacement of the grips, thus avoiding any drift in image observation, was kept for the micro tensile system described here. The preparation of submicron thickness, self standing, tensile-test specimens of metallic or dielectric thin films can be accomplished with the use of standard silicon technology processes.

EXPERIMENTAL DETAILS

The micro tensile testing machine is shown in Figure 1. The grips are moved symmetrically by the rotation of a central endless screw, actuated by a DC micro motor. A laser displacement sensor measures the relative displacement of the two movable grips. Forces are measured using miniature piezo resistive load cells, with maximum ranges of 40N (for metal on polymer samples, of about 50μm thick) or 1.5N (for self standing aluminum samples, of 1μm and less thick). Two piezoelectric transducers, inserted in exchangeable grips, allow cyclic deformation. The system can be installed in a scanning electron microscope in order to perform in-situ tensile testing of micro-scale specimens. A fixed displacement rate is used for testing.

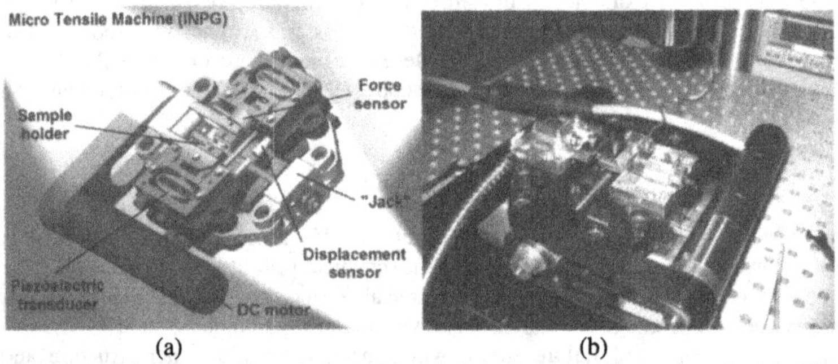

(a) (b)

Figure 1. (a) Schematic representation of the micro tensile testing device. The term "Jack" indicates the elements that will transfer the displacements induced by the rotation of the associate endless screw. (b) View of the device installed on a testing bench.

Our specimens were designed to deal with the problem that stress concentrations always occur near the end of the gauge length of a tensile sample, which is generally located near the clamping zone. These stress concentrations very often produce a premature fracture of the specimen. To avoid or diminish them, and to distribute homogeneously the deformation along the gauge length of the sample, the samples must be designed with rounded ends, which progressively increase the sample section towards the clamping zone. This is a well known approach based on earlier macroscopic tensile tests.

The circular shoulders give to the tensile samples the so-called "dog-bone" shape. However, even though the design (curvature radius), is well established and defined for macroscopic samples, for very thin samples it becomes a critical problem. During our first experiments we noticed that the final rupture of the samples was observed regularly at these shoulders and close to the sample's clamping, demonstrating the local stress concentrations, and making the validity of the test doubtful. Next, the samples were designed with a single circular shoulder. This was changed after analyzing by 3D Finite Element Modeling (FEM) several shapes for the tensile samples: a double curvature transition was chosen for the transition zone (see Figure 2).

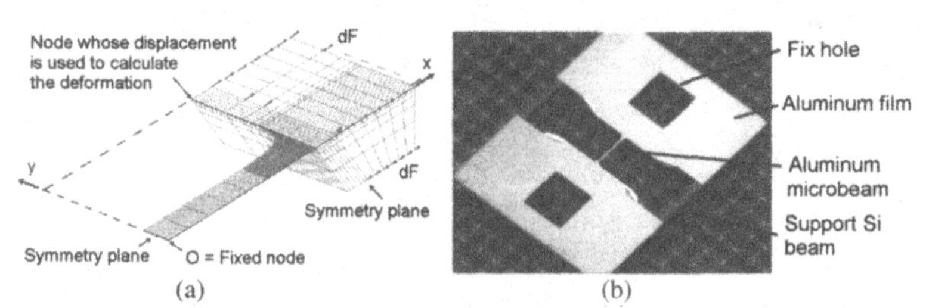

Figure 2.(a) Finite element mesh, boundary conditions and nodes of applied loads on a thin film sample.(b) self standing sample (each little white square has 1mm per side).

Two main results for the sample design were obtained from the FEM analysis. First, the maximum stress generated at a sample shoulder was reduced by more than 10 %, and second: the stress distribution towards the uniform part of the sample was improved [7,8] With this new sample shape, the micro machined specimens are prepared on double side polished, p-type <100> silicon wafers using a two-mask fabrication process. On one side of the wafer (front side), 0.5μm to 1μm thick aluminum is sputtered. The metallic film is patterned by wet etching to define the shape of the tensile specimens. By dry etching the backside of the wafer, the windows to obtain the self standing part of the sample are opened. The free-standing beams are released by a final two-sided etching process [7]. Deep trenches etched around every sample allow an easy separation of the specimens from their wafer. Each complete die, supporting a detached beam, is 14mm wide, 18mm or 23.5mm long and 400μm thick. A detached sample can be mounted on the movable grips by clamping it with screws through the holes that have been etched through the silicon frame. The detached parts of the aluminum films were: 3 mm long (2mm gage length), about 200μm up to 500μm wide, and 0.5 μm to 1μm thick. The silicon frame insures the rigidity of the specimen during the process of installing and fixing the sample on the grips. A temporary tension can then be applied to the frame, checking the machine stiffness and verifying the alignment of the clamping of the sample. The silicon frame is then cut away with a dentistry drill, and the sample is ready to be pulled.

DISCUSSION

We consider first some true stress vs strain curves, which vary with the applied strain rate (Figure 3a), or the presence of thought thickness defects (Figure 3b). The sample thickness, from 800nm to 1000 nm, had no significant effect on the stress vs strain curve.

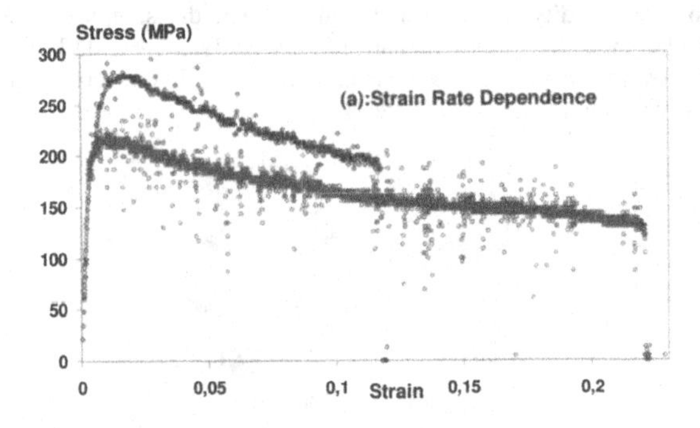

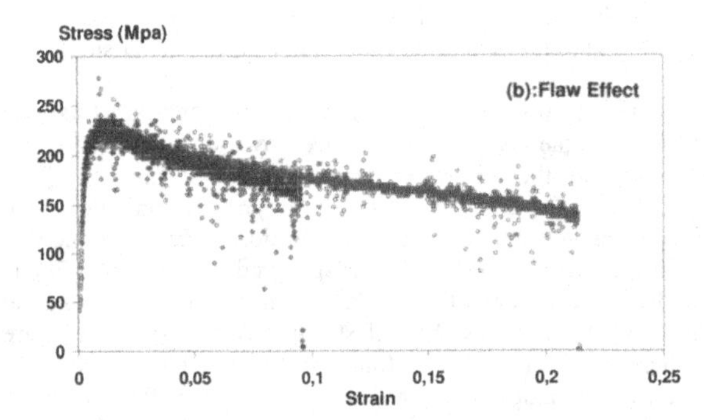

Figure 3. Stress vs strain curves showing: a strain rate effect (a). The shorter elongation but higher stresses correspond to an experiment performed at 1 10^{-3} sec^{-1}, the other curve was obtained with a strain rate of 1, 6 10^{-4} sec^{-1}. On (b), the shorter curve corresponds to a sample with a visible through thickness cavity of about 7 μm diameter.

As may be seen in the tensile curves, a faster strain rate or the presence of a critical defect (through thickness circular cavity) reduces the total elongation of the sample; but flow and rupture strength increase with increasing strain rate. Our results show several main characteristics about the tensile response of our films:

- The corresponding strain at maximum stress never exceeds 1, 5% to 1, 8%.
- After reaching the maximum stress, the stress vs. strain curves show a progressive decrease with sample strain of 20% and more (samples without optically observed defects, and with an applied deformation rate of 0,5 μm/sec).
This particular behavior can be understood in terms of the micro structural evolution, as discussed below.
- The mechanical parameters derived from our curves, for instance the Young's Modulus, ranged from 58 GPa to 70 Gpa, while the flow stress was in almost all cases about 155+/-5 MPa, and the maximum stress about 220 MPa, with the exception of 280 MPa reached at a higher strain rate.
-Our samples showed a mean grain size of about 150nm (Figures 4).

The Young's modulus values we obtained agree with the order of magnitude of bulk aluminum values (see for example [9]). When comparing the flow stress values, we note that with respect to previous works on Al thin films: the higher flow stresses, are associated to microstructures with the smaller grain sizes. Indeed, the flow stress for our samples with a mean grain size of 150 nm never exceeded 160 MPa. These values are higher with respect to results reported for thin samples with grains ranging from 300nm to 1000nm: maximum 100 MPa, [10,11], but lower, compared to the flow stress of samples with a mean grain sizes ranging from 50 nm to 20 nm: which present flow stresses increasing from 200 MPa up to 310 MPa [12,13] . These results, point out a grain size effect, predictable through a Hall Petch type relation [14, 15].

With respect to the deduced strains, the total strains for our samples could reach 20%, when applying strain rates on the order of 10^{-4} sec^{-1}. This high tensile elongation includes a short initial strain hardening stage: never exceeding 2%. This short stage may be controlled by slip and dislocation interactions, in favorably oriented grains. Following this stage, an extended and continuous strain softening stage develops, characterized by little deformation bands appearing progressively on the sample's surface with increasing strain (Figures 4 and 5).

The TEM observations (Figure 4), showed mainly two populations of grains, oriented <001> and <111> with respect to sample's surface. For these grains, the direction of the tensile axis aligns parallel to a <110> in plane type direction. The other <110> type directions at 45°, correspond to the directions of the observed deformation bands (Figure 5). Thus, depending on the grain orientations (mainly <001> or <111>), two shear mechanisms may act:
-A limited intra granular slip of dislocations activated on <110> (111) slip systems favorably oriented with respect to the tensile axis; then,
-the inter granular sliding, which appears determined along the maximum resolved shear directions of the sample (deformation bands).

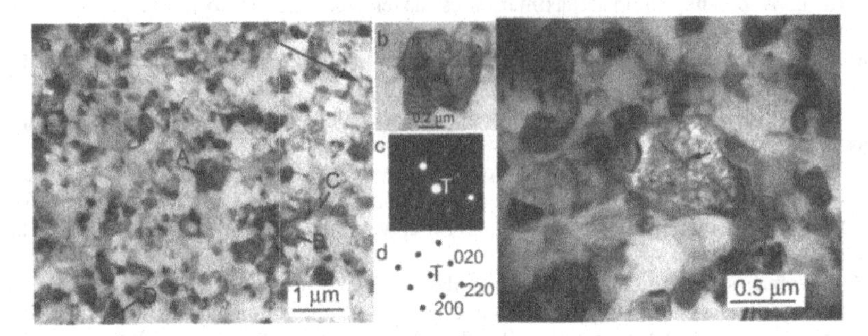

Figure 4. TEM micrographs from a deformed sample. The black arrow on the left micrograph corresponds to the tensile axis, which is parallel to a <110> direction in the <001> oriented grains. The letters A, B, C and D correspond to diffracting grains, oriented <001>, with respect to the sample's surface. The T letters on the diffraction patterns correspond to the incident beam (e⁻s//<001>). The right micrograph shows a dislocation in the grain A pointed out on the left micrograph.

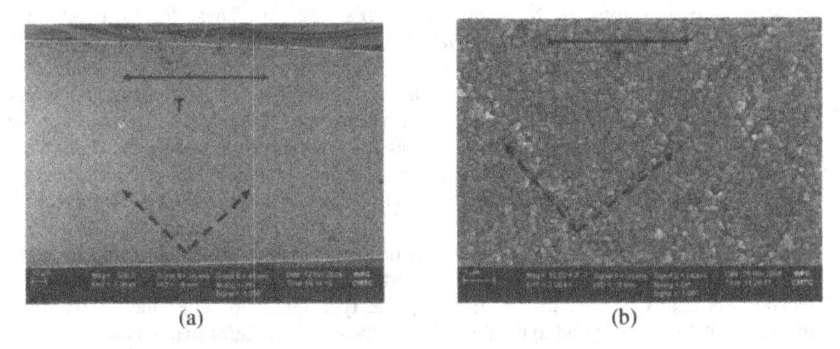

(a) (b)

Figure 5. FEG SEM, showing at lower magnification (a) then at higher magnification (b), the observed in-situ deformation bands (dashed lines); the tensile axis is indicated by the black T arrows. The SEM observations were performed after the complete tensile deformation of a sample, at a strain rate of 1, 6 10^{-4} sec^{-1} (see stress vs strain curve on Figure 3a).

CONCLUSIONS

After a crucial change in the design of the samples to avoid stress concentrations near the extremities, we developed a useful new experimental procedure. The parameters deduced from our experiments: Young's modulus, flow and rupture stress, are in agreement with values reported in the literature for thin aluminum samples. These parameters appear to be dependent on the mean grain size presented by the microstructure. With respect to the deformation mechanisms which drive the tensile behavior, the observed short hardening stage is associated to limited dislocation slip while the long softening stage, to grain boundary sliding, activated along the resolved shear directions of the sample.

ACKNOWLEDGMENTS

The authors would like to thank Professor Paul A. Flinn for discussion and critical review of the manuscript.

The research developed on thin films was initially supported by Intel Components Research. The results presented here have been funded by the French National Research Agency (ANR).

REFERENCES

1. J.A.Schweitz,*MRS Bulletin* 17 (7),34,(1992)
2. W.N.Sharpe, *Mater.Res.Soc.Symp.Proc.*Vol.1052 (2008)
3. *Reliability of MEMS, Advanced Micro and Nanosystems,*O.Tabata and T.Tsuchiya Editors, Wiley Verlag GmbH,Vol 6 (2008)
4. M.Ignat , "Stresses and mechanical stability of CVD thin films" Chapter 3 in *Surface Engineering Series* Ed. J.H. Park. ASM International Series Vol 2, (2001) pp 45-80
5. M.Ignat.*Key Eng. Mat.*, Vol.116-117,. 290 (1996)
6. M.Ignat,T.Marieb,H.Fujimoto,P.A.Flinn,*Thin Solid Films* 353,.201 (1999)
7. C.Malhaire,M.Ignat,K.Dogeche,S.Brida,C.Josserond and L.Debove. *Proceedings of Transducers'07 and Eurosensors XXI*,Symp.2. Vol.1,623 ,(2007)
8. C.Malhaire,C.Seguineau,M.Ignat,C.Josserond,L.Debove,S.Brida,J.M.Desmarres, X.Lafontan,*Review of Scientific Instruments* (2009)(in press)
9. M.F. Asby. *Material Selector* Granta Design. Cambridge University.(2005)
10. G.Cornella "Monotonic and cyclic testing of film materials for MEMS applications" PhD, Stanford University (1999).
11. D.Read, Y.Cheng,R.Keller and JD McColskey, *Scripta Materialia* 45, 583 (2001)
12. M.A.Haque,M.T.Saif, *Scripta Materialia*,47,863 (2002)
13. D.S. Gianola,S.Van Petegem,M.Legros,S.Brandstetter, H.Swygenhoven, KJ Hemker, *Acta Materialia* 54, .2253 (2006)
14. A.H.Choski,A.Rosen, J.Karch,H.Gleiter, *Scripta Materialia*,23, 1679 (1989).
15. G.E.Fougere,J.R.Weertman,R.W.Siegel,S.Kim, *Scripta Materialia*,26,1879 (1992)

Mater. Res. Soc. Symp. Proc. Vol. 1139 © 2009 Materials Research Society 1139-GG04-09

Strength and Fatigue Life of Nanocrystalline Titanium/Platinum Multilayer Membranes for Implantable MEMS Reservoir Array Devices

Karl B. Yoder[1], John M. Maloney[2], and Jonathan Coppeta[1]
[1]MicroCHIPS, Inc., 6-B Preston Court Bedford, Massachusetts, 01730, U.S.A.
[2]Materials Science and Engineering, Massachusetts Institute of Technology, Cambridge, Massachusetts, U.S.A.

ABSTRACT

Thin-film metal membranes that can be electrically opened when desired are used in MEMS reservoir array devices designed for long-term medical implants in order to protect and selectively expose drugs or biosensors; one such device is an implantable continuous glucose monitor. The membranes' unique role as reservoir caps directly impacts the safety, efficacy, and potential lifetime of such medical implants. Consequently, these sub-micron metal films are evaluated by their mechanical behavior in addition to their electrical properties and biocompatibility. We have investigated the microstructure and mechanical properties of sputtered titanium and platinum multilayer membranes of thickness 380 nm and 680 nm with TEM, rupture, and nanoindentation tests. Additionally, a novel fatigue-testing apparatus was designed and used with over one thousand membranes for 100 million cycles in simulated *in vivo* conditions without membrane failure.

INTRODUCTION

One difficulty in creating long-term implantable biosensor or drug-delivery devices is to protect the drug or biosensor from the *in vivo* environment, yet preserve functionality for extended periods of time. One successful approach was the creation of a device with multiple, protected reservoirs that can individually be exposed to the body environment whenever desired [1-4]. MicroCHIPS, Inc. produces both biosensor and drug-delivery devices, which are currently being qualified for human clinical trials.

In addition to electronics, telemetry, and a power source, the devices consist of MEMS modules containing sensors or drug within hermetically sealed reservoirs. The membranes sealing the reservoirs are thin metal films, which can be opened by passing a current through them. The activation process that melts the metal, takes less than 10 μs. Electrical activation and biocompatibility considerations have led to the selection of platinum-passivated sub-micron titanium films as the material system of choice; Ti for mechanical and electrical properties and Pt for passivation. However, in order to demonstrate the requisite device robustness to serve as implantable medical devices, it is important to characterize the mechanical strength and integrity of the membranes.

In this report, we present results of our investigations into the mechanical behavior of sputtered titanium/platinum membranes of total thickness 380 nm and 680 nm along with the microstructure of the films as characterized with SEM, TEM and X-ray diffraction.

EXPERIMENT

Low stress LPCVD nitride films (0.2 μm) were grown on 200 μm thick [100] silicon wafers with 1.5 μm thermal oxide. This process was followed by sputtering a 20 nm Ti adhesion layer, 40 nm platinum, 300 or 600 nm Ti, and finally a 40 nm platinum as top passivation layer. A series of backside RIE and DRIE steps were used to etch a single large cavity beneath silicon lattice containing an array of windows covered with metal membranes. Most of the MEMS chip samples studied in this report consisted of a single cavity measuring 1500 μm by 830 μm with 18 membrane-covered windows, each 100 μm square, set in a 50 μm thick silicon scaffold (Figure 1). Unpatterned Pt/Ti/Pt films atop LPCVD nitride and thermal oxide were produced during the same sputter deposition runs and used for nanoindentation testing and microstructural characterization.

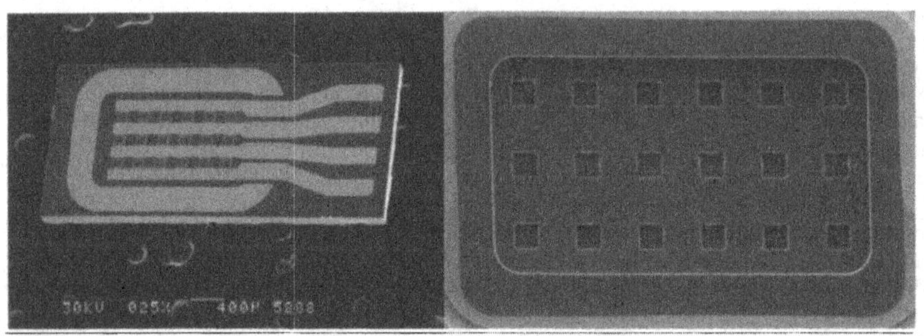

Figure 1. (Left) Top view of a 200 μm thick, 18-membrane MEMS chip; gold traces that extend to the edge of the chip are used to activate the membranes which are seen as 3 rows of 6 membranes between the traces. (Right) Rear view of chip; a cavity for containing a biosensor or drug payload lies under a silicon lattice of membrane-covered windows.

The strength of the suspended membranes was determined by rupture testing entire arrays. A fixture was built to apply high pressure N_2 to the top side of an array of membranes suspended over a single cavity (i.e., pressure was applied opposite the cavity). Images of the membranes were recorded as the pressure was increased until all membranes failed, and the failure pressure for each noted. The pressure was ramped at 5 psi/s.

Nanoindentation was performed on a TriboIndenter (Hysitron, Inc.) located at the MIT NanoMechanical Technology Laboratory. Testing was performed with a Berkovitch tip using load ranges between 100 and 10,000 μN at a loading of 100 μN/s. Data were analyzed with both a modified Doerner-Nix approach [5-6] and the Oliver-Pharr method [7].

A unique fatigue apparatus was developed and built to allow low-stress, high-cycle fatigue testing of the membranes in a simulated *in vivo* environment. In this procedure, 24 MEMS chips were mounted via o-rings at the bottom of a stainless steel chamber that was filled with phosphate buffered saline (PBS) and maintained at 37±1 °C. A second sealed cavity below each chip contained a glass rod serving as a light-pipe. A tracer dye (fluorescein) was added to the PBS, so that the failure of any membranes would fill the lower cavity with dye. A UV light source illuminated the array of light-pipes, and a camera recorded the output, to allow *in situ* determination of any membrane failure during the test. A PID-controlled high-load voice coil was used to apply a sinusoidal pressure profile to the chamber through a neoprene diaphragm. A Lab View program running on a PC operated the system and recorded camera images at 1 Hz.

The mean pressure applied to the chamber during the fatigue testing was 170 kPa with a stress amplitude 100 KPa at a frequency of 20 Hz; applied pressure was significantly greater than anticipated *in vivo*.

RESULTS AND DISCUSSION

Microstructure

Cross-sectional TEM and X-ray diffraction and reflectivity were used to determine the grain size and microstructure of the films. Focused Ion Beam (FIB) cuts were made at the edge of the membranes to make samples suitable for TEM (Figure 2). Roughness in the film arises from specific columnar Ti grains, and the film roughness of the thicker Ti layer was noticeably greater. X-ray diffraction and reflectivity data show the microstructure of Pt layers for both Ti thicknesses were the same: FCC with a very strong <111> orientation, greater than 95% crystallinity, compressive stress about 350 MPa and columnar crystallites about 1 nm in plane and 23 nm out of plane. The microstructure of Ti for both thicknesses was also similar: HCP with a very strong <110> orientation, greater than 95% crystallinity, and columnar crystallites about 125 Å in plane and 33 nm (thin) or 35 nm (thick) out of plane. Residual stress in the Ti could not be determined. However, wafer bow measurements indicate total residual stress for the entire stack to be about 120 MPa compressive, so the Ti can be inferred to be more tensile than the compressive Pt.

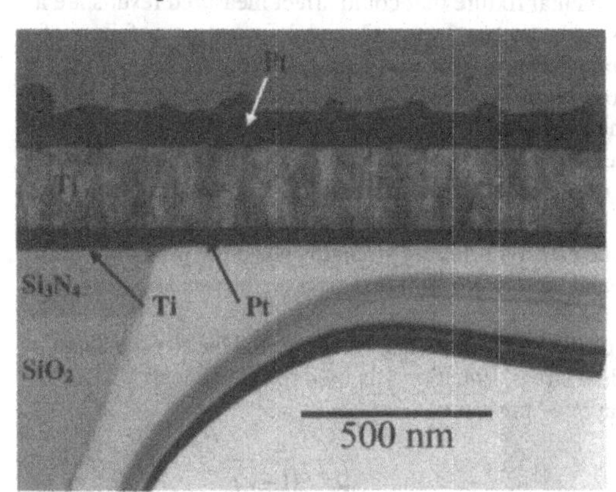

Figure 2. TEM micrograph of FIB-cut cross section of a suspended 380 nm thick membrane stack (prior to the FIB cut, an additional Pt layer was added). The stack layers and Ti grain structure are visible along with the geometry of the corner on the left. No DRIE or RIE undercutting is evident (the angle of the cut is not quite vertical against the membrane). The lighter curved layers near the center of the image are epoxy used during sample preparation.

Rupture Testing

Multiple chips from six wafers were tested on the rupture tester, three each for membranes with 380 or 680 nm metal stacks. The results from rupturing several thousand membranes are presented in Figure 3. The data were analyzed as fraction surviving, and fit to a two-parameter Weibull model, also shown. The metal stack with the thinner Ti layer is both stronger and more consistent from wafer to wafer than the thicker film: 95% of all thin film membranes survived a pressure of 440 psi, while thick films from different wafers had between 0 % and 70% surviving the same pressure.

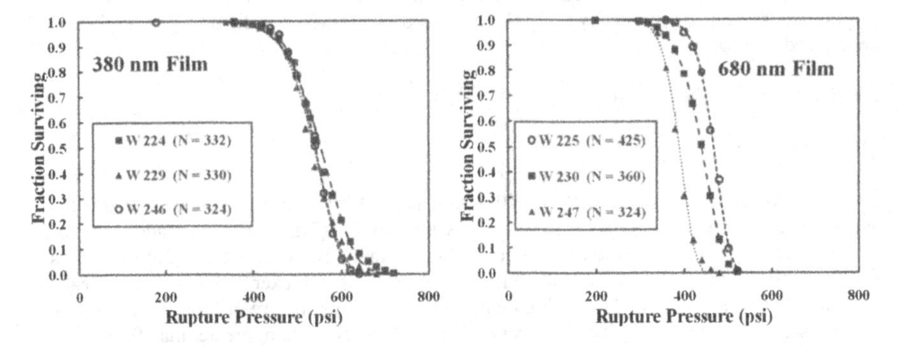

Figure 3. Survival plots from three different wafers each for the thin and thick Ti film stacks (data points represent average survival fraction for each 20 psi). The thicker Ti layer (right) exhibts lower mean and maximum rupture stress, as well as more scatter in the data. The lines represent fits to a two-parameter Weibull model for each wafer, where the number of individual ruptured membranes is indicated in the legend.

Potential problems with the experimental fixture that could affect measured results are a pressure drop in the cavity as membranes in the array start to fail, and deformation or failure of the silicon lattice supporting the array of membranes. Because the cross-sectional area of the supply line was much greater than the cross-sectional area of the windows as membranes failed, the cavity pressure was at least 95% of the measured line pressure even after 17 of 18 membranes had failed. The silicon lattices do deform elastically during rupture testing, but an analysis showed no location-dependence of the order of membrane failure in the array which would be expected if elastic deformation of the lattice significantly affected the rupture stress of the membranes. Additionally, there was no effect on rupture strengths when the pressure ramp rate was changed between 1 and 10 psi/s and there were no transient creep effects as evidenced by pressure hold experiments.

The approximate failure strength of the membranes can be calculated based on a bulge testing analysis [8-10]. Neglecting residual stress, Equation 1 is used to calculate an approximate deflection at the maximum applied pressure:

$$P = \frac{t}{a^2} d \left[C_1 \sigma_0 + \frac{f(\nu)}{a^2} \frac{E}{1-\nu} d^2 \right] \quad \text{or} \quad d = \sqrt[3]{\frac{a^4 P(1-\nu)}{tEf(\nu)}} \qquad (1)$$

P is the applied pressure, d is the deflection at the center of the membrane, ν is Poisson's ratio, E is the Young's modulus, σ_0 is residual stress, C_1 and $f(\nu)$ are geometric constants which for a square membrane are 3.45 and 1.994(1-0.271ν) [9], and $2a$ is the width and t the thickness of the membrane. Based upon a volume average, the elastic constants of the film stacks are E = 128, ν=0.35 and E = 123 and ν=0.35 for the 380 nm and 680 nm membranes, respectively. Finally, using the analysis from Chalekian et al for a square membrane [10], a fracture stress is calculated from deflection:

$$\sigma = \frac{0.872E}{\left(1-\nu^2\right)}\frac{d^2}{a^2} \qquad (2)$$

Nanoindentation

The hardness is generally interpreted as about 3 times the flow stress at 8% strain. Flow stress data (from nanoindentation hardness data) for thick and thin films are reported in Figure 4. Data are plotted as a function of the normalized contact area ($\sqrt{A}/h$) where h is the film thickness, allowing for comparison of data between films of different thickness. Surface roughness caused some scatter in the data, especially at shallow indents. Hall-Petch data are also presented, with reference data for commercially pure Ti [11]: grain size used is the crystallite size measured with x-ray diffraction, while flow stress data from rupture experiments were calculated (Eq. 2) from the pressure causing 90% of membranes to fail.

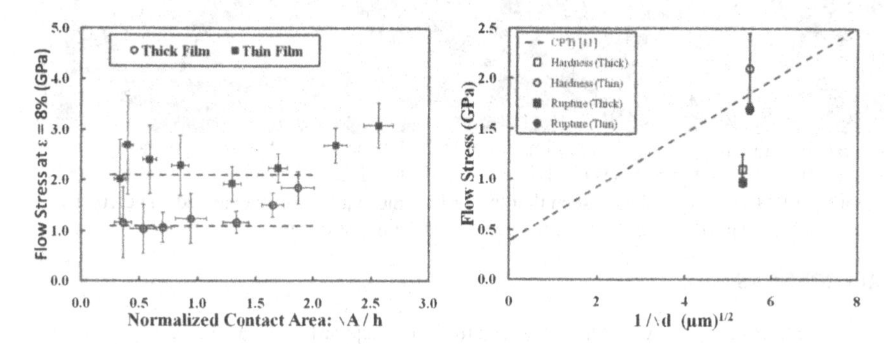

Figure 4. Indentation hardness data plotted as a flow stress versus normalized contact area (left). A Hall-Petch relationship for CP Ti (right) with stress data from the membranes from either average rupture stress (solid markers) or indentation hardness (open markers).

Fatigue Testing

Typical membrane deflections at the maximum applied pressure of 270 kPa were ~2 μm, so from Eq. (2) the maximum fatigue stress was about 255 MPa. Though over 1000 membranes (of both thicknesses) have been tested to 100 million cycles, there have been zero failures and as can be seen in Figure 5, the surface morphology of the membranes remains unchanged. The lack of evidence of stress-corrosion cracking indicates the Pt passivation layer also remains robust.

CONCLUSIONS

Thin Pt-passivated Ti membranes have high strength and fatigue resistance, with thinner Ti layers stronger than thicker layers. The strength increase, seen in both hardness data and maximum rupture stress, does not correlate with the size of coherently-diffracting domains that give rise to the crystallite-size measurements in the x-ray analysis. However, it is possible that grain sizes in the thick films are actually larger (which is somewhat evidenced by the rougher surface), or point defect density rather than crystallite size controls the dislocation motion; future studies of the defect density of the two film thicknesses are anticipated.

Fatigue testing results are consistent with fatigue behavior of bulk Ti with its high fatigue limit. However, the results are somewhat surprising, since sputtered nanocrystalline films should have a higher surface defect density and it could be expected that the fatigue limit would be lower. Further studies are directed to testing membranes at higher fatigue stress levels.

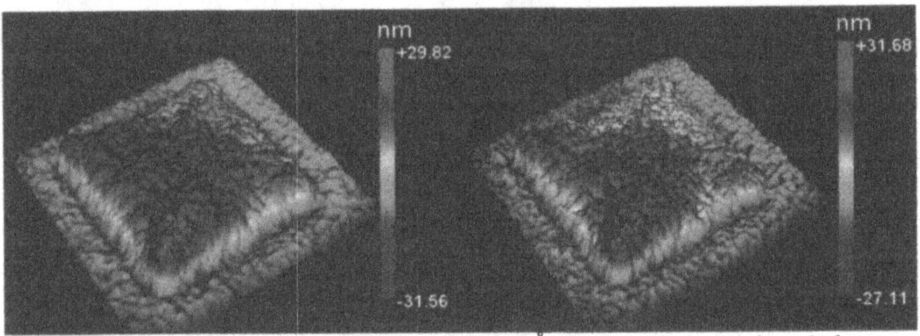

Figure 5. White light interferometry images of a 6800 Å membrane before (left) and after (right) 100 million fatigue cycles (20 Hz) in 37 °C buffered saline (mean stress 170 kPa, stress amplitude 100 kPa). It is clearly seen that the surface morphology is unchanged. The 100 μm square of the suspended membrane is the central region in the image.

REFERENCES

1. J.M. Maloney, S. A. Uhland, B. F. Polito, N. F. Sheppard Jr., C. M. Pelta, and J. T. Santini Jr., *J. Controlled Release* **109**, 244-255 (2005).
2. J. H. Prescott, S. Lipka, S. Baldwin, N. F. Sheppard Jr., J. M. Maloney, J. Coppeta, B. Yomtov, M. A. Staples, and J. T. Santini Jr., *Nature Biotechnology* **24** (4), 437-438 (2006).
3. J. H. Prescott, T. J. Krieger, S. Lipka, and M. A. Staples, *Pharmaceutical Res.* **24** (7), 1252-1261 (2007).
4. E. R. Proos, J. H. Prescott, and M. A. Staples, *Pharmaceutical Res.* **25** (6), 1387-1395 (2008).
5. D. S. Stone and K. B. Yoder, *J. Mater. Res.* **9**, 2524 (1994).
6. M. F. Doerner and W. D. Nix, *J. Mater. Res.* **1**, 601 (1986).
7. W. C. Oliver and G. M. Pharr, *J. Mater. Res.* **7**, 1564 (1992).
8. J. S. Mitchell, C. A. Zorman, T. Kicher, S. Roy, and M. Mehregany, *J. Aerospace Engineering* **16** (2), 46-54 (2003).
9. D. Maier-Schneider, J. Maibach and E. Obermeier, *J. Microelectromech. Syst.* **4** (4), 238-241 (1995).
10. A. J. Chalekian, R. L. Engelstad, and E. G. Lovell, *Mat. Res. Soc. Symp. Proc.* **795** U11.34 (2004).
11. M. J. Donachie, *Titanium: A Technical Guide, 2nd Ed.*, (ASM International, 2000), p. 144.

MEMS Reliability and Tribology

Mater. Res. Soc. Symp. Proc. Vol. 1139 © 2009 Materials Research Society 1139-GG05-04

Investigation of gold sputter coated vertically aligned multi-walled carbon nanotubes for RF MEMS contact surfaces.

E.M. Yunus, S.M. Spearing and J.W. McBride
School of Engineering Sciences, University of Southampton, SO17 1BJ, UK

ABSTRACT

A novel approach has been developed for the long-standing problem of achieving durable contact surfaces for Radio Frequency MEMS switches. The performance of gold-coated vertically aligned multi-walled carbon nanotubes is investigated under the low force electrical contact conditions typical of MEMS switches. The carbon nanotubes provide a compliant support for the conducting gold surface layer, minimizing the degradation, and adhesion between the contact surfaces which is the principal source of wear in such contacts. Greatly increased lifetimes have been obtained over monolithic gold contact surfaces.

INTRODUCTION

MEMS RF switches have been viewed as a promising technology for nearly two decades. The promise has not been fully realized, in part due to the lack of sufficiently durable contact materials. There are a number of materials commonly used for this application including noble metals such as gold [1]. The weakness of such materials is that they are relatively soft and wear easily. Carbon nanotube (CNT) coated surfaces have previously been investigated for electrical contacts [2,3]. In recent experiments [4], Au sputter-coated (5μm thick) vertically aligned multi-walled carbon nanotubes (MWCNT) 100μm long were investigated and found to perform comparably to gold alloys. In our own previous work, [5,6] low numbers of load cycles (up to 1000) were applied without current. In the present study, tests were conducted in "dry-circuit" and "hot-switched" conditions out to higher numbers of cycles with the aim of examining the stability of Au/MWCNT composite contacts under realistic MEMS contact conditions.

MATERIALS AND TEST METHODOLOGY

The test geometry used in the present work consisted of one flat surface contacting a 2mm diameter hemisphere. In all cases the hemisphere consisted of a stainless steel core, sputter coated with Au, ~500 nm thick, the total surface roughness $R_a \approx 400$nm. In the Au to Au case the flat surface was a silicon substrate (~2mm by 7mm), sputter coated with Au ~500 nm, with a surface roughness $R_a \approx 30$nm. For the Au to Au/MWCNT case, a "forest" of MWCNTs was grown on the Si wafer using thermal CVD. The catalyst used was sputter-deposited Fe and the gaseous carbon source was ethylene. The growth temperature and time were 875°C and 3 minutes respectively to produce a dense forest of vertically aligned MWCNT of an average length of ~50μm. The MWCNT forest was then sputter-coated with Au to produce composite coatings as shown in Fig 1. The Au penetrated the MWCNT surface to a depth of 700nm-4 μm.

The performance of the surfaces was determined using a PZT actuator to apply cyclic contact loading as shown in Fig. 2. This experimental setup is described more fully in [6]. A signal generator with voltage amplification was used to drive the PZT actuator. Typically frequencies of 10 Hz were applied. The dynamic force is measured using a high-resolution load cell situated under the Au-coated hemisphere.

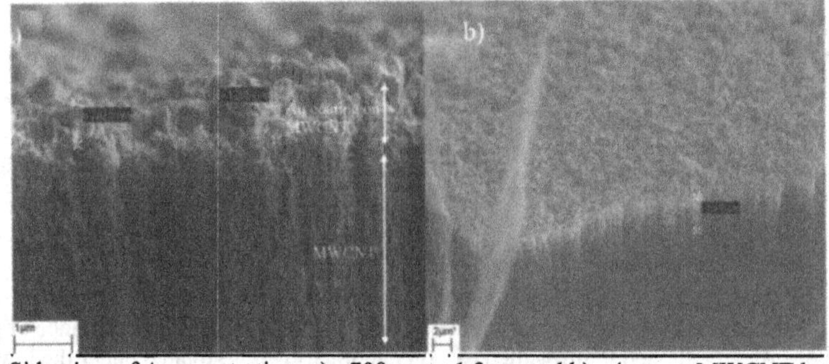

Fig. 1. Side view of Au penetration a) ~700nm to 1.3μm and b) ~4μm on MWCNT by sputtering.

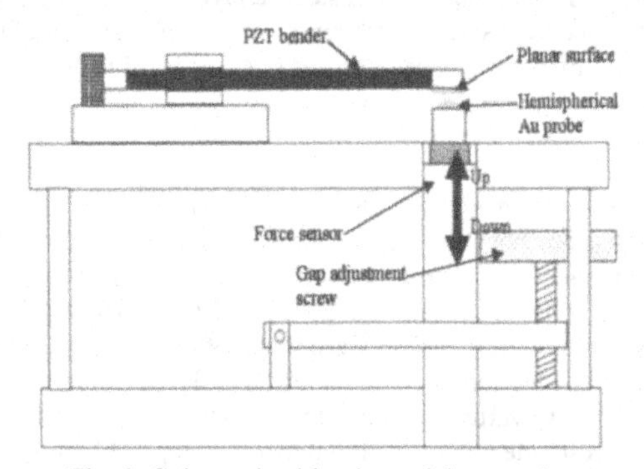

Fig. 2: Schematic side view of the test rig.

In the "dry" (no current) tests load levels of 1 mN and 3mN were employed. After predetermined numbers of cycles quasi-static cycles were applied so that the contact resistance, R_c could be determined using a 4 wire-measurement. The DC current source across the contact surfaces was set at 1mA and 20mV using a micro-ohmmeter. During testing the apparatus was enclosed and held at ambient air and room temperature. Fig. 3 shows an example of the load history. Fig. 4 shows the variation in R_c with cycles.

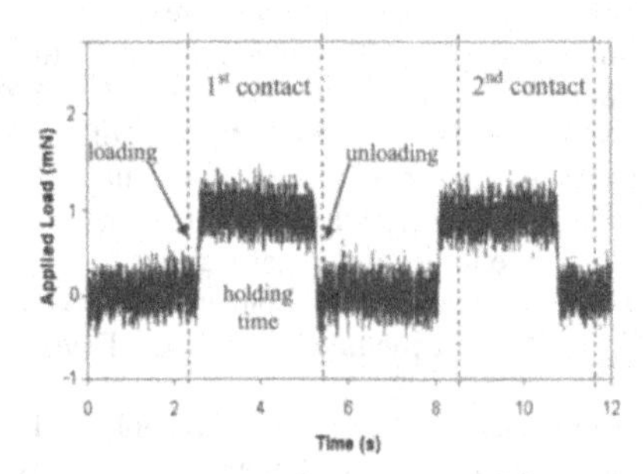

Fig. 3. Example of load cycles (0.2 Hz) load cycling for Au-Au pair.

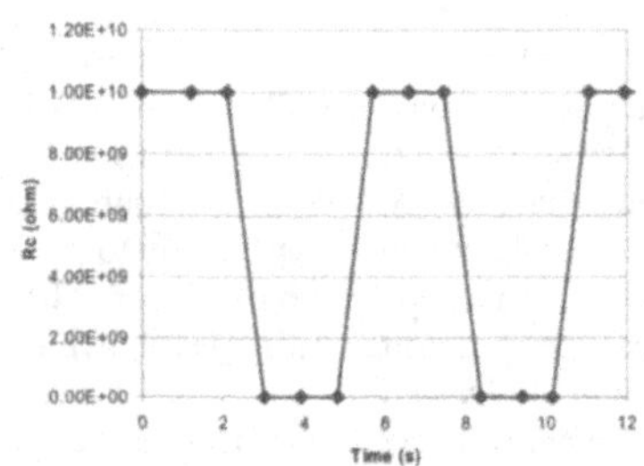

Fig. 4. Example of contact resistance during Cycling for an Au-Au pair at 1mN.

The same methodology was used for the hot switching tests, in which a constant applied force of 1mN was used. Currents of 1mA or 10 mA at 4V were applied. 4V, 1 mA and 1 mN are typical maximum values for RF-MEMS switches. 3 mN and 10 mA represent higher values than would be used in practice [7,8,9], but allowed for accelerated testing.

RESULTS AND DISCUSSION

For "dry" tests on Au-Au contact pairs [5] it was found that the initial contact resistance was ~0.2-0.5Ω, which increased rapidly to 4-6Ω at ~430 cycles. This is due to the purely mechanical deterioration of the Au-Au pair surfaces, reflecting the recognized problems of using soft metals for electrical contacts. It is thought that the mechanism consists of initial smoothing of the Au surfaces which leads to increased adhesion and then material transfer [7]. The smoothing is the result of the repeated impacts and time-dependent deformation of the Au [10]. Figure 5 shows the corresponding behavior for the contact resistance of a Au-Au/MWCNT pair. The initial contact resistance is higher, but it remains essentially unchanged over nearly 2 million load cycles (vs 430 cycles for Au-Au). The compliance of the MWCNTs helps to reduce the deterioration. It is also likely that the increased surface roughness of the Au/MWCNT contact surface is responsible for the high initial contact resistance as well as playing some role in the greatly increased lifetime.

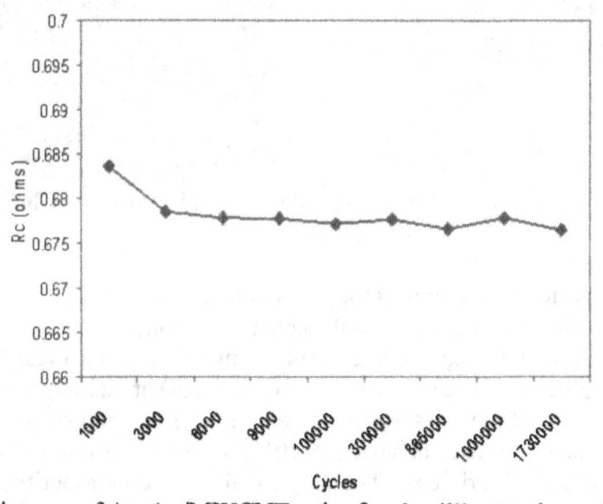

Fig. 5. Contact resistance of Au-Au/MWCNT pair after 1 million cycles at 1mN.

A scanning confocal laser image of the Au/MWCNT contact surface is shown in Fig. 6. This reveals permanent deformation of the composite surface, consistent with the shape of the hemisphere. Further cycling at higher load levels showed signs of deterioration. Figure 7. shows two corresponding SEM images of the Au hemisphere surface and the Au/MWCNT surface after 20 million cycles at a maximum applied force of 3mN. Tearing/cracking of the Au was observed consistent with adhesion and/or fatigue of the Au. This deterioration had not yet resulted in any measurable change in the contact resistance. No failures were observed at 1mN dry switching up to 2 million cycles.

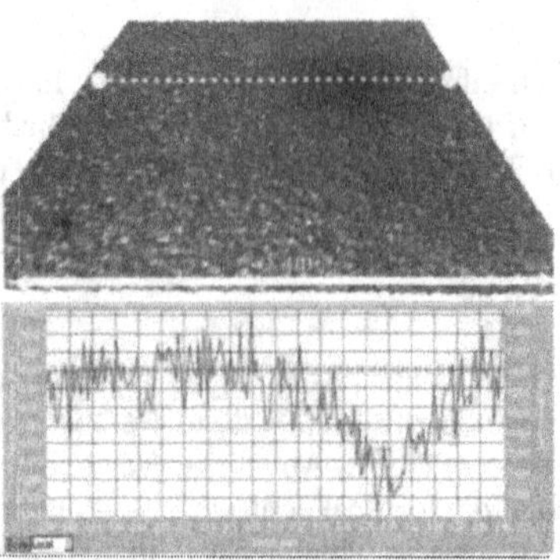

Fig. 6. Confocal laser scanned image of an Au/MWCNT planar for Au-Au/MWCNT pair (201x201/0.4mmx0.4mm) after 10^6 cycles at 1mN.

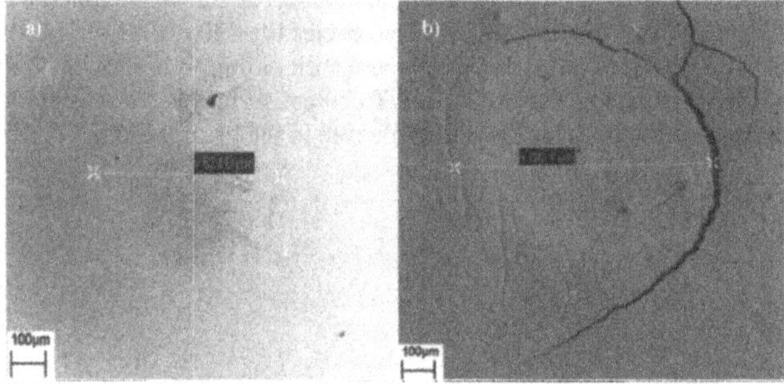

Fig. 7. SEM images of a) Au hemispherical probe and b) Au/MWCNT surface after 20 million cycles at 3mN.

Under hot switching conditions the Au-Au and Au-composite contact surfaces exhibited similar behaviors. Fig 8 shows the contact resistance of an Au-Au pair over 1000 cycles (quasi static) at a maximum applied load of 1mN. The contact resistance of the Au-Au pair is initially ~0.58Ω and increases dramatically to 4-10Ω at ~220 cycles. The reason for the sharp increase in R_c of the Au-Au pair is due to the melting and smoothing of the Au surfaces which leads to increased adhesion [8,10]. The melting is due to the supply and current load used in this experiment. The theoretical voltage for asperity melting of Au contacts is ~0.43V [11]. Fig. 9 (a) and (b) show the delaminated Au hemispherical probe surface and Au adhered on the Au planar surface for the Au-Au pair respectively. Fig 8 also shows the contact resistance of Au-Au/MWCNT pairs with current loads of 1mA and 10mA at 4V. Fig. 10 shows the contact resistance variation over one million cycles. The contact resistance is stable for both current (1mA and 10mA) conditions. SEM imaging indicates some Au adhesion on the Au hemispherical probe as shown in Fig 11. The material transfer is much less than for the Au-Au pair. It is surmised that the combination of MWCNT compliance, higher surface roughness and possibly greater heat dissipation contribute to improving the cyclic contact lifetimes. No observable damage on the Au/MWCNT planar composite surfaces can be detected, further suggesting that the CNT under-layer has improved the mechanical integrity of the gold surface.

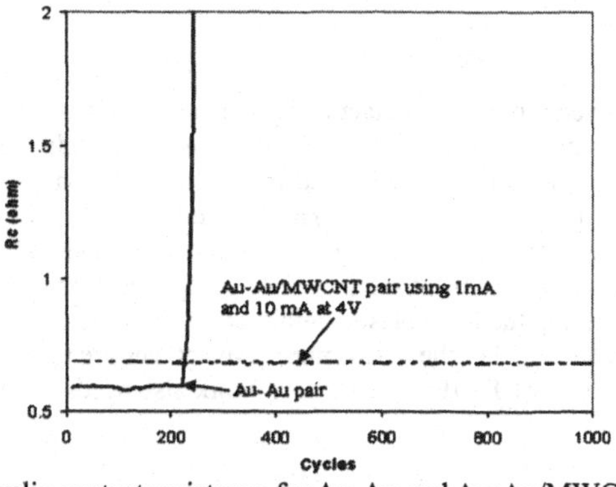

Fig. 8 Cyclic contact resistance for Au-Au and Au-Au/MWCNT pair.

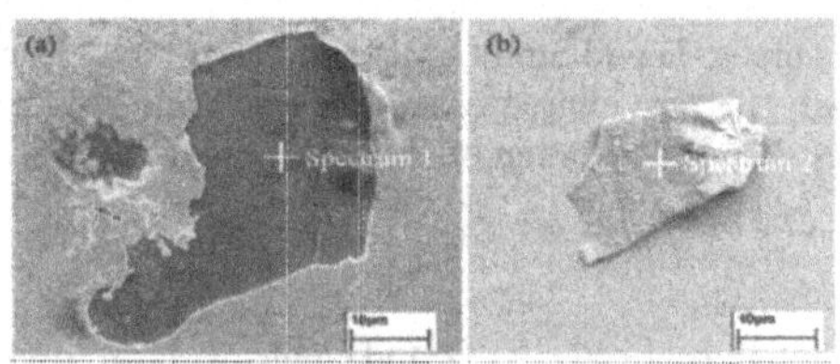

Figure 9: (a) SEM image of Au hemispherical probe degradation (b) SEM image of Au planar with Au debris.

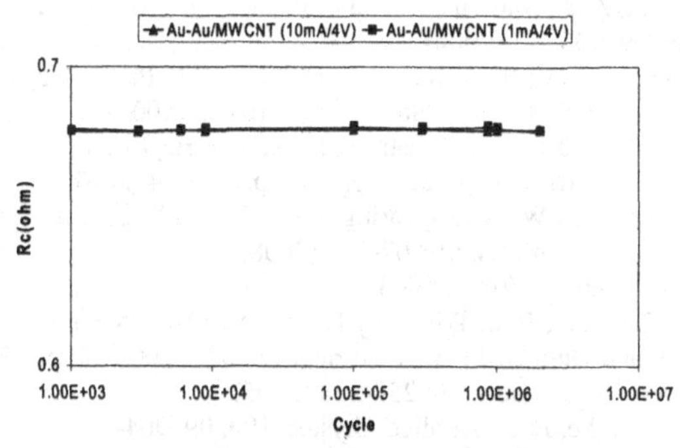

Fig. 10. Contact resistance after a million cycles for Au-Au/MWCNT pair at 1mA and 10mA.

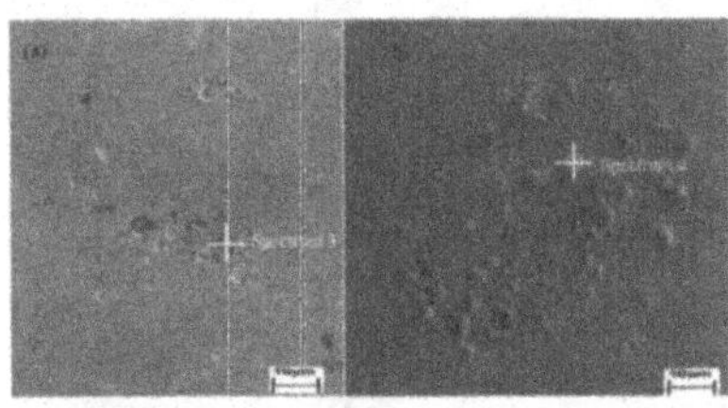

Fig. 11: SEM image of (a) Au hemispherical probe for Au-Au/MWCNT pair after more than 1 million cycles at current load 1mA, 4V and (b) Au hemispherical probe for Au-Au/MWCNT pair at current load 10mA, 4V.

CONCLUSION

The applied cyclic load and contact resistance between Au-Au and Au-Au/MWCNT contact pairs was investigated using electrical and mechanical loading conditions typical of MEMS switches. The Au-Au pair showed catastrophic degradation over ~430 (dry circuit condition) and ~200 (hot-switched) cycles. The Au-Au/MWCNT pair demonstrated a much more stable contact resistance beyond one million load cycles under all conditions. This improvement is believed to be due to the increased compliance of the Au/MWCNT composite surface limiting the local plastic deformation and thus adhesion. Further work is underway to investigate whether the performance advantages are realized at the much higher cycle numbers required for RF MEMS switches. Work also needs to be done on fabrication process integration.

ACKNOWLEDGEMENT

The authors thank Dr. David Smith of the School of Physics and Astronomy at Southampton University, for providing the facilities for developing the MWCNT deposition, and to Mr. Tim Hartley for his dedication in designing and fabricating the test apparatus.

REFERENCES

1. Coutu, R. A., Kladitis, P. E., Leedy, K. D. and Crane, R. L., J. Micromechanics and Microengineering, (14) pp. 1157-1164 (2004).
2. Tzeng, Y., Chen, Y. and Liu, C., Diamond and Related Materials, 12, pp. 774-779 (2003).
3. Yaglioglu, O., Hart, A. J., Martens, R. and Slocum, A. H., Review of Scientific Instruments, 77, pp 095105/1-3 (2006).
4. Bult J, Sawyer W G, Voevodin A, Muratore C, Dickrell P, Pal S, Ajayan P and Schadler L., Proceedings of Material Research Society, vol. 1085E (2008).
5. Yunus, E.M., McBride, J.W., and Spearing, S.M., Electrical Contact, Proc. 53rd IEEE Holm Conference on Electrical Contacts, Vol.6.4, pp.167-174 (2007).
6. Yunus, E.M., McBride, J.W., and Spearing, S.M., Proceedings of the 24th International Conference on Electrical Contacts, pp.507-513 (2008).
7. Rebeiz, G.M., New Jersey, Wiley (2003).
8. Patton, S.T. and Zabinski, J. S., Tribology Letters, Vol 18, No.2, pp. 215-230 (2005).
9. Dickrell III, D. J. and Dugger, M. T., Electrical Contacts 2005 – Proc. 51st IEEE Holm Conference on Electrical Contacts, pp. 255-258 (2005).
10. Gregori, G. and Clarke, D. J. Applied Physics, 100, 094904-1-10 (2006).
11. Slade, P. New York, Basel, Marcel Dekker, Inc (1999).

Mater. Res. Soc. Symp. Proc. Vol. 1139 © 2009 Materials Research Society 1139-GG05-05

An Improved Nanotribological System for Hard Disk Drives

Xuan Li, James Economy
University of Illinois at Urbana-Champaign, Department of Materials Science and Engineering, Urbana, Illinois, 61801, U.S.A.

ABSTRACT

A new family of aliphatic polyester lubricants has been developed for use in micro-electromechanical systems (MEMS) specifically for hard disk drives that operate at high spindle speeds (>15000rpm). Our program was initiated to address current problems with spin-off of the perfluoroether (PFPE) lubricants. The new polyester lubricant appears to alleviate spin-off problems and at the same time improves the chemical and thermal stability. This new system provides a low cost alternative to PFPE along with improved adhesion to the substrates. In addition, it displays a much lower viscosity which may be of importance to stiction related problems. The synthetic route is readily scalable in case additional interest emerges in other areas including small motors.

1. INTRODUCTION

Future progress in the hard disc drive industry depends on achieving higher storage densities and faster accessing of data. This requires that the head-disk interface (HDI) spacing be decreased to below 5nm and the spindle speed increased from 7,000 beyond 15,000 rpm. [1] To achieve these spacings and spindle speeds, it is essential to come up with a new greatly improved lubricant. [2], [3] Perfluoropolyethers (PFPE) are almost universally employed by the magnetic recording industry as hard disk lubricants. This is due to several advantages of PFPE including relatively low volatility, high thermal stability, low surface energy, and excellent lubricating ability. [4] However, the performance and lifetime of PFPE lubricants is proving to be inadequate for next generation HDI, which is mainly limited by insufficient chemical stability and their tendency to display problems of stiction and adhesion problems. [5] Recent research has focused on trying to improve the tribology of the HDI by modifying PFPE lubricant end groups, but none has led to molecularly uniform thick lubricant films with superior thermal and tribochemical stability. [6]

In this study, a new family of sterically hindered aliphatic polyester lubricants was designed with the goal of overcoming some of the drawbacks of PFPE lubricants for hard disk drive applications. Branching can greatly reduce the glass transition temperature and crystallinity, while improving thermal and hydrolytic stability for more severe HDI conditions. The polar ester groups in the polyester main chain should provide sufficient dipole interactions between the lubricant and the substrate minimizing the tendency for spin-off and dewetting. [7] In addition, polar lubricants may form a solid-like boundary where the film thickness is under 5nm range and possibly act to reduce stiction problems. [8] The alkyl side groups will minimize the surface energy and improve the interfacial interactions leading to a stable uniform ultra-thin film. The polar nature of esters also makes the polyester lubricant more miscible with a wide range of solvents and dispersants. This allows the polyesters to dissolve easily and redistribute on the carbon overcoat. In comparsion PFPE lubricants are relatively insoluble, and require use of ozone-damaging solvents such as Freon during processing. The PFPE lubricants also exhibit autophobicity which can lead to formation of capillary wave patterns. [9]

2. EXPERIMENTS

2.1. Materials

2M Lithium diisopropylamide in heptane /tetrahydrofuran /ethylbenzene, isobutyric acid, 2-methyl butyric acid, 1, 3-dibromopropane, 1, 5-dibromopentane, antimony oxide, triphenyl phosphate, trimethylacetyl chloride, 2,2,6,6-tetramethylheptanedioic acid, 2,2-diethyl-1,3-dipropanediol, pyridine, chloroform and hexane were all purchased from Aldrich. 2-ethyl-2-methyl-1, 3-propanediol was purchased from Lancaster. The following materials were dried for24 hours under vacuum at room temperature prior to use : 2,2,6,6-tetramethylheptanedioic acid, 2,2-diethyl-1,3-dipropanediol and 2-ethyl-2-methyl-1,3-propanediol.

2.2. Equipments

FT-IR scans were acquired using a Nexus 670 FT-IR E.S.P. (Thermo Nicolet) with 64 scans at 4 cm^{-1} resolution. Crystalline samples for FT-IR analysis were prepared by forming pellets with KBr powder. Liquid samples for FT-IR used chloroform as solvent.

^{1}H-NMR measurements were carried out using a Varian Unity 400NB NMR system with $CDCl_3$ as solvent.

Gel permeation chromatography (GPC) measurements were performed in tetrahydrofuran at 25°C with a Waters 515 HPLC pump, a Viscotek TDA model 300 triple detector, and a series of three Viscogel 7.8 × 300 mm columns (2 × GMHXL16141 and 1 × G3000HXL16136). Molecular weights were determined using Viscotek's TriSEC software. The light scattering, mass, and viscosity constants were determined from a single 96 kDa narrow polystyrene standard and checked against other known polystyrene standards for accuracy. The column exclusion limit was 1.0×10^7 Da, and the flow rate was 1.0 mL/min.

Differential scanning calorimetry (DSC) measurements were obtained with a Mettler Toledo Star System with a heating rate of 10°C/min in a nitrogen atmosphere from -80°C to 80°C. Thermal gravimetric measurements were made with a TA Instruments TGA 2950 with high resolution option under a constant stream of nitrogen or air (100mL/min).

2.3 Materials Synthesis

2.3.1. Monomer Synthesis

Diacid monomers, 2, 2, 6, 6-tetramethylheptanedioic acid (I), 2, 6-diethyl-2, 6-dimethylheptanedioic acid (II), and 2, 8-diethyl-2, 8-dimethylnonanedioic acid (III) were synthesized according to the following protocols. (Scheme 1)

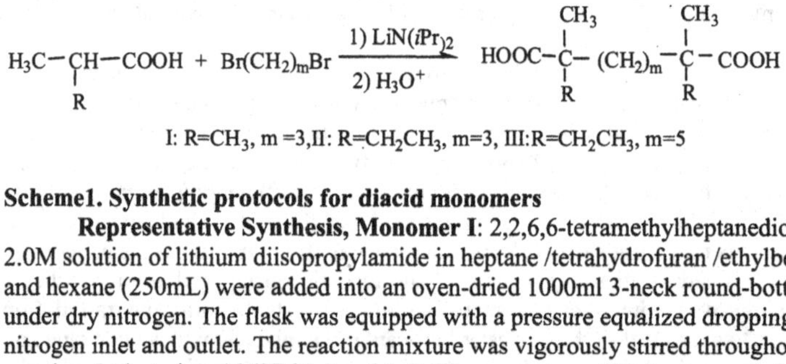

$$H_3C-\underset{R}{CH}-COOH + Br(CH_2)_mBr \xrightarrow[2)\ H_3O^+]{1)\ LiN(iPr)_2} HOOC-\underset{R}{\overset{CH_3}{C}}-(CH_2)_m-\underset{R}{\overset{CH_3}{C}}-COOH$$

I: R=CH_3, m =3,II: R=CH_2CH_3, m=3, III:R=CH_2CH_3, m=5

Scheme1. Synthetic protocols for diacid monomers

Representative Synthesis, Monomer I: 2,2,6,6-tetramethylheptanedioic acid. 2.0M solution of lithium diisopropylamide in heptane /tetrahydrofuran /ethylbenzene (444.2mL) and hexane (250mL) were added into an oven-dried 1000ml 3-neck round-bottomed flask at 0°C under dry nitrogen. The flask was equipped with a pressure equalized dropping funnel and nitrogen inlet and outlet. The reaction mixture was vigorously stirred throughout the entire reaction. Isobutyric acid (37.00g) and 1, 3-dibromopropane (25g) were cooled to 0°C and then added sequentially dropwise over 30 min. The solution temperature was kept at 35°C for 1hr. The solution was washed with HCl, deionized water, and extracted by chloroform. The chloroform layer was washed with NaCl-saturated water solution and then dried over sodium sulfate. Evaporation of chloroform yielded 31g (72%) of monomer I. The monomer I was further purified by recrystallization from acetone/water mixture (1:1 volume ratio).

Monomer I Characterization: FT-IR analysis of monomer I showed three characteristic bands (caused by COOH functional group). The broad band centered in the range 2700-3000cm^{-1} is caused by the presence of OH, as well as CH stretch. Hydrogen bonding and resonance weaken the C=O bond, resulting in absorption at a lower frequency than the monomer. The band at 1701cm^{-1} is due to the C=O double bond stretch. Two bands arising from C-O stretching and O-H bending appear in the spectra of carboxylic acids near 1320-1210 and 1440-1395 cm^{-1}, respectively.

^{1}H-NMR of monomer I displayed four sets of functional groups: 1.18 (s, 12H, **CH_3**), 1.18-1.23 (m, 2H, CH_2-**CH_2**-CH_2), 1.49 (t, 4H, **CH_2**-CH_2-**CH_2**), 12.53 (s, 2H, CO**OH**). These results confirmed that 2, 2, 6, 6-tetramethylheptanedioic acid was successfully synthesized and purified.

2.3.2 Polymer Synthesis

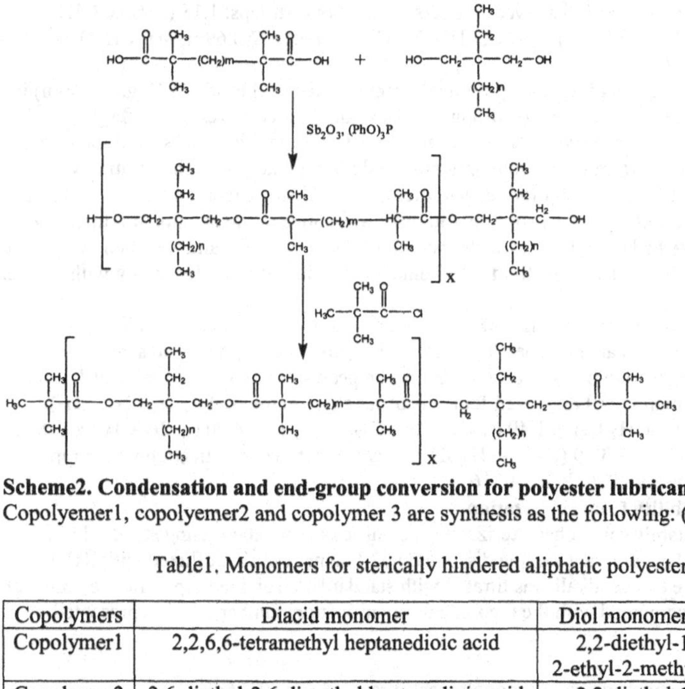

Sb_2O_3, $(PhO)_3P$

Scheme2. Condensation and end-group conversion for polyester lubricants

Copolyemer1, copolyemer2 and copolymer 3 are synthesis as the following: (Table 1)

Table1. Monomers for sterically hindered aliphatic polyester lubricants

Copolymers	Diacid monomer	Diol monomers(1:1 molar ratio)
Copolymer1	2,2,6,6-tetramethyl heptanedioic acid	2,2-diethyl-1,3-propanediol, 2-ethyl-2-methyl-1,3-propanediol
Copolymer2	2,6-diethyl-2,6-dimethyl heptanedioic acid	2,2-diethyl-1,3-propanediol, 2-ethyl-2-methyl-1,3-propanediol
Copolymer3	2,6-diethyl-2,6-dimethyl heptanedioic acid, 2,8-diethyl-2,8-dimethyl nonanedioic acid mixture(1:1 molar ratio)	2,2-diethyl-1,3-propanediol, 2-ethyl-2-methyl-1,3-propanediol

Representative Polymerization, Copolymer1: 2,2,6,6-tetramethylheptanedioic acid (I)(28.85g ,150.26mmol), 2,2-diethyl-1,3-dipropanediol(11.92g,90.16mmol), and 2-ethyl-2-methyl-1,3-propanediol (10.66g ,90.16mmol) were melt polymerized in the presence of antimony oxide (0.5g) and triphenyl phosphate(0.25g). The reaction mixture was held at 180°C at atmospheric pressure for 15hr, and then at 190°C for 26hr. To complete the reaction, the pressure was reduced to 1mm Hg with a vacuum pump over 2hr. The viscous fluid was cooled down to room temperature, dissolved in chloroform, and then filtered to remove the antimony oxide. A viscous transparent liquid lubricant was obtained by evaporation of chloroform under vacuum at 70°C.

Coploymer1 Characterization: the reaction was monitored by FT-IR. The strong broad OH stretch due to COOH (2700-3000 cm^{-1}) disappeared. Carbonyl C=O double bond stretch in diacid monomers had sharp absorption at 1701cm^{-1}. As the polymerization reaction progressed, the absorption peak shifted to a higher value (1717cm^{-1}). The sharp adsorption peak at 1216cm^{-1} was due to the ester group formation.

^{1}H-NMR of monomer I displayed five sets of functional groups: 1.18 (s, 12H, $\mathbf{CH_3}$), 1.18-1.23 (m, 2H, CH_2-$\mathbf{CH_2}$-CH_2), 1.49 (t, 4H, $\mathbf{CH_2}$-CH_2-$\mathbf{CH_2}$), 3.57-3.69 (t, 4H -$\mathbf{CH_2}$OH), 3.82-4.199 (d, 2H-$\mathbf{CH_2}$OCO)

Representative Capping of Copolymer1: trimethylacetyl chloride (3.62 g, 30mmol) in chloroform (100mL) solvent was added dropwise to a solution of copolyester I (6.83g) in chloroform (200mL) and pyridine (2.77 g, 35mmol) at 5°C. The mixture was stirred for 1.5hr allowing the solution to warm up to room temperature during the last hour; the solution was refluxed for 2 hr. The pyridinium chloride was filtered off after the reaction mixture was cooled down to room temperature. 5% HCl solution (300mL) was added to the reaction mixture. The organic layer was washed several times with water until it was acid free and was then dried with sodium sulfate. A yellowish viscous fluid was isolated from chloroform after drying with a water aspiration at 80°C.

Capped Copolymer1 Characterization: the reaction was monitored by FT-IR. The peak at 2870-2980 cm^{-1} became sharper. The carbonyl group in copolymer1 had sharp absorption at 1717cm^{-1}, as esterification progressed, the absorption peak shifted to higher value at 1729cm^{-1}.

^{1}H-NMR of monomer I displayed four sets of functional groups: 1.18 (s, 12H, $\mathbf{CH_3}$), 1.18-1.23 (m, 2H, CH_2-$\mathbf{CH_2}$-CH_2), 1.49 (t, 4H, $\mathbf{CH_2}$-CH_2-$\mathbf{CH_2}$), 3.82-4.20 (d, 2H-CH_2OCO). The disappearance of peak 3.57-3.69 (t, 4H -CH_2OH) indicated that the esterification was completed.

GPC (g/mol): Mw=820, PDI= 4.316

2.4. Hydrolytic Stability Characterization

Hydrolytic stability was characterized by saponification numbers using standard test method ASTM-94. The lubricant sample dissolved in 2-butane and was refluxed with KOH-ethanol solution. The excess alkali was titrated with standard acid and the saponification number was calculated and compared with the theoretical saponification number which represented 100% hydrolysis.

3. RESULTS

DSC measurements on capped copolymer1, prepared from 2,2,6,6-tetramethylheptanedioic acid (I), 2,2-dimethyl-1,3-propanediol and 2-ethyl-2-methyl-1,3-propanediol (molar ratio 5:3:3) are shown in (Figure 1). As can be seen there is a glass transition at -47.35°C and a large melting endotherm at 42°C.

Introduction of ethyl groups in the diacid monomers will introduce further irregularities into the main chain and therefore could lower the crystallinity of the polymers. Capped

Copolyester2, which was prepared from 2,6,-diethyl-2,6-dimethylheptanedioic acid, 2,2-dimethyl-1, 3-propanediol and 2-ethyl-2-methyl-1, 3-propanediol (molar ratio 5:3:3), displayed a lower T_g at -54.04°C and no melting endotherm was observed.

Additional asymmetry in the molecular structure could be further achieved by copolymerizing a mixture of monomers from copolymers1 and 2. Thus, capped copolymer3, which was prepared from 2,6,-diethyl-2,6-dimethylheptanedioic acid(II), 2,8-diethyl-2,8-dimethylnonanedioic acid(III), 2,2-dimethyl-1,3-propanediol and 2-ethyl-2-methyl-1,3-propanediol (molar ratio 1:1:1.2:1.2), the T_g was lowered to -67.47°C and no melting endotherm was observed (Figure 1).

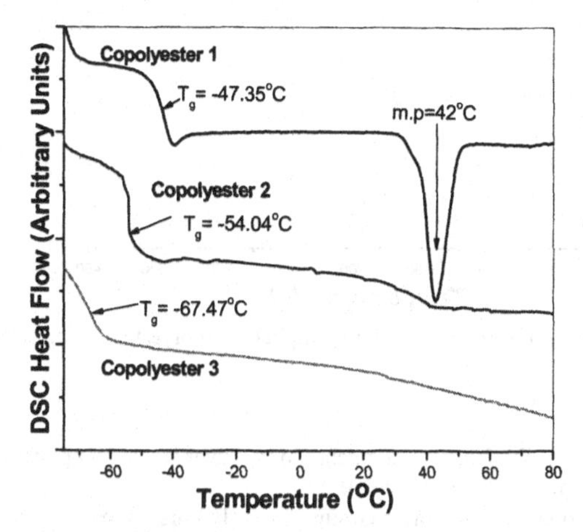

Figure1. T_g of sterically hindered aliphatic copolyesters using a 5°C/min heating rate

The thermal degradation behavior of the sterically hindered copolyester3 in an Al_2O_3 crucible shows no weight loss up to 230°C under N_2 and no degradation up to 185°C in air. Isothermal measurements in air at four different temperatures for copolyester3 demonstrate excellent long-term thermal stability from 100-150°C (Figure 2). In comparison, PFPE has a relatively weak physical absorption on the carbon overcoat and shows up to 96% weight loss after 1hr at 200°C. [10]

The hydrolytic stability was carried out by refluxing 30 min in 0.5 N alcoholic KOH, and copolyester1, copolyester2 and copolyester3 showed 20.7%, 5.7% and 4.9% hydrolysis, respectively. In comparison, conventional polyesters would hydrolyze completely under the test conditions.

4. CONCLUSIONS

In conclusion, we have successfully synthesized sterically hindered polyester lubricants with improved thermal stability and oxidation resistance, low glass transition temperature and crystallinity, and good hydrolytic stability. These new materials are a promising alternative to PFPE for hard disk drive lubricant. We expect these aliphatic polyester lubricants have much stronger bonding to carbon overcoat than PEFE, which will effectively alleviate the spin-off and dewetting problem for hard disk drive industry

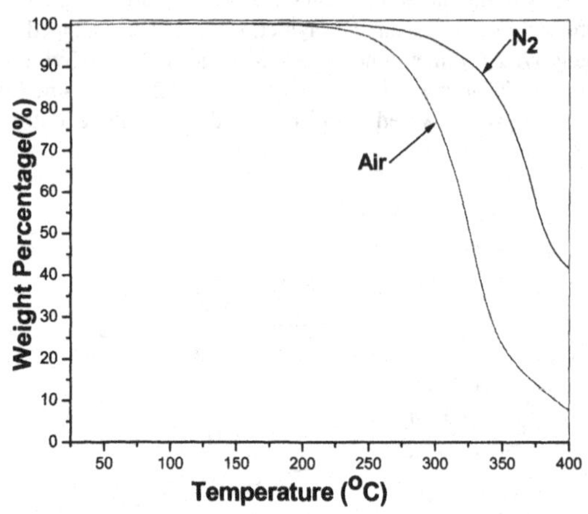

Figure2. Thermal stability of sterically hindered aliphatic copolyester1 using a 10°C/min heating rate in air and in N_2.

5. REFERENCES

1. Juang, J. and Bogy, D. (2005) 'Nanotechnology Advances and Applications Information Storage', *Microsyst Technol* 11(8): 950-57.
2. Economy, J. Professor Economy's Group Currently Pursues the Following Areas of Research on Advanced Materials', ULR: http://economy.mse.uiuc.edu/research.htm
3. Hitachi Global Storage Technologies, 'Storage Technology', URL: http://www.hitachigst.com/hdd/research/storage/adt/index.html
4. Chen, C. Bogy, D. and Singh, B. (2001) 'Effects of Backbone and Endgroup on the Decomposition Mechanisms of PFPE Lubricants and their Tribological Performance at the Head-Disk Interface', *Transaction of the ASM* 123(2): 364-67.
5. Homola, A.M. (1996) 'Lubrication Issues in Magnetic Disk Storage Devices', *IEEE Transactions on Magnetics* 32(3):1812–18.
6. Xiao, W. (2005) *New Materials Systems for Advanced Tribological and Environmental Applications*, University of Illinois at Urbana-Champaign
7. Hatco Corporation, About Ester Chemistry Web Site, URL: www.hatcocorporation.com/pages/syntheticlubes/aboutesters.htm
8. Fabio, B. (1997) *The Application of Advanced Materials to Four Unique Problems Involving Surface Interfaces*, University of Illinois at Urbana-Champaign.
9. Karis, T. Kim, W. Jhon, M.S. (2005) *Tribology Letter* 18(1): 27-41.
10. Kasai, P. H. Tang, W. T.Wheeler, P. (1991) *Applied Surface Science* 51(3-4):201-11.

AUTHOR INDEX

SUBJECT INDEX

Printed in the United States
By Bookmasters